Second Edition

Fundamentals of Industrial Ergonomics

B. Mustafa Pulat

Lucent Technologies, Inc.
Global Provisioning Center
and
School of Industrial Engineering
University of Oklahoma

WAVELAND

PRESS, INC.

Prospect Heights, Illinois

For information about this book, write or call:
    Waveland Press, Inc.
    P.O. Box 400
    Prospect Heights, Illinois 60070
    847/634-0081

To my wife, Simin, and my sons, Özgür and Önder

Printed in the United States of America

7 6 5 4 3 2

Contents

Contents

Preface to the Second Edition

Although the first edition of this book is dated 1992, most of the development was complete by June 1990. Hence, a few important issues were not included in the first edition. For example, the Revised NIOSH Equation (1991), which was published in a scientific journal in 1993, is now a part of this edition. This equation appears in Chapter 3 on Physical Egonomics, along with an application. Also in Chapter 3 we present the Krebs Cycle and supporting systems to the musculoskeletal system, including respiration, nutrition, and circulation. Furthermore, there is a presentation on The Americans with Disabilities Act of 1990 in Chapter 8. The final major addition is in Chapter 12, which now features discussion of research techniques in ergonomics—including basic and applied research. Other changes span the entire book to delete outdated material and refresh various topics by new research results. We maintained several early applications and case studies. Hence, any reference to AT&T is a result of the material included in the book at an earlier date.

I express my appreciation to all reviewers, Waveland Press, Inc., and my family for their support in the development of the second edition. Finally, much appreciation is due Jim Styring of Lucent Technologies, Inc., who encourages out-of-the-box thinking and promotes scholarly work.

B. Mustafa Pulat
Lucent Technologies, Inc.
Oklahoma City

Preface

Ergonomics has come a long way. Many years have passed since the heavy ergonomics attention to military equipment during and immediately after World War II. Practitioners and ergonomics students in engineering, psychology, and design are realizing that manufacturing and service systems that are designed for human use do not function as well as they can because in many instances the human element has been taken for granted. Much of the ergonomics information, theory, and data developed for the military setting can be applied to the industrial setting for such systems to work more effectively with the fewest accidents and the least human suffering.

The Reason for This Book

I wrote this book because I felt the need for a textbook that presents and links the theory and practice of industrial ergonomics for engineering students, especially industrial engineering students and practicing professionals. Having served as a full-time assistant/ associate professor of industrial engineering for six years (and continuing as adjunct professor at the University of Oklahoma), I always had a hard time finding a textbook that would appeal to these groups. In all cases I resorted to a book that had been in existence for a long time and was oriented toward ergonomics in general rather than industrial ergonomics.

Text Organization

This book is intended to cover the basics of ergonomics as applied to industrial environments. We start with *introductory concepts* (see the chapter flow diagram). The first chapter in this section covers general introductory notes such as the assumptions, definitions, related organizations, and historical notes. The second chapter is more focused. It looks at the ergonomics function in an industrial concern, starting with planning for deployment and continuing with functional organization and organizing for program continuity.

The three chapters that follow present the *underlying theory* of industrial ergonomics. First, the physical factors are covered, including the basics of human activity, fatigue, strength/endurance, biomechanics, and the widely researched area of manual material handling. After establishing some structure in work physiology and biomechanics, Chapter 4 goes into the sensory and information-processing mechanisms and their underlying structure. The final section in this chapter focuses on the contemporary issues in human–computer interaction. Finally, chapter 5 focuses on the human anthropometry, both structural and functional, and the related issues.

The mission of each chapter in the theory section is to present human capability/ limitation information on which design decisions need to be based. Chapters 6 to 11 build on these concepts. The first four of these chapters are concerned with the *design of work systems*. Chapter 6 provides a lead-in, presenting the basic elements of human–machine

CHAPTER FLOW

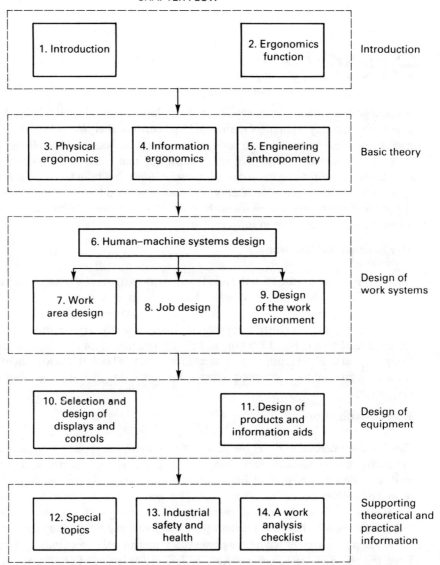

systems design. Then Chapter 7 focuses on the underlying issues in workstation and work-space design. Chapter 8 expands these ideas to cover issues in job design. Finally, Chapter 9 focuses on the missing element, work environment design, which covers the four most common physical environment elements in industrial establishments: the visual, the auditory, the vibratory, and the thermal environments. Next, we focus on the *design of equipment*. After an inside-out discussion of work systems design in Chapters 7 to 9, the two chapters in this section (Chapters 10 and 11) provide an outside-in look at the issues of display, control, product, and information aids design.

The two chapters that follow present specialized information along the lines of human error, selection and training, job performance aids, shift work, design for the elderly, advanced manufacturing, and safety and health, all important to industrial ergonomics. Finally, Chapter 14, a work analysis checklist, provides a systematic work evaluation methodology. These last three chapters *support the core issues* presented in the earlier chapters.

Although the appendices are not displayed in the flowchart, they support the relevant chapters with extensive design data. Appendix A focuses on the design parameters of common controls. Appendix B provides extensive anthropometric data. Appendix C presents research citations along several major dimensions of industrial ergonomics published within the last decade. The objective here is to offer quick help to a focused reader who wishes to have additional information.

There are two types of end-of-chapter exercises contained in the text. The first type helps the student to review the text material. The second type provides more project-oriented problems/cases for future research or applications. Furthermore, Appendix C provides extensive leads to a person with an inquisitive mind who wishes to do additional reading in the various topics of industrial ergonomics published in the most recent past. In addition to many spot examples in the text, there are nine *case studies* that support the theory. The case studies have been taken directly from applications at the factory where I work. They serve as a reality check to the reader.

A project of this magnitude could not be completed without the support of many. First, I am most pleased to have the opportunity of working with Prentice Hall editor Elizabeth Kaster and with Joe Mize and Walter J. Fabrycky, co-editors, Prentice Hall's International Series in Industrial and Systems Engineering. Their encouragement and never-ending support were important factors in timely completion of the manuscript.

I also wish to thank William C. Moor of Arizona State University, James K. Hennigan of Texas A&M University and others for reviewing the manuscript. As a result of their valuable suggestions, we have a more comprehensive text.

I am indebted to the management of AT&T Oklahoma City Works for their encouragement during the preparation of this book. Special acknowledgment is due to P. A. Gannon, R. E. Wagner, R. J. Vidmar, and A. L. Arms.

Recognition is also extended to the early supporters of my interest in ergonomics: Professors Ömer Saatçioğlu, Ünver Çinar, and Halim Doğrusöz of the Middle East Technical University, Ankara, Turkey, and to the Scientific and Technical Research Council of Turkey for funding my advanced studies at North Carolina State University, Raleigh, North Carolina, with Professors R. G. Pearson and M. A. Ayoub.

Finally, and most important, this text would not exist without continuous backing from my family. They tolerated months of neglect during the preparation of this volume.

B. Mustafa Pulat

CHAPTER 1

Introduction

1.1 PROLOGUE

People play an important role in the functioning of socioeconomic systems. For example, they develop equipment, machinery, workstations, and objects. Most of the things that we use in our daily lives are human-made, and we are either direct users of these or are involved in maintaining them.

Ergonomics or *human factors* is an interdisciplinary science that deals with the interaction of people with the objects they use. In many cases, human beings take for granted the inefficiencies that may exist in such an interface. It is when significant ill effects accrue that a person realizes the problem and takes a corrective action. The effects may also develop instantaneously as single-incident trauma. A production worker may lacerate a thumb while opening a carton that is placed on a high work bench that limits vision. This is a single-incident trauma case caused by a use problem. Garden shears that place concentrated stress on the soft tissues of the hand, causing blisters and other ailments, is an example of use problems that may develop over time. Some interfaces simply lead to comfort problems. A spray device that requires a bent posture for lawn weed control, due to a short handle-nozzle mechanism, is a good example. In many cases, poor design of workplaces leads to a drop in production efficiency. An assembly

1

worker develops muscle fatigue around the shoulder and neck area when the work surface is only 5 centimeters too high. In such a case, the worker will have to raise the shoulders and/or abduct the arms, leading to fatigue that must be balanced by short periods of idle time.

In this chapter we first present the assumptions and costs of ignoring ergonomics in workplace and work methods design. Then a number of definitions and related concepts are presented, including accident statistics and descriptions of government and other organizations active in the field. A brief history follows, along with the role that an ergonomist plays in most organizations and the types of problems with which the ergonomist is concerned. Finally, two industry examples are cited.

1.2 ASSUMPTIONS

As with many branches of scientific study, ergonomics makes several assumptions when focusing on the functioning of human–object systems (Chapter 6). Time after time, these assumptions have been shown to hold via experimental and applied studies, two of which are discussed at the end of this chapter. One of the assumptions is that we can link the efficiency of person–object systems to the efficiency of human functions in those systems. Hence if people cannot function effectively in a system, the system's performance may be degraded. Another assumption relates to motivation. It states that people achieve more if they are properly motivated. Designers must study characteristics of the work environment that motivate people, such as meaningful work and opportunity for advancement, and build these into organizations in which people play important roles (see Chapter 8).

Perhaps the most important assumption of ergonomics is that equipment, objects, and environmental characteristics influence human performance and thus total human–object system performance. Thus if products, equipment, workstations, and work methods are designed keeping human capabilities and limitations in mind, the performance of the resulting system will be better than otherwise. Conversely, if ergonomics is ignored during design, one should be ready to accept the consequences.

1.3 COSTS OF IGNORING ERGONOMICS

Experience shows that there are many potential ill effects that can be expected if ergonomics is ignored during design:

1. Less production output
2. Increased lost time
3. Higher medical costs
4. Higher material costs
5. Increased absenteeism
6. Low-quality work
7. Injuries, strains
8. Increased probability of accidents and errors

9. Increased labor turnover

10. Less spare capacity to deal with emergencies

Potential gains of ergonomics are the reverse of the above. In addition to the tangible gains, there are many intangible benefits. One can count increased job satisfaction and work acceptance as two important intangible gains of considering the human being when undertaking system design.

1.4 ERGONOMICS DEFINED

Over the past several decades, many definitions of ergonomics have been offered. In simple terms, one can describe it as the *study of the interaction between human beings and the objects they use and the environments in which they function.* This definition covers the most important elements: human beings, objects, environments, and the complex interactions among them. In the literature, the first use of the term seems to have occurred during the mid-nineteenth century. A Polish educator and scientist, Wojciech Jastrzebowski, introduced the term combining two Greek words: *ergon,* meaning work, and *nomos,* meaning laws [1]. The literature also suggests that the word was independently formed in the same manner in 1949 by a British scientist, K. F. H. Murrell. *Human factors* or *human factors engineering* are the equivalent expressions used primarily in the United States. Ergonomists frequently use the term *human engineered* to describe a design that conforms with human expectations, or which people use without undue stress.

Design for human use is another practical definition of ergonomics. Again, the emphasis is on the use of human-made objects, equipment, machines, and systems. Sanders and McCormick [2] provide a comprehensive definition of the discipline. Briefly, the authors offer a three-pronged approach: the central focus, the objectives, and the central approach. The *central focus* is the consideration of human beings in the design of objects, machinery, and environments. The *objectives* are to increase the effectiveness of the resulting human–machine system while maintaining human well-being. The *central approach* is systematic application of available data on human characteristics (capabilities, limitations) to the design of such systems or procedures. Thus, as indicated in Figure 1.1, there are three major objectives of ergonomics [3].

The central theme of ergonomics is *fitting the task to the person.* This means that designers must design for human use, keeping in mind at all times what people can and will do. There is a wealth of information in the literature concerning human capabilities and limitations that is available for the designer's use. Such data are generated through basic research with human subjects. They are then published in handbooks, periodicals, textbooks, manuals, and the like [4–7]. The opposite of the central theme is exercised during personnel selection. Here the job requirements are defined, and then the job applicants are assigned to jobs based on the best match between skills possessed and required. Although this is the case, the challenge to the job and work area designer is to human-engineer those aspects of work so that the need for selection is minimized.

Another point regarding ergonomics is taken from general systems theory. For any system to function effectively, the two major prerequisites are: (1) the components must have been designed properly, and (2) the components must function together cohesively

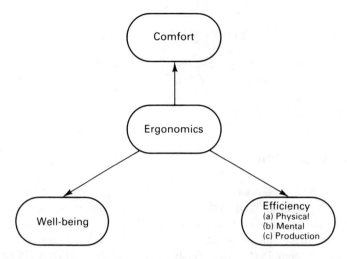

Figure 1.1 Objectives of ergonomics. (Reprinted from *Industrial Ergonomics: A Practitioner's Guide,* 1985. © Institute of Industrial Engineers, 25 Technology Park, Norcross, GA 30092.)

toward a common goal. If we wanted to develop a superior car engine and selected the best-built engine components (carburetor, engine block, crankshaft, etc.) on the market, based on individual performance, and assembled these components, the resulting engine would not work. No matter how well the individual parts have been designed and manufactured, they will not fit and function together. Similarly, the user is an integral part of any human–machine system. If he or she cannot function effectively, the performance of the total system will be affected and major mishaps will sometimes occur.

Frequently, we hear that human error was the cause of a major accident. However, detailed investigations reveal that the real reasons are often design deficiencies that lead to a human error. Case in point: the Three Mile Island mishap in 1979. One of the major reasons for late discovery of the low reactor coolant level was that the status of a valve draining the coolant was erroneously indicated as closed on the instrument panel in the control room, when in fact, the valve was open. Naturally, under a wrong assumption, control room operators looked for the problem elsewhere. After several hours, the problem was discovered, but before then, a considerable amount of the coolant had been lost. Detailed investigations concluded that the design of the valve control system was at fault. Rather than checking the actual condition of the valve, the control system monitored whether or not the valve was powered. The designer did not take into consideration that even though it is powered, the valve may still fail to close due to mechanical failures, which was the case here.

1.4.1 Accident and Injury Statistics

Data compiled in the United States indicate that ergonomics-related problems capture five of the top 10 leading work-related diseases and injuries [8]. Table 1.1 lists those ailments. Items 2, 4, 5, 8 and 10 are ergonomics related. In 1994, employers

TABLE 1.1 TEN LEADING WORK-RELATED DISEASES AND INJURIES IN THE UNITED STATES

1. Occupational lung disease
2. Musculoskeletal injuries
3. Occupational cancers
4. Amputations, fractures, eye loss, lacerations, and traumatic deaths
5. Cardiovascular diseases
6. Diseases of reproduction
7. Neurotoxic disorders
8. Noise-induced loss of hearing
9. Dermatologic conditions
10. Psychologic disorders

Source: After Ref. [8].

reported 6.3 million work injuries and 515,000 cases of occupational illnesses. In that year injuries alone cost $121 billion in lost wages and lost productivity, administrative expenses, health care and other costs. With 125 million workers in the United States—almost one of every two in the entire population—each day, an average of 137 individuals die from work-related diseases, and an additional 16 die from injuries on the job [9].

In October 1988, the meat-packing firm of John Morrell and Company was fined $4.33 million for a series of workplace injuries termed *cumulative trauma disorders* (CTDs). Industrial ergonomics was involved in identifying CTDs—the injury of the 1990s. The cost of one case of carpal tunnel syndrome, a common form of cumulative trauma disorder, can reach $100,000, with an average overall cost of about $15,000. In comparison, job and workplace design with built-in ergonomics may cost only several hundred dollars extra. The Bureau of Labor Statistics indicates that the number of injuries due to cumulative trauma increased tenfold, to 332,000, from 1984 to 1994. Soft-tissue disorders, which include both motion and back injuries, now account for a significant portion of all workers' compensation claims according to National Institute for Occupational Safety and Health (NIOSH). OSHA (Occupational Safety and Health Administration) predicts that this will reach 50% by the year 2000 [10].

1.5 HUMAN-INTEGRATED DESIGN AND MANUFACTURING

In earlier sections we saw that the primary objective of ergonomics is designing objects, equipment, and machinery for effective use by human beings. *Design for human use* requires a central and consistent strategy to be applied to all stages of design. HIX is a term coined by Pulat and Alexander [11] to symbolize attention to the human being from start to finish in the design cycle. "X" is a variable that is replaced by either "design," "manufacturing," "test," or other. "HI" stands for "human integrated." *Human-integrated design* (HID) refers to systems designed with human capabilities and limitations in mind so that the resultant design can be used effectively by people (Figure 1.2). In manufacturing

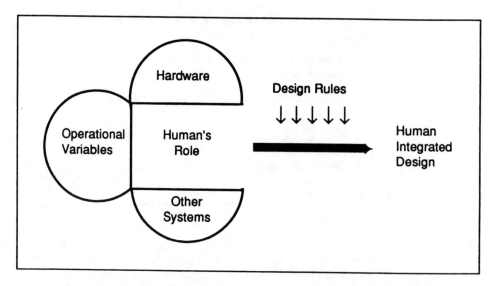

Figure 1.2 The various components of human integrated design. (Reprinted from *Industrial Ergonomics: Case Studies,* 1990. © Institute of Industrial Engineers, 25 Technology Park, Norcross, GA 30092.)

concerns the concept can be symbolized by HIM (*human-integrated manufacturing*). One of the principal concerns in *design for manufacturability* is whether or not human operators can work effectively with the design. The concept extends to cover all instructions, design documents, and other design support issues. For example, design documents from product design organizations can be so difficult to read that material planners may order incorrect parts or incorrect quantities of the correct parts. Another example is that of misinserted components on a printed wiring board assembly line due to human operators not being able to read engineering information correctly from a video display terminal with which glare and orientation and resolution problems are associated. Both cases lead to human errors, which in turn lead to thousands and sometimes millions of wasted dollars. *Consider HIM when designing for manufacture.*

Prevention versus correction.   Many ergonomics projects are undertaken as a corrective measure for an existing problem. This practice is especially prevalent under limited resources: labor, money, and so on. Naturally, the preferred application mode is *preventive.* A system is most effective when ergonomics is designed into it. A process is most effective when ergonomics is built into the process. Both are cases for HIX.

1.6 ORGANIZATIONS

For maximum effectiveness, ergonomics must be supported by research and development organizations. Furthermore, data and information developed as a result of research must be applied. Certain information affecting safety and health may have to be compiled into laws, which then must be enforced. Several government and other organizations that support and/or enforce application of ergonomics and safety in the United States were mentioned in the preceding section, and several more are discussed in Section 1.7. OSHA is the government organization responsible for generating safety standards for industry and for inspecting and enforcing the application of these standards in organizations engaged in interstate trade. Another government agency, NIOSH, conducts research to support this activity. MSHA (the Mine Safety and Health Administration) is OSHA's counterpart in the mining industry.

Several other professional societies that are of relevance to ergonomics in the United States are the Human Factors and Ergonomics Society (HFES); Institute of Industrial Engineers (IIE), Ergonomics Division; American Society of Safety Engineers; American Industrial Hygiene Association; Industrial Designers Society of America; Systems, Man, and Cybernetics of the Institute of Electrical and Electronics Engineers; and Division 21 of the American Psychological Association. All of these bodies are actively involved with organizing conferences, promoting focused study and lobbying on behalf of effective functioning of human beings in socioeconomic systems.

Other professional groups in other parts of the world include the Ergonomics Society (ES), ILO (International Labor Organization), WHO (World Health Organization), and the affiliated societies of the IEA (International Ergonomics Association). Parallel societies exist in the Soviet Union, the emerging democracies of eastern Europe, and Japan. The Human Factors Society in the United States and the Ergonomics Society of the United Kingdom are the two flagship organizations promoting the research and practice of ergonomics in the world. More will be said about these organizations in the next section.

Many universities offer short courses and advanced degrees in ergonomics. As of 1996, very few universities in the world offered undergraduate study in the field. Industrial engineering or psychology programs are more apt to offer formal study at advanced levels. Many members of the Human Factors and Ergonomics Society have undergraduate degrees in engineering or science. Professional licensing is available through the Board of Certification for Professional Ergonomists (BCPE) in Bellingham, Washington.

1.7 BRIEF HISTORY

Since the beginnings of the human race, our ancestors have always looked for ways to make life easier and more enjoyable. Many different designs of tools, weapons, and food containers excavated provide the reference point. Perhaps the first major thrust for worker safety and health came during and after the *industrial revolution*, which took place primarily in Europe between the mid-eighteenth and early nineteenth centuries. During this time, many machines were invented, primarily in the textiles industry, which brought the factory and power age into being. These devices optimized production rather than considering the human interface. Naturally, many worker injuries and fatalities followed. Facing a major social problem, the countries involved (primarily England, France, and the United States) passed laws to protect the workers. Although these laws helped to reduce industrial accidents, they fell short of making significant inroads with regard to preventing them.

Research into the effective use of human capabilities and limitations in systems and equipment design started almost simultaneously in various countries. As early as during

World War I, both the United States and Great Britain exercised selection and training, thus *fitting the human to the task.* Documents show that these countries also continued research in these areas after the war. It was not until World War II that ergonomics made its impact on design. During this time it was observed that many complex military equipment could not be operated effectively or safely. Close to 400 aircraft and many lives were lost in the United States alone due to use problems. As a result, work focused on improving designs with human capabilities and limitations in mind.

The time during and right after World War II is known as the "knob and dial era." Much research was conducted during this time to develop optimal design parameters for displays and controls. The first textbook dealing primarily with ergonomics was published in the United States in 1949. Since 1950, more people, including academicians, have gotten involved with the discipline. It was during this time that nonmilitary applications were developed in the areas of manufacturing, communications, and transportation. The Human Factors Society (HFS) of the United States was formed in 1957 in Tulsa, Oklahoma. HFS changed its name to Human Factors and Ergonomics Society in 1992. This society, now headquartered in Santa Monica, California, leads the research and application of ergonomics in the United States. A second U.S. body, the International Foundation of Industrial Ergonomics and Safety Research, formed in March 1986, also aims at supporting research and application projects in ergonomics worldwide.

In Great Britain, as early as 1915, a body known as the Industrial Fatigue Research Board was set up by the government, later changing its name to the Industrial Health Research Board. These bodies carried out practical research in the areas of environmental factors and performance. Between the disbanding of the Industrial Health Research Board in 1920 and the beginning of World War II, the amount of work-related research carried out was insignificant. From 1942 on, organized groups of interdisciplinary researchers representing areas such as anatomy, physiology, psychology, industrial medicine, design engineering, architecture, and illumination engineering, met routinely to identify and work on ergonomics problems. In 1949, the Ergonomics Research Society was formed, which later changed its name to the Ergonomics Society (ES).

Ergonomics research started in the Scandinavian countries at about the same time, with heavy emphasis on work physiology and human physical resources. After many years, psychological parameters were studied. ERGOLAB, founded in 1971, oriented its research efforts primarily to human psychology. As in many other European countries, research in the Scandinavian countries during the 1970s assumed a more practical role, dealing increasingly with real-life situations. Strong unions in Europe were quite influential in ergonomics practice at the workplace.

Ergonomics development in Japan took place much later than World War II due to strong Japanese traditions of loyalty and devotion and a strong spirit of overcoming all obstacles. The Japanese soldiers were expected to adapt themselves to machines. After the war, the need to rapidly rebuild industry in Japan precluded any major role the unions may have played in ergonomics applications [12].

The umbrella association that encourages international activity in this field is the IEA (International Ergonomics Association), which held its first meeting in 1961 in Stockholm, Sweden. The association now has at least 15 affiliated societies from several continents.

1.8 THE ROLE OF THE ERGONOMIST

Organizations function with people carrying various responsibilities. A person whose primary responsibility is to deal with ergonomics problems in an organization is known as an *ergonomist* or *human factors specialist.* There may or may not be a full-time ergonomist in a factory or service establishment. A general expectation is that all engineers must possess some ergonomics background. Supporting them may be a full-time ergonomist attending to complex problems. Unfortunately, not all engineers who have design responsibility take formal training in this field. If there is a safety department in an organization, the ergonomist may be included in this group.

A survey of members of the ergonomics division of IIE indicates that in manufacturing and service industries one can find full-time ergonomists when the size of the organization reaches around 1500. Naturally, those industries where there is significant potential for such problems fund an ergonomics position at much lower head counts [13]. According to the survey, subjects that attract most project work are VDT (video display terminal) ergonomics, office furniture and layout, general workspace design and improvement, manual material handling and biomechanics, and hand tool design. Other subjects of interest are software design, seating, training, physiological and psychological stress, environmental factors, job/task design, job evaluation, and equipment design.

1.9 TYPES OF ERGONOMICS PROBLEMS

The discussion above leads one into categorizing the ergonomics problems into various groups, depending on the specific area(s) of the body affected [14]:

1. *Anthropometric.* These relate to the dimensional conflict between functional space geometry and the human body. Anthropometry deals with the measurement of linear body dimensions, including weight and volume. Reach distances, sitting eye height, and buttock-to-knee length are several examples (see Chapter 5). Anthropometric problems manifest themselves as lack of fit between these dimensions and the design of the workspace. The solution is to modify the design and establish the compatibility.

2. *Cognitive.* Cognitive problems arise when there is either information overload or underload under information-processing requirements. Both the short-term and the long-term memory may be strained. On the other hand, these functions may not be sufficiently utilized for maintenance of optimum state of arousal. The solution is to complement human functions with machine functions for enhanced performance (see Chapter 4), as well as job enrichment.

3. *Musculoskeletal.* Problems that strain the muscular and skeletal systems are in this category. They can induce single-incident or cumulative-effect trauma. A slipped disk is an example. Solutions of these problems lie in providing job performance aids or redesigning the job to keep the requirements within the bounds of human capability (see Chapter 3).

4. *Cardiovascular.* These problems place stress on the circulatory system, including the heart. The result is that the heart pumps more blood to the muscles to meet the elevated oxygen requirements. Paced muscular work and work under heat stress are examples. Remedies lie primarily in the redesign of the job to protect the worker and job rotation (see Chapter 3).

5. *Psychomotor.* Problems that strain the psychomotor systems can best be dealt with by redefining the job requirements to suit human capabilities and providing job performance aids. Paced manipulative work with significant visual demands is an example of such tasks (see Chapters 4 and 8).

In many cases, a problem task may impose stress on multiple areas. Stress on each area must be analyzed first independently, and then considering interactions. In this way, one can mold a remedy package that resolves most of the problems. Education and training help, in general, to extend human capabilities and create awareness with respect to limitations. This is particularly important in avoiding obvious problem cases. For example, many back injuries can be prevented simply by educating workers in correct lifting methods.

1.10 ILLUSTRATIONS

Two examples will help illustrate the point for ergonomics in industrial environments. The first case is concerned with the programming of an automated guided vehicle (AGV) used for material handling at a textile plant. The vehicle followed a magnetic tape on the floor. There were 30 possible stop points on the path of the AGV. The stop-and-go action was controlled by 30 small toggle switches positioned next to each other on a control panel. The AGV stopped at a point on its path if the corresponding switch was set at the "up" position. The down position indicated "go." The plant was experiencing many inefficiencies in this handling operation. The problem was traced to programming errors with the switches (seemingly, a human error). The two underlying causes, however, were (1) the short lever arm of the toggle switch, and (2) poor color and brightness contrast between the switches and the metal plate on which they were mounted. Both the switches and the plate were silver in color. For the two reasons stated above, the programmer could not visually identify switch positions with perfect accuracy. Under the current incentive system, he was not motivated to feel the position of each switch either. Thus at any given run of the AGV, several switch positions did not necessarily correspond to the intended stop-and-go action. The plant experienced at least a 30% productivity increase in this particular operation after a simple ergonomic modification. All it took was the placement of small pieces of dark adhesive tape on the tips of switch lever arms, for better contrast with the background and improved visual discrimination of switch positions. Thus the operator could, at a glance, spot the switch positions. Cost of the modification was within $50, including the engineer's time. The benefits could be measured in the thousands of dollars.

The second illustration involves a riveting operation at a flashlight and lantern plant. An expensive machine was designed and custom built to automate the label-riveting operation on flashlight tubes. The task involved the placing of oval-shaped plastic labels (1 in. major axis length, $\frac{1}{4}$ in. minor axis length) on a rotating table (1 ft diameter) with very close fit reqirements. Labels were picked up from the table by suction and positioned on a tube delivered to the riveting station by a conveyer mechanism. Instantaneously, the label was riveted to the tube at both ends. The plant was experiencing many rejects on this operation. Again, the problem was traced to the task of the operator. Most of the rejects had the label missing. Thus it was evident that either the suction mechanism did not work properly, or the worker could not keep up with the speed of the rotating table. Visual observations led to the conclusion that the second alternative was true. It was simply impossible for the operator to keep up with the speed of the rotating table. Three adjustments helped turn what was an extremely costly operation into a profitable one. First, the angular speed of the turn table was reduced. Second, an operator who possessed better finger dexterity was assigned to the job. Third, the vendor was contracted to deliver labels packaged in proper alignment, eliminating many finger movements for orientation. After changes, the operation achieved close to a 50% output increase, with significant reductions in the reject rate.

The foregoing examples show how significant gains in productivity can be achieved via simple ergonomic modifications. They may also seem to be commonsense solutions to simple problems; however, it is simple interfaces that most designers overlook. Ergonomics is most effective if exercised during design. Frequently, modifications made later are much more costly and not as effective.

1.11 SUMMARY

Industrial ergonomics focuses on the working relationship between human beings and the equipment, machinery, and tools that they use in the industrial setting. Both manufacturing and service industries are included in the work domain. The purpose is to maximize the effectiveness of human functioning while minimizing the risk of injury and discomfort. An ergonomist or an engineer (mostly industrial engineer) deals with these concerns. Many government and other bodies participate in the work domain of the ergonomist. It is important that their objectives and control spans be clearly understood.

QUESTIONS

1. Why should one be concerned with ergonomics?
2. What is the other term for ergonomics that is commonly used in the United States?
3. Define *ergonomics*.
4. The term *ergonomics* was coined by two scientists. Who are they?
5. What is the central theme of ergonomics?
6. Ergonomics has several objectives. List and discuss.
7. What is the fundamental assumption of ergonomics?
8. List and discuss five major costs of ignoring ergonomics.
9. What significance did the industrial revolution play in the emergence of concern for people?
10. Discuss the differential emphasis on ergonomics research in the Scandinavian countries, the United Kingdom and the United States.
11. Define IEA and its objectives.
12. What is the general expectation from all designers of equipment and systems for human use?
13. List and briefly discuss three types of ergonomics problems.
14. Discuss HIX, HID and HIM.

15. Is there any accident/injury related justification for ergonomics projects? Discuss.

16. What are OSHA, NIOSH, MSHA, AIHA, APA, ILO, and WHO?

EXERCISES

1. Give an example from everyday life where designers have fit the task to the person. Give an example of where they have not. Discuss each in detail.

2. Write a five-page article on OSHA, NIOSH, and their interrelationships.

3. Briefly discuss the activities of the Human Factors Society since its inception in 1957. List the current technical groups and their activities.

4. Select one of the several types of ergonomics problems discussed in Section 1.9 and relate it to an everyday problem. Propose solutions.

REFERENCES

1. Polish Ergonomics Society. 1979. 7th International Ergonomic Association Congress. *Ergonomia*, 2(1).
2. Sanders, M. S., and McCormick, E. J. 1993. *Human Factors in Engineering and Design*, 7th ed. McGraw-Hill, New York.
3. Alexander, D. C., and Pulat, B. M. 1985. *Industrial Ergonomics: A Practitioner's Guide*. Industrial Engineering and Management Press, Atlanta, GA.
4. *AFSC Design Handbook*. 1977. *DH1-3*. Personnel Subsystem, January. Air Force Systems Command. Andrews Air Force Base, DC.
5. Diffrient, N., Tilley, A. R., and Harman, D. 1983. *Humanscale 1/2/3, 4/5/6, 7/8/9*. MIT Press, Cambridge, MA.
6. Van Cott, H. P., and Kinkade, R. G. (Eds.). 1972. *Human Engineering Guide to Equipment Design*. U.S. Government Printing Office, Washington, DC.
7. Woodson, W. E. 1981. *Human Factors Design Handbook*. McGraw-Hill, New York.
8. Centers for Disease Control. 1983. Leading Work-Related Diseases and Injuries—United States. *Morbidity and Mortality Weekly Report*, 32, pp. 24–26.
9. *National Occupational Research Agenda*, NIOSH. April 1996.
10. *Industrial Engineering*. 1989. 21(8), p. 82.
11. Pulat, B. M. and Alexander, D. C. 1991. *Industrial Ergonomics: Case Studies*. Industrial Engineering and Management Press, Atlanta, GA.
12. Eberts, R., and Eberts, C. 1995. *The Myths of Japanese Quality*. Prentice Hall PTR, New Jersey.
13. Pulat, B. M. 1984. Unpublished report. North Carolina A&T State University, Greensboro, NC.
14. Alexander, D. C. 1986. *Practice and Management of Industrial Ergonomics*. Prentice Hall, Englewood Cliffs, NJ.

CHAPTER 2

The Ergonomics Function

In this chapter we discuss business functional divisions and provide a framework of ergonomics organization. Various underlying concepts of ergonomics, such as physical work and effects on the body, information input and processing, anthropometry, stress, and the like, will be more readily accepted by readers who appreciate how these concepts cluster to make meaningful projects and who the important players are that need to get exposed to these projects. Equipped with this information, one can enhance his or her communication effectiveness with others in an organization.

2.1 BUSINESS FUNCTIONAL DIVISIONS

Before discussing various aspects of ergonomics involvement, it is beneficial to review the general functional structure of any business. Our review concentrates primarily on manufacturing businesses, since most cases relevant to industrial ergonomics work arise in such environments. However, the topics discussed are equally applicable to other types of industries, such as mining, health care, service, and the like.

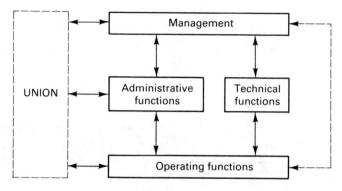

Figure 2.1 General functional structure in industry.

Figure 2.1 gives the general structure of major functions. Although variations are possible, in general, management, administrative and technical functions, and operating functions form a stack of three layers [1,2]. On top of the stack is *management*. Its primary function is to set business objectives consistent with company policy and market conditions. Management teams also develop macro plans to achieve these objectives with time reference. With these two functions, management provides leadership for both technical and administrative and operating functions. A third major task of management is to provide the manufacturing resources to meet the business objectives. Management does this by selective funding of strategic projects and hiring/layoff decisions. At this stage it should be clear to the reader that the *driving force in any business is based on the requirements and needs of the business.*

In the middle of the stack are the *technical and administrative* resources. The latter provide nontechnical support to the business plans as well as to the daily functioning of the production activity. Examples are human resources, accounting, payroll, and so on. On the other hand, business activity moves primarily along the technical backbone provided by the technical work force. These groups develop technical production plans consistent with the business plan. They acquire the necessary facilities, make process plans that make optimal use of the existing resources, and resolve daily production issues.

The *operating* groups are responsible for executing the daily functions of the production facilities. They function under the guidance of the technical plans and processes developed by the technical work force. Operating shops also target the meeting of weekly and monthly production goals and are responsible for making effective and nondestructive use of the production facilities.

The primary group of interest to the industrial ergonomics effort is the labor force functioning within the operating group. Although many production activities are automated, many others are still driven by human effort. Even automated functions require human involvement in the form of maintenance activity. The ergonomist's primary target is to design operating activities (tasks, functions, jobs) and physical facilities such that operators can function effectively with them and without undue stress. Perhaps the greatest help to an ergonomist comes from the labor union, if there is one. The union represents employee concerns at all levels of the organization, including management. Effective cooperation between union and management may bring about the implementation of many ergonomics projects that help the organization as a whole.

2.2 SELLING THE IDEA TO MANAGEMENT

Just as any other project would, ergonomics activity requires that the idea first be sold to management. There is no single most effective method of doing this, however. If there is no ongoing ergonomics activity in the organization, the formation of an ergonomics group or task force and obtaining management's backing need to be carried out in a step-by-step manner. The first step will concentrate on a detailed definition of business objectives and identifying the important players. The second step will define the objectives of the ergonomics group and the compatibility between its objectives and business objectives. As the third step, data could be collected as to operations being performed, available technical expertise, economic conditions, existence and strength of a union, and so on [3]. The fourth step is the correlation of the results of steps 1 and 2 as they relate to data collected in step 3. Here the objective is to prepare the most effective presentation to the right parties, at the right time, with the right concerns to be addressed.

Even if there is already ongoing ergonomics activity in the organization, careful analysis is necessary to implement its major projects. The frequent reason for unsuccessful or incomplete implementation is lack of acceptance [4]. This may come from management, peer professionals, or end users.

Justifying ergonomics projects.   No matter what the specific situation, any ergonomics project could be justified on the basis of the following [5]:

1. *Productivity increase.* This should be linked to cost reduction or increase in output with a constant cost structure. If low productivity is the current situation, state how ergonomics will improve worker efficiency. If significant injury cost is the case, cite how ergonomics involvement will notably reduce accident potential, thus injuries, and thus costs. If operators are committing too many errors, discuss how the restructured workplace for the job will reduce error potential, thus rejects, and hence costs. If workers are unproductive or frequently absent, describe how they will enjoy working in the new environment.

2. *Social Responsibility.* This could span a range from appeals to the social consciousness of the firm to the legal responsibilities. Potential penalties associated with safety violations or product liability suit potential may also be mentioned [6]. Such expenses will undoubtedly result in increased operating costs.

Planning the presentation.   Once the strategy for management conviction is determined, the next step is to plan the presentation. This step involves gathering the tools of conviction, including literature, studies conducted at other facilities (friendly and unfriendly), empirical evidence from similar applications already implemented in the organization, consultant views, and so on [7]. A very strong point can be made if ergonomics is already working in the plant with significant results. If no significant work has been conducted before, several pilot studies will help. Results of these studies could be of benefit during the presentation. Again, do not overlook applied studies reported in the literature. It is impossible to cite all of this work in this chapter; however, the interested reader may refer to [8–10] as a starting point.

Making the presentation.   If a thorough preparation has been done, things should go smoothly during the presentation. Make sure to define ergonomics in very simple terms as the first step. After presenting the background, applications at other facilities, and the proposed application(s), make sure that the various concerns of each level of management are addressed. For example, show the expected impact of the implementation(s) on long-term disability payments to upper management. It is important to invite the key people to the presentation. Make sure that management representatives are present. Their commitment is extremely important.

2.3 FORMING A GROUP

An ergonomics group may either be formally assembled by management or may start from a grass roots movement. The group could assume a permanent or temporary role. An ergonomics task force within the industrial engineering department is an example of a temporary formation. Such groups will investigate a specific problem with a planned attack, meeting regularly, and making a final presentation after the project objectives have been met. An ergonomics task force assembled for a temporary role very frequently will turn into a permanent one. Naturally, the most important factor is the severity of the problems. A task force may also be organized across various departments, including medical, safety, manufacturing, management, engineering, and wage representatives. Frequently, a union (if there is one) representative will also be a part of the group [11]. Such multidisciplinary groups get together routinely to discuss ergonomics problems and develop solutions.

A grass-roots movement to organize and support a group will face more hurdles to be overcome. It will have to go through all steps of "selling the idea to management." Although it may take longer for the movement to be recognized, skillful handling, good preparation, and severity of problems to be tackled will speed up the process.

Size of the group.   Group size depends on many factors, such as degree of management support, severity of problems, and the like. Unless a formal multidisciplinary approach has been taken, few groups will start with more than one professional. Frequently, the ergonomics responsibility is given to one professional on a part-time basis. As awareness is elevated in the firm, there may be internal or external expansion of the effort. Internal expansion refers to adding personnel to the function. External expansion refers to educating other professionals in the field, especially designers and engineers. Probably the most effective method is external expansion [12].

Types of work.   The group concerns itself with two types of work. First and most prevalent will be the corrective work. At any given point in time, many cases will demand attention. Ergonomics professionals will rank order these in terms of severity and other criteria, and will develop and implement corrective processes or work areas. Second and less prevalent is preventive work. Routine review of existing or planned designs may reveal modification requirements. Although industry spends less time on this type of work, it is the more important type and should be the most prevalent. The ergonomics group should not only evaluate existing designs, it should carry out ergonomic reviews on all planned workspace, work methods, and product and process designs.

Responsibilities. The group will have jurisdiction over several areas. First is the safety and health of the workers. It is widely recognized that maintaining the well-being of employees is the primary function of an ergonomist. By the same token, the group will place a primary emphasis on the avoidance of accidents and injuries. Second in line are performance issues. Poor job and workplace design results in degraded worker performance, if not in accidents and injuries. Management will be very receptive to ideas that enhance human and thus system performance, reducing costs and improving quality. Third in line will be the projects that span quality of work life. Although this type of effort will produce mostly intangible results, satisfied and motivated workers will be greatly noted and appreciated by management, especially in a unionized environment.

Available help. The group may obtain help from various resources in carrying out its work. First and foremost is the available literature. University and/or company libraries may be excellent sources. A small ergonomics library may also be established containing primarily design handbooks [13–15]. Advisors, including company professionals who have had significant experience with the types of problems at hand and outside consultants (from universities or consulting firms), are secondary sources of help. They could be contracted for specialized work. A third source of help is the available equipment. Data-gathering and data-analysis equipment such as audiometers, portable physiograph systems, tape measures, photometers, sound-level meters, timers, and personal computers will help the group perform investigations with confidence and objectivity [16].

2.4 SUSTAINING THE EFFORT

The ergonomics effort must be placed into a long-term program posture, not into a fighting for survival mode. This requires careful management of resources, developing directions for growth, and building information networks. It is important to emphasize "management," not simply "overseeing" the activity. A passive role such as just waiting in the office for a phone call is asking for the program to die out. Keep in mind that many similar programs are being pushed simultaneously with the ergonomics program. One has to maintain pressure to keep the attention focused on the ergonomics program.

Several prerequisites of sustaining the effort are discussed below. These will propel the program, after which careful management must be exercised.

Develop a background. This step will involve three major elements. First, one must establish credibility. Users of the products (new designs, improved designs, critical evaluations) of the ergonomics group must feel that there really is something to be gained.

The work should be based on sound scientific principles with as much objectivity as possible. Occasionally, basic research results that have been used in the ergonomic analyses may be shared with the users. In addition, occasional use of consultants will indicate that the group is serious about use of the best available resources to attack problems. Second, education of others is a definite requirement. Not only should awareness be created, but a level must be reached where equipment and workplace designers *accept* ergonomics as part of their daily work. However, one must be careful here. The designer should not be stressed to the point of frustration. Ergonomics should not be forced upon anybody. It should be diffused into the culture. The designer should feel that the resulting design will not be complete without a thorough ergonomic review. Third, an information network must be created. Collection of accident and injury data, employee concerns, management concerns, and other relevant data; processing these; and passing results and suggestions to the concerned require that a reliable and effective network be established and maintained. This will establish the communication medium, the lack of which will isolate the program, leading to its eradication.

Value return on investment. It is a fact that ergonomists follow their convictions, with the primary objective of maintaining and improving human well-being. Many results of this effort may be intangible. However, keep in mind that profitability of the business is a primary concern to management. Maintaining top management's support requires that the tangible gains be documented. Cost reduction through reduced disability payments, insurance premiums, and liability settlements should be sought. Cost avoidance through prevention of accidents, human suffering, and property damage should be documented. Estimated training costs for employee turnover replacements is another category to be pursued as cost avoidance. Productivity enhancement via improved output is relatively easy to pinpoint. This may take several forms, such as shortened duty cycle time, fewer rejects due to fewer errors, and more time spent on the job due to lower absenteeism and turnover. Time spent on documenting savings will pay off later when ergonomics projects are funded by management.

Establish customer satisfaction. Everybody will agree that a satisfied end user will be of notable help to an ergonomics group in maintaining and expanding a program. A material handler who is no longer feeling a sore back, and a clerk who no longer feels a strange numbness in the fingers, will tell the story to other workers, union representatives, and supervisors. There simply is no more effective way to establish credibility than to have the customer testifying to it. On the other hand, a worker who feels the same discomfort after ergonomic intervention will not help the cause. *Success stories are the best means to sustain a program.* A method to document such stories may be to conduct opinion surveys before and after modifications. Such surveys will also help manage the program effectively with objective data. Videotapes of a job before and after an ergonomic redesign with the affected employee taking part in it and telling a success story can be taken to meetings, funding conferences, and other gatherings for publicity purposes.

Assume a design reviewer role. As we mentioned earlier, prevention-oriented work is much more effective than corrective approaches. Although at the initial stages of an ergonomics program the ergonomist will be involved primarily with carrying out corrective work, this activity will establish the credibility needed as well as develop the necessary information network. The next step, though, is to get into a posture such that major designs are routinely routed through the ergonomics group to establish their concurrence. Naturally, if the education step has been successful, the designer will have already exercised ergonomics prior to the formal group review. However, it is not unwarranted to have the design reviewed again by the specialists. Although this seems to lengthen the design cycle time, a project that makes significant contributions to factory operations will need to be done properly the first time. On smaller-scale designs, the ergonomist may assume an advisory role with occasional audits.

Ergonomics efforts prosper when designers feel that they constitute an integral part of their work and management feels that they are worthy of support. It is up to the ergonomist(s) to build such efforts into the culture.

2.5 INDICATORS OF NEED FOR ERGONOMICS ATTENTION

Ergonomics is most effective if applied during the design stage. However, routine audits are effective in identifying jobs that need ergonomics attention. Over the years, several tables have been published that list symptoms of ergonomic problems [17,18]. These may be a good starting point for a comprehensive ergonomics review program. The following list provides a summary of indicators for ergonomic review:

1. There is an unacceptably large number of rejects on the operation.
2. Operators are making frequent mistakes on the job.
3. There is high material waste.
4. Labor turnover on the job is excessive.
5. Employees frequently complain about job requirements.
6. Production output is unacceptably low.
7. Absenteeism is unacceptably high.
8. There are accidents or near-accidents on the job.
9. Employees frequently visit the infirmary.
10. Product quality is low.
11. Training time is unacceptably long.
12. Employees take frequent rest periods.
13. Personal and fatigue allowances are too high.
14. Operator assignment to the job is limited by size, sex, age, or physique.
15. There is interference with verbal communication.
16. There are requests for transfers to other jobs.
17. Employees cannot meet production standards.
18. Employees are frequently away from the workplace.
19. Breaks seem excessively long.

Significant involvement with modifications is necessary when:

1. Employees are being injured.
2. Production volume is high.
3. Labor turnover is high.
4. The workplace is utilized for more than one shift.

2.6 CASE ILLUSTRATION 1

As an illustration, Lucent Technologies, Inc. supports both central and distributed philosophies in ergonomics focus. The corporate safety and health group specializes in overall support to all facilities of Lucent, including international joint venture locations. Training of all designers and factory engineers is another responsibility of the corporate group. From time to time, the corporate personnel may develop position statements on various topics of interest, such as VDTs and cumulative trauma disorders (see Chapter 3). All organizations of the company function under applicable policy statements (see example in Figure 2.2).

The distributed working groups provide specific factory or function support. For example, handling of hazardous material including waste may be the responsibility of several field units, while factory ergonomists and safety and health engineers provide specific support to local operations. Activities of these groups and results of programs are routinely monitored by the corporate group for consistency, detail, and modification if necessary.

2.7 SUMMARY

The ergonomics function has benefited many organizations, while others have not implemented relevant programs. Usually, the first time a facility gets involved in an ergonomics program is when collective injuries or complaints are forwarded to management. It is important to begin preventive work before major mishaps occur. A grass-roots movement may be successful if several good applications are developed and benefits clearly shown. Although reactive (after-the-fact) programs are common in industry today, the best approach is proactive (before the occurrence). Every designer (workplace, work methods, products) must exercise ergonomics as an integral part of the design activity.

QUESTIONS

1. What are the three major layers of a business organization?
2. What is the primary function of management?
3. What role do unions play in the application of ergonomics projects in an organization?
4. On which bases could ergonomics projects be justified?
5. What types of work does an ergonomist or ergonomics group get engaged in?

Lucent Technologies

environmental, health and safety policy

Lucent Technologies is committed to protecting the environment and the health and safety of our people, our customers, and the communities where we operate. Meeting this commitment is a primary management objective and the individual and collective responsibility of all Lucent employees worldwide. To that end, we shall:

> **comply** with all applicable environmental, health and safety laws, regulations and Lucent's Global EH&S standards

> **establish** management systems for environment, health and safety based on recognized standards, and set company-wide goals for continual improvement

> **integrate** environment, health and safety into our business plans and decisions -- including in the design, production, distribution and support of our products and services

> **ensure** that our products are safe, and work with suppliers and customers to promote responsible use throughout their life cycle

> **reduce** environmental impact and conserve natural resources by minimizing waste and emissions, reusing and recycling material, and responsibly managing energy use

> **motivate and prepare** all employees to take personal accountability for protecting the environment and creating a safe and healthy workplace

> **be a leader** in deploying and promoting innovative, cost-effective environmental, health and safety technologies and procedures both within and outside the company.

We will regularly review and improve this policy, communicate it to all employees, and make it available to all stakeholders.

Henry Schacht
Chairman and
Chief Executive Officer

Richard McGinn
President and
Chief Operating Officer

John Pittman
Vice President, Chief Quality,
Environment, Health and Safety Officer

Figure 2.2   Environmental, health and safety policy. Used with permission.

6. List and discuss the kinds of help available in carrying out ergonomics projects.

7. What are some of the steps suggested for focusing attention on ergonomics?

8. List and discuss seven indicators for ergonomics attention.

EXERCISES

1. Discuss various conflicts between the major parties (management, union, etc.) of an organization with respect to ergonomics projects.

2. How can you get a production supervisor's attention for an improvement project in his or her area? Assume that the supervisor is not familiar with ergonomics.

3. If you were to present a 3-hour general awareness course to all supervisors at a textile plant on ergonomics, what would you cover?

4. You have been given the go-ahead to form a two-person team to spearhead ergonomics programs in a Fortune 500 company specializing in wholesale grocery distribution. How would you allocate functions to the two members? What would you do to maintain your programs?

5. Assume that a project requires skills that do not exist within your company for the completion of an ergonomics project. How and from where would you get help?

REFERENCES

1. Amos, J. M., and Sarchet, B. R. 1981. *Management for Engineers.* Prentice Hall, Englewood Cliffs, NJ.

2. Peters, T. J., and Waterman, R. H., Jr. 1982. *In Search of Excellence.* Harper & Row, New York.

3. Doxie, F. T. 1988. Committee Approach to Ergonomic Cases. In *Trends in Ergonomics/Human Factors V.* Aghazadeh, F. (Ed.). Elsevier, Amsterdam, pp. 1009–1016.

4. Wick, J. L. 1988. Implementing Ergonomics Projects. In *Trends in Ergonomics/Human Factors V.* Aghazadeh, F. (Ed.). Elsevier, Amsterdam, pp. 1017–1021.

5. Smith, L. A., and Smith, J. L. 1982. How Can an IE Justify a Human Factors Activities Program to Management? *Industrial Engineering,* 36. February, pp. 39–43.

6. Smith, L. A. 1985. Selling the Idea to Management. In *Industrial Ergonomics: A Practitioner's Guide.* Alexander, D. C., and Pulat, B. M. (Eds.), Industrial Engineering and Management Press, Atlanta, GA.

7. Alexander, D. C. 1986. *The Practice and Management of Industrial Ergonomics.* Prentice Hall, Englewood Cliffs, NJ.

8. McAtee, F. L., Jr. 1988. Practical Ergonomic Applications. In *Trends in Ergonomics/Human Factors V.* Aghazadeh, F. (Ed.). Elsevier, Amsterdam, pp. 1023–1026.

9. Klym, M. P. 1988. Practical Applications in Ergonomics. In *Trends in Ergonomics/Human Factors V.* Aghazadeh, F. (Ed.). Elsevier, Amsterdam, pp. 1037–1043.

10. Zimmermann, R. L., and Shaull, J. E. 1989. Hand Tool Ergonomics: The Operator's Story. In *Advances in Industrial Ergonomics and Safety.* Mital, A. (Ed.). Taylor & Francis, London, pp. 257–266.

11. Longmate, A. R., and Welker, C. 1985. Components of an Industrial Ergonomics Program: The Johnson and Johnson Experience. In *Industrial Ergonomics: A Practitioner's Guide*. Alexander, D. C., and Pulat, B. M. (Eds.). Industrial Engineering and Management Press, Atlanta, GA.

12. Alexander, D. C. 1985. Ergonomic Group Organization. In *Industrial Ergonomics: A Practitioner's Guide*. Alexander, D. C., and Pulat, B. M. (Eds.). Industrial Engineering and Management Press, Atlanta, GA.

13. *AFSC Design Handbook*. 1977. *DH1-3*. Personnel Subsystem, January. Air Force Systems Command. Andrews Air Force Base, DC.

14. Salvendy, G. (Ed.). 1987. *Handbook of Human Factors*. Wiley-Interscience, New York.

15. Woodson, W. E. 1981. *Human Factors Design Handbook*. McGraw-Hill, New York.

16. Pulat, B. M. 1985. Ergonomics Equipment. In *Industrial Ergonomics: A Practitioner's Guide*. Alexander, D. C., and Pulat, B. M. (Eds.). Industrial Engineering and Management Press, Atlanta, GA.

17. Pulat, B. M. 1985. Summary. In *Industrial Ergonomics: A Practitioner's Guide*. Alexander, D. C., and Pulat, B. M. (Eds.). Industrial Engineering and Management Press, Atlanta, GA.

18. Alexander, D. C. 1986. *The Practice and Management of Industrial Ergonomics*. Prentice Hall, Englewood Cliffs, NJ, pp. 57–61.

CHAPTER 3

Physical Ergonomics

After an introductory chapter on the general concepts of ergonomics and a chapter on the position of ergonomics in an industrial concern, in this chapter and Chapters 4 and 5 we provide the theoretical bases of the discipline. Discussions that follow in Chapters 6 through 11 are all based on the foundations established in Chapters 3 through 5.

3.1 HUMAN ACTIVITY CONTINUUM

Human beings are engaged in many different types of activity (work related, leisure, sports, etc.) every day. As a result of these, the body undergoes various changes: physiological, psychological, cognitive, and so on. The terms used to distinguish between the outside effector and the resulting change in the body are *stress* and *strain,* respectively [1]. The intensity of the activity determines whether or not the strain is within acceptable limits. In general, there is no long-term wear and tear on the body if the work intensity is within acceptable limits. On the other hand, one can expect long-term effects—if not an instantaneous effect—if stress is outside this range. An example of a single-incident case is electrical shock that results in immediate incapacitation or death.

A human activity continuum can be developed for the purpose of categorizing and quantifying strain. This is important, since decisions as to whether or not a task is stressful can most effectively be based on data collected from the body. The effects on the body are more important to an ergonomist than what is taking place around us. Figure 3.1 gives one such continuum. On the left of this range are the physical activities. Here the primary emphasis is on muscular functioning. Although other elements (psychomotor, cognitive) are not totally absent, their contributions (or use) to the task are on a greatly reduced scale. A good example of dynamic muscular activity is exercise with a bicycle ergometer. Moving toward the right on the continuum, one encounters static muscle loading. An example is that of holding a piece part on an assembly line while applying pressure on it. At the center of the scale are psychomotor activities which involve light muscular work with loading on the sensory systems. An example is light assembly work. At the extreme right, one finds mental tasks that primarily load cognitive functions with secondary loading on sensory mechanisms. An example is a clerk's task involving mental arithmetic. Again, the muscular element in the activity is not absent. Its contribution is minimal.

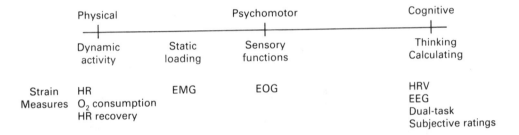

Figure 3.1 Human activity continuum and the primary measures of strain.

Any industrial task will fall somewhere on this continuum with respect to its demands on the body. It is also possible to have an activity sampling from various points on this scale. That is, it is possible to have a given activity broken into its constituting elements, with several elements classified primarily as muscular, several as psychomotor, and several as cognitive. In these cases, each element's requirements must be evaluated first independently and then collectively with other task elements. Overload periods on a time scale can thus be determined with overall activity evaluation. Job performance aids may help bring overload conditions to within acceptable limits.

Strain measures listed under each activity type in Figure 3.1 are discussed later. These have been researched over decades and are the best available methods of evaluating stress objectively in industrial environments. Some are more easily evaluated in the field than others. An activity that cannot be evaluated effectively in the field can be simulated in the laboratory and evaluated there.

Stress and performance. It should be mentioned here that stress does not always lead to ill effects. Research [2,3] shows that there is an inverted U-shaped relationship between it and performance (Figure 3.2). Human productivity is low at low levels of stress. That is, when there is little stress, performance will be less than optimum. This is evident in the fact that one tends to go to sleep when there is little to do. Moving right on the horizontal axis, performance will get better with increasing levels of stress. At some point, performance will be optimum. This point depends on many factors, including the type of stress agent and the existence or lack of other stressors. For any given situation, the optimal point can be found experimentally.

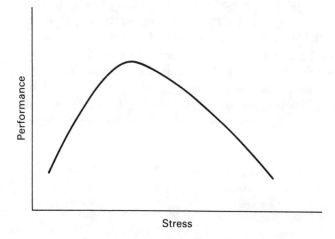

Figure 3.2 Relationship between stress and performance.

Figure 3.2 Relationship between stress and performance.

Still moving right, one encounters increasing levels of stress, which result in drops in efficiency. This is the overload condition. There simply is too much for the person to do or to attend to. Forgetting to attend a meeting or displaying suboptimal performance on one project when there are too many projects at hand are examples. The conclusion from all this discussion is that *human beings always need some stress* to be proficient on the job. Primary sources of job stress are:

1. Arrangement of the workplace (dimensions, relative locations, etc.)

2. Work content (speed and accuracy requirements, etc.)

3. Work organization (work-rest cycles, start-stop times, etc.)

4. Work environment (co-workers, supervisors, heat, cold, etc.)

Naturally, the total stress acting on the body during work is a combination of stress due to work and residual stress due to non-work-related activities (problems with children, finances, and the like).

3.2 MUSCULAR ACTIVITY

The main thrust of this chapter is to provide the underlying concepts of muscular activity. Sensory and mental work are discussed in Chapter 4. After a general discussion of human activity (Section 3.1), we look at general muscular activity and its evaluation. Sections 3.3, 3.4, and 3.5 focus on specific physical work categories/groups.

3.2.1 The Musculoskeletal System

The musculoskeletal system provides the primary components for muscular activity. It is composed of the muscles, bones, and connective tissue. Metabolism is needed to provide the necessary energy to the musculoskeletal system.

Bones. There are 206 bones that establish the framework of human structure. If there were no bones in the body, it would simply be a mass of flesh. Bones give the body its structure, and together with the muscles, they provide its mobility, an ingredient necessary for human activity. Some bones simply provide cover to vital organs, such as the skull for the brain and the rib cage for the lungs. Others provide the basis for activity, such as the bones of the upper and lower extremities. Bones are connected to each other via ligaments. Muscles are connected to bones via tendons. Neither the bones nor the ligaments can be stretched or contracted. Bones are connected to each other at joints. There are several types of joints: hinge joints (wrist), pivot joints (elbow), and ball-and-socket joints (shoulder, hip). Much greater forces can be applied through ball-and-socket joints than through hinge joints.

Mobility at the joints is limited by their design, by cartilage, by muscles, and by gender. Staff [4] researched the voluntary (unforced) mobility in 32 major body joints. The subjects in this study were 100 women, and the study was carefully controlled to resemble an earlier study on 100 men. An electric bubble goniometer was attached to each joint on each subject to measure the relevant motion. The subjects were to move their limbs "as far as comfortably possible" using the dominant limb. A total of 24 measurements showed significantly greater mobility by females than by males. Men showed superior mobility only in ankle flexion and wrist abduction. This finding is consistent with earlier findings of motion-range superiority exhibited by women.

Bones change their shape, size, and structure depending on the mechanical demands placed on them over time. In 1892, Wolff [5] expressed the relationship between such changes and mechanical loads. According to *Wolff's law,* the bone is deposited where needed and resorbed where not needed.

Muscles. Muscles are among the major prerequisites of human activity. Among the several different types of muscles, the skeletal (or voluntary) muscles are of most concern to an ergonomist. Muscles are composed of bundled muscle fibers. The larger the muscle is (in terms of its cross section), the larger the forces that can be applied with it. For mechanical leverage action, the muscle applies forces on the bone(s) to which it is attached. The muscle contracts in order to generate forces. It can generate maximum force when it is in an extended state. A contracted muscle can apply little force.

A muscle generates mechanical work by converting chemical energy into mechanical energy. It is believed that the thin filaments (composed of actin, troponin, and tropomyosin B proteins) and the thick filaments (myosin) slide over to bring about contraction (Figure 3.3). Motor nerves regulate muscular activity, a subject dealt with in Chapter 4.

Metabolism. Metabolism supplies the necessary energy to the musculoskeletal system. It is the chemical process of conversion of food into mechanical work and heat.

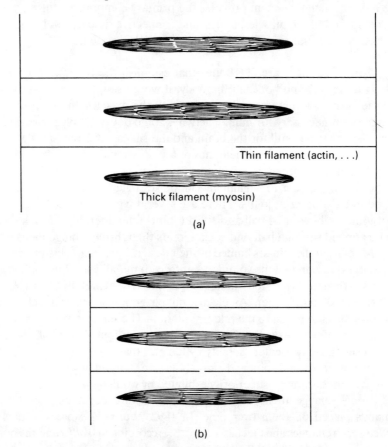

Thin filament (actin, . . .)

Thick filament (myosin)

(a)

(b)

Figure 3.3 Two states of muscle fibers: (a) stretched; (b) contracted.

Some mechanical work is consumed internal to the body, and some is consumed for physical activity. Excess heat is dissipated.

The basic source of energy for the contraction of muscle is glycogen or glucose, which is abundant in blood. However, this is not the initial source of energy. At the beginning of muscular activity (the first 3 to 5 seconds), adenosine triphosphate (ATP), a high-energy phosphate compound available in the muscle tissue, is mobilized (Table 3.1). Its breakdown into adenosine diphosphate (ADP) releases significant amounts of energy.

To continue the muscular activity, the ATP must be regenerated (i.e., the battery must be recharged). The first source of energy for recreation of ATP is creatine phosphate, another high-energy phosphate compound, already present in the muscle tissue in small amounts. Its reaction with the free ADP releases creatine and ATP. After depletion of creatine phosphate (15 seconds), if the activity is continuing, the blood glucose or glycogen is mobilized. Glucose, an important blood sugar, is circulated to the muscle tissue through the permeable capillary structure. It is converted by various stages first into pyruvic acid [6]. Further breakdown may take two possible routes:

1. *Anaerobic work.* If sufficient oxygen is not being supplied to the muscle, pyruvic acid is converted into lactic acid while ATP is regenerated. Lactic acid accumulation between muscle fibers causes muscular fatigue and pain to develop.

2. *Aerobic work.* Under sufficient oxygen supply, the pyruvic acid is broken down into water and carbon dioxide, releasing large amounts of ATP. This is a more efficient reaction than that in anaerobic work.

TABLE 3.1 ENERGY EXCHANGE IN THE MUSCLE

$ATP \rightleftharpoons ADP + P + $ free energy	Generate energy
creatine phosphate $+ ADP \rightleftharpoons$ creatine $+ ATP$	Re-create ATP
glucose $+ 2$ phosphate $+ 2 ADP \rightarrow 2$ lactate $+ 2 ATP$	anaerobic work
glucose $+ 38$ phosphate $+ 38 ADP + 6 O_2 \rightarrow 6 CO_2 + 44 H_2O + 38 ATP$	aerobic work

Source: Ref. [6].

The oxidation of pyruvic acid in aerobic work is a very complex process involving enzymes, co-enzymes, and fatty acids. This process made of a cycle of events is known as the Krebs Cycle [7]. The stage is set with pyruvic acid converting into an intermediate compound acetyl CoA which then combines with oxaloacetic acid to form citric acid (Figure 3.4), the same six-carbon compound found in citrus fruits. The citric acid converts to ketoglutaric acid to succinic acid and back to oxaloacetic acid with many byproducts. As a result of this process, two molecules of carbon dioxide and two molecules of water are formed from each molecule of pyruvic acid.

Oxygen is the key to efficient work. Its supply to the muscle fibers requires that more blood be pumped to the muscle per unit time as well as heavier breathing in order to oxygenate more blood through the respiratory system.

The most common measure of energy requirement for physical activity is the *kilocalorie* (kcal). Typically, the resting energy required to maintain the molecular structure and electrical potentials in the muscle is 0.3 kcal per minute for a man weighing about 70 kg (154 lb) [8]. Other functions of the body also require energy. A resting person lying down with no digestive activity will require about 1700 (man of 70 kg) or 1400 [woman of 60 kg (132 lb)] kcal per day (24-hour period). This minimum steady energy consumption to maintain involuntary activity (circulation, etc.) is known as the *basal metabolism*. *Total metabolism* is the energy consumption of an active person as an algebraic sum of basal metabolism, digestive metabolism (energy consumed by the digestive system), and activity metabolism. Digestive metabolism can be estimated at 10% of the sum of basal and activity metabolism. Activity metabolism is discussed in Section 3.2.3.

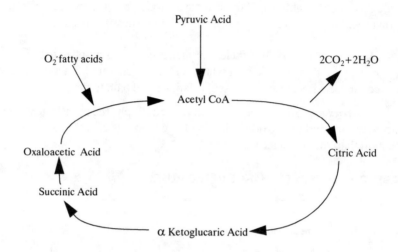

Figure 3.4   Krebs Cycle (simplified)

3.2.2 Supporting Systems

The musculoskeletal system is supported by other systems in the body during work and rest. The nervous system is described in the next chapter. Other important support systems are the respiratory, nutritional, and circulatory systems.

Respiration.   The respiratory system is composed of many organs, including the nose and mouth, pharynx, larynx, and trachea; and the lungs, consisting of bronchi, bronchioles, alveolar ducts, alveolar sacs, and alveoli. The primary function of the respiratory system is to expose a large volume of water-saturated air to a large volume of blood to facilitate rapid exchange of gasses between blood and the inhaled air. It is inherently linked to the circulatory system, which carries gasses, water, electrolytes, etc. to tissues and cells. The term respiration denotes interrelated processes: absorption of oxygen and elimination of carbon dioxide, transportation of oxygen and carbon dioxide, use of oxygen and release of carbon dioxide by cells and the interstitial fluid, and lastly it denotes the roles of these two gasses in the metabolic processes.

The inspired air passes through two bronchi, the primary passages to each lung. By this time, the inspired air temperature is adjusted to body temperature and also moistened in order not to injure the delicate lung structures. The airways end at alveolar sacks and alveoli. There are more than 300 million alveoli in the lungs, and they provide the vital surface for gas exchange; 70 to 90 m^2 for average adult. Each alveolus is covered by a capillary network of around 2,000 segments. The gas exchange between pulmonary air and the blood in capillaries is achieved by diffusion. At rest, 250 ml of oxygen is absorbed per minute through the alveoli and 200 ml of carbon dioxide is expelled. The diffusion is made possible by partial pressure difference in oxygen and carbon dioxide between the blood and the air. In the lungs and tissues, gas molecules move from an area of higher pressure to an area of lower pressure. The venous blood contains oxygen at a partial pressure of 40 mm Hg and carbon dioxide at 46 mm Hg. The average alveolar oxygen partial pressure is 103 mm Hg and the carbon dioxide partial pressure is 39 mm Hg. This partial pressure differential causes the carbon dioxide to diffuse from the blood to the pulmonary air and oxygen to diffuse and dissolve through the alveolar membrane into the blood.

Nutrition. The process of nutrition involves ingestion, digestion, absorption, and utilization of nutrient substances. Proper nutrition is the foundation of physical activity providing the necessary fuel and the chemicals. Proteins, carbohydrates, and fat are the main nutrients that supply energy for muscular work, although their contribution to the energy liberation process in the muscle cell is not equal. Certain vitamins and inorganic compounds are also necessary for nutritional balance and efficient energy transfer.

The body cannot make use of the ingested food directly. It has to be converted into a usable form by digestion. This takes the form of a combination of mechanical and chemical processes. In the oral cavity, the food is broken and mixed with saliva, which splits starch into maltose by the action of enzymes present in it. The tongue pushes the chewed food to the stomach through the pharynx and the esophagus. The stomach glands in the mucosa secrete gastric juices which dissolve the food and break the proteins down as well as cause the emulsification of fats. The small intestines in turn digest the food through the action of enzymes, which break down the carbohydrates, fat, and proteins into molecules that can be absorbed into the body fluids (blood or lymph) through the intestinal cells. These nutrients are carried to distant tissues via the vast vascular supply to the gastrointestinal (GI) tract. This supply also delivers oxygen and nutrients to the GI tract, using some of the nutrients within the intestinal lymphatic system. Most of the secretions entering the small intestines are produced by the intestinal mucosa; however, the secretions of the liver and the pancreas also enter the small intestine and play an important role in digestion. The length of the small intestines is about 6 meters. Normally 18 to 24 hours are required for material to pass through the large intestine, whereas 3 to 5 hours are sufficient for movement in the small intestines. The large intestine primarily converts chyme into feces to be expelled from the body through the anus. This process involves many microorganisms, mucus and the absorption of water and salt [9].

Circulation. The circulatory system carries oxygen from the lungs and nutrients from the GI tract to the cells, muscles, etc. It also returns metabolic byproducts (carbon dioxide, heat, water, and waste) to be dissipated. The heart acts as the double pump for the system. Blood vessels outside the heart can be categorized into two classes: (1) the pulmonary vessels, which transport blood from the right ventricle of the heart through the lungs and back to the left atrium of the heart, and (2) the systemic vessels, which transport blood from the left ventricle of the heart to all other parts of the body, including the head and the lower body, back to the right atrium. The pulmonary circulation and the systemic circulation together make up the peripheral circulation. The kidneys remove waste products, many of which are toxic, from the blood. Although kidneys are the primary organs of excretion, liver, lungs, the skin, and intestines also eliminate waste.

Arteries are vessels that carry blood away from the heart. Arteries branch repeatedly to form smaller arteries, and the arterioles. Blood flows from the arterioles into capillaries. It is at the capillaries that the exchange occurs between the blood and tissue fluid (interstitial fluid). Capillaries have thinner walls, and blood flows through them more slowly. There are far more capillaries than any other blood vessel type.

Within the capillary, the net hydrostatic pressure at the arteriole side may be 40 mm Hg. This pressure drops gradually to 10 mm Hg on the venous side. The net colloid-osmotic pressure (on the capillary wall from outside-interstitial fluid) around the capillary is about 25 mm Hg. Hence, the net filtration pressure on the arteriole side is 15 mm Hg (40 minus 25) giving an outflow (proximal) of vascular fluid into the interstitial and cell areas. On the venule side, the net pressure is 15 mm Hg (25 minus 10) giving an inflow or suction of fluids back into the capillary. These pressure differentials allow the exchange of gasses, nutrients, and waste products between the capillaries and the surrounding areas [7].

From the capillaries, blood flows into venules, which combine to form veins. Veins carry blood toward the heart. As veins get closer to the heart, they increase in diameter and decrease in number. Valves prevent the backflow of blood in the veins.

Blood flow through a tissue is usually proportional to the metabolic needs of the tissue. The nervous system is responsible for routing the flow of blood except in the capillaries and precapillary sphincters. The nervous system is also responsible for maintaining blood pressure. The sympathetic nervous system controls blood vessel diameter. There is also some local control of arterial vessel diameter as a function of metabolite concentration in the muscle.

Approximately 30 liters of fluid pass from blood capillaries into the interstitial areas each day. The corresponding figure from interstitial spaces back into the blood capillaries is about 27 liters. The remaining three liters of fluid enter the lymphatic capillaries. This fluid is called lymph, and it passes through the lymph vessels to be emptied into the blood through the right and the left subclavian veins [9].The lymphatic system also absorbs fats and other substances from the gastrointestinal tract.

3.2.3 Types of Muscular Effort

Ergonomists differentiate between two types of muscular effort for the purpose of correct evaluation of physical demands of work from the body:

1. *Dynamic effort.* This type is characterized by rhythmic contraction and relaxation of the muscles involved. An example is turning a handwheel to open a valve. Alternated tension and relaxation allow more blood to be circulated through the muscle than when in the resting state. Thus both oxygen needs and waste removal needs are met effectively.

2. *Static effort.* In comparison, static effort is characterized by a prolonged state of contraction, which restricts blood flow to the muscle tissue. Neither oxygen supply nor waste removal needs can be met. Examples are: holding a box in static posture and applying continuous pressure on a part to maintain its position. The muscle bulk experiencing static loading will quickly deplete the reserve ATP and creatine phosphate. Since no oxygen or glucose are being received, this type of activity will not last long. Severe muscular pain will also develop due to waste products, including lactic acid, accumulating in the muscle tissue.

Compared to dynamic loading, static effort will require longer rest periods. It will also result in more employee complaints and turnover. Designers must make sure to minimize job elements in a process that requires static loading.

3.2.4 Energy Cost of Work

With the onset of physical work, energy demands increase. The amount of increase depends on several factors, such as physical conditioning of the body, intensity of activity, sex, and body weight. To provide the needed oxygen and remove the accumulated waste, vascular activity in the muscles increases. More blood is pumped to the affected areas. Oxygen consumption increases. Breathing accelerates to oxygenate more blood per unit time in the lungs. Heart rate increases since more blood is processed in this organ per unit time. There is increased heat production in parallel to elevated energy generation for muscular activity. Sweating occurs to dissipate the excess heat. Blood pressure increases. To an ergonomist these are the most important changes in the body in reaction to work [10]. Thus stress due to a physical activity can be evaluated using one or preferably a combination of measures of these changes. Oxygen consumption and heart rate are the indices most often used to measure physical strain that the body is undergoing (Figure 3.1). Portable devices exist to measure these changes unobtrusively during work. Both static effort and dynamic load on limited muscle groups can be analyzed with the help of electromyogram (EMG). EMG measures electric muscle action potential. Although in absolute terms there is no clear-cut definition of muscular work limit with local loading, subjective feelings of fatigue may be valued to make such judgments. EMG may be a valuable tool in making relative judgments. In general, the activity that generates EMG tracings with higher amplitude and frequency is the more stressful (tiring, fatiguing).

Heart rate, oxygen consumption, and work. Let us now analyze the behavior of heart rate (HR) before, at the onset, during, and after submaximal (work demand less than work capacity) physical work (Figure 3.5) [11]. During rest, the HR averages around some value between 60 and 85, with individual variances. At the onset of work, the HR starts to rise. For some time it lags the energy cost of work, using locally available high-energy phosphates, while developing oxygen debt. At some point it levels off, indicating that the amount of blood supply to the muscles is sufficient to meet the demand (steady state). If work is stopped suddenly, a gradual decline in HR occurs until it again levels off at the resting rate. Notice that the HR does not immediately drop back to resting level, indicating that the oxygen debt initially developed is being paid back, possibly with interest. The difference between HR_R and HR_{T_1} is a measure of the intensity of physical stress. Thus task T_2 is more stressful than T_1. Notice also that HR recovery time is longer in more stressful work. Thus HR recovery time may also be used as a measure of task intensity. With physical conditioning, one can attain lower elevated heart rates than otherwise. This is why trained athletes can endure more physical stress than can others. There is a maximum attainable heart rate for any person, the primary determinant being age. At the age of 65, one can attain only about 60% of the heart rate at the age 20.

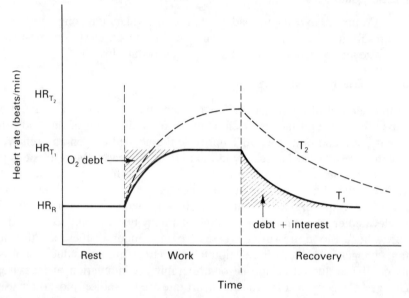

Figure 3.5   Heart rate with two tasks under submaximal work.

There appears to be a linear relationship between heart rate and oxygen consumption under submaximal work. Astrand and Rodahl [12] showed this relationship using a bicycle ergometer under laboratory conditions. Here the subjects were asked to cycle at a steady workload for a constant period while the heart rate and oxygen consumption were being recorded. Then the workload was increased in a stepwise fashion while cycling continued for constant periods. Figure 3.6 gives the plot of heart rate versus oxygen consumption during the experiment.

The linear relationship between heart rate and oxygen consumption has also been demonstrated in field studies. The work of Rodahl [13] is an example. In this study the researcher made measurements of heart rate and oxygen consumption in a 21-year-old coastal fisherman while he was engaged in net fishing. Figure 3.7 gives the results.

Several work situations violate the linear relationship between oxygen uptake and heart rate. An example is working with the arms overhead, as in the case of hammering

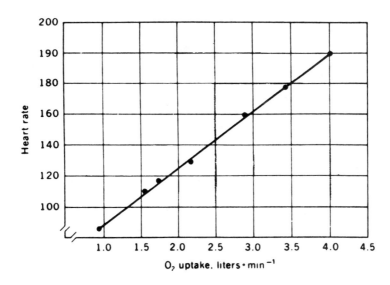

Figure 3.6   The relationship between heart rate and O_2 consumption on the cycle ergometer. (From *Textbook of Work Physiology, 3/E*, P.O. Astrand and K. Rodahl, 1986, used with permission of McGraw-Hill, Inc.)

nails into a ceiling. In this case the heart rate is exceptionally high. Another example is working in hot environments in which the heart rate is elevated significantly. Emotional stress also elevates the heart rate disproportionately.

All of this discussion leads one to conclude that oxygen uptake can be estimated from recorded heart rate, with some error. In trying to quantify this error, Rodahl et al. [14] made direct measurements of oxygen uptake using the Douglas bag method and compared the results with oxygen uptake calculated from simultaneously recorded heart rate in six commercial fishermen during commercial fishing. Figure 3.8 gives the results. It is evident that estimating workload on the basis of recorded heart rate gives good results. In the experiment above, the difference between calculated and empirical data was within ± 15%.

Figure 3.9 demonstrates that mental and emotional stress elevates heart rate [13]. Here a managing director is engaged in intense discussions with a small group of interested, nonopposing listeners. Heart rate also fluctuates between different work elements of the same job, depending on the amount of stress due to each element. Rodahl and Vokac [15] measured average heart rate on the crew of two long-line fishing vessels. Figure 3.10 gives the results. It is clear that in both cases the "unhooking fish" element poses most stress.

In situations where the work demand exceeds one's capacity, there is continuously growing oxygen deficit and blood lactate content. In this situation there is no steady state, and, once energy expenditure reaches the worker's capacity, work is voluntarily stopped and recovery starts.

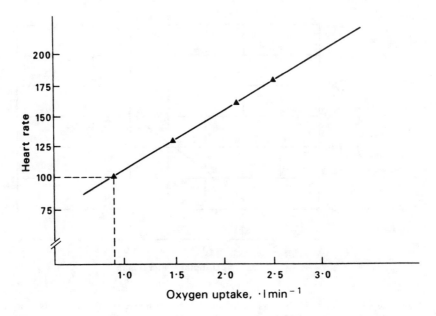

Figure 3.7 O_2 uptake equivalents of recorded heart rates in a 21-year-old fisherman. 100 bt/min = 0.91 1/min O_2. (From *The Physiology of Work*, K. Rodahl, 1989, Taylor & Francis, Ltd., London, used with permission.)

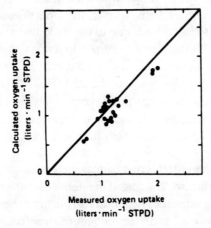

Figure 3.8 Relationship between O_2 consumption measured by the Douglas bag method and calculated from recorded heart rate. (From Rodahl et al. [14]).

Physical work capacity. A worker's capacity for energy output is referred to as the physical work capacity, which is a function of the energy available to the person in the form of food, oxygen, and the sum of the energy provided by the aerobic and anaerobic processes. Age also places a limit on the maximum increase in heart rate and oxygen

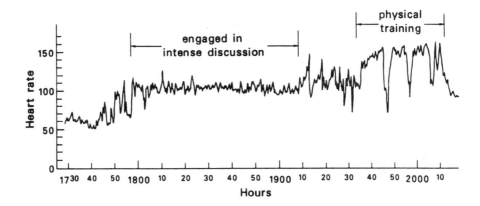

Figure 3.9 Heart rate versus management activity. (From *The Physiology of Work*, K. Rodahl, 1989, Taylor & Francis, Ltd., London, used with permission.)

consumption. Working at more than 30 to 40% of one's maximum aerobic power in an eight-hour shift causes notable muscular fatigue [10].

Energy cost of work on specific activities.   Energy consumption in humans is measured in kilocalories. Since there is a good relationship between the two, energy consumption of work can be measured indirectly by oxygen consumption. For every liter of oxygen consumed, there is an average of 4.8 kcal of energy expended. As discussed in Section 3.2.1, every activity requires an activity metabolism. Much research has been carried out to estimate the calorific value of various activities [16–19]. Table 3.2 gives a sample list of industrial activities and their average activity calorie requirements.

Several variables that affect the amount of calories expended on a task are age, sex, posture, body weight, and the intensity of activity. The values given in Table 3.2 assume average parameters for each variable. In terms of posture, a task that affects the center of gravity least is also the least energy consuming. As the rate of activity increases, there is elevated energy cost. With increasing body weight, there is increasing energy cost. A sedentary life leaves many parts of the body unused. It is estimated that a healthy occupation should require 3000 to 3500 kcal for a man and 2500 to 3000 kcal for a woman each day. As is obvious from these data, in general, women expend fewer calories than men and thus require fewer calories. At the age of 65, one can spend 75% of the calories that can be spent at age 25 [20].

Daily energy expenditure due to specific activities is the sum of the product of time spent on each activity and the energy cost of work for that activity, summed across all activities. Naturally, the total daily energy requirement must also include the needs for basal and digestive metabolism.

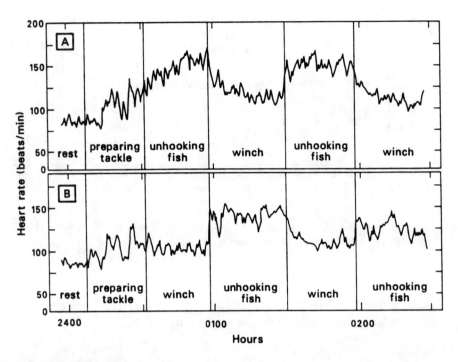

Figure 3.10 Heart rate versus fishing activity. (From *The Physiology of Work*, K. Rodahl, 1989, Taylor & Francis, Ltd., London, used with permission.)

TABLE 3.2 ENERGY REQUIREMENT FOR VARIOUS INDUSTRIAL JOBS

Activity	Average energy requirement (kcal/min)
Screwdriving (horizontal)	0.5
Light engineering work	2.0
Light assembly (seated)	2.2
Printing	2.3
Laboratory work	2.5
Bench soldering	2.5
Walking on level (3 km/h)	2.8
Cleaning windows	3.1
Shoe repair and manufacturing	3.8
Walking on level (4.5 km/h)	4.0
Bricklaying	4.0
Tractor plowing	4.2
Agricultural work	4.8
Milking by hand	4.9
Heavy washing (housework)	5.0
Weeding	5.1

Walking on level (6 km/h)	5.2
Pushing wheelbarrow	6.0
Binding	6.2
Handsawing wood	6.8
Chopping wood	8.0
Mowing	8.3
Shoveling (7-kg weight)	8.5
in front of heat	10.2
Tending the heat furnace	11.5

Source: Adapted from Refs. [18] and [19].

Keeping energy cost of work within acceptable limits. Designers, engineers, and managers must know the acceptable limits of energy expenditure in order to keep task requirements within bounds. Research [21,22] led to the following rules of thumb for average age:

1. For men:

· A maximum time-weighted average of 5 kcal/min due to activity energy cost of work

· A maximum time-weighted heart rate average of 100 beats/min.

2. For women:

· A maximum time-weighted average of 4 kcal/min due to activity energy cost of work

· A maximum time-weighted heart rate average of 90 beats/min.

The figures above are absolute work standards. Relative comparisons can be made by comparing two tasks with respect to HR or oxygen consumption. The calorific values of 5 and 4 can be estimated via measured oxygen consumption at approximately 1.04 and 0.83 liters/min, respectively. Data are also available with respect to categorization of work based on energy requirements [23]. Table 3.3 gives one such data set. More conservative (about 10% less) estimates also exist [24].

3.2.5 Fatigue

A natural result of work that is physically and mentally demanding is fatigue. Feelings of fatigue occur during or after work. Such feelings manifest themselves in forms ranging from slight tiredness to complete exhaustion. Fatigue due to mentally demanding tasks is discussed in Chapter 4. In this section we focus on fatigue due to physical stress. Many scientists have attempted to form a relationship between subjective feelings of fatigue and objective measures such as accumulation of lactate in the blood. Although such relationships have been well established for strenuous muscular effort such as athletic events, they are not clearly present in prolonged light or moderate work.

TABLE 3.3 WORK CLASSIFICATION BY ENERGY COST

Work grade	Energy expenditure (kcal/min)	O_2 consumption (liters/min)
Severe	> 12.5	> 2.5
Very heavy	10.0–12.5	2–2.5
Heavy	7.5–10.0	1.5–2
Moderate	5.0–7.5	1.0–1.5
Light	2.5–5.0	0.5–1.0

Source: Ref. [23].

Volle et al. [25] examined the fatigue effects of a compressed workweek (40 hours in 4 days) compared with the usual schedule of 40 hours in 5 days. Two groups of subjects in industry participated in the study, working at two different plants engaged in making similar products: one group following a 40-hour, 4-day schedule, and the other a 40-hour, 5-day schedule. Certain physiological data (heart rate, blood pressure, body temperature, oxygen consumption, CO_2 output, etc.) collected before and after the work-week did not reveal significant differences between the two groups. However, the critical flicker fusion (see Chapter 4, Mental Workload) frequency and right-hand strength showed significant deterioration in the 4-day group. Although this was the case, the extent of the higher level of fatigue indicated by these parameters can be questioned.

Astrand [26] observed a rise in heart rate in subjects who worked at loads corresponding to about 50% of the individual's maximum oxygen uptake during a period of about 8 hours. However, since the research was carried out throughout a workday, the elevation in heart rate may have been due to factors not controlled, such as the circadian rhythms and others.

Research with strenuous lifting tasks indicate that lactate production may be a good predictor of fatigue and exhaustion. Yates et al. [27] had three male subjects perform six progressive lifting tasks to exhaustion. Three tasks used a constant-weight, increasing-frequency format. The other three tasks used a constant-frequency, increasing-weight format. Blood lactate and heart rate were monitored throughout the experiment. Heart rate increased monotonically with increase in workload. Lactate production was similar to that seen in treadmill or cycle ergometer tests. The results also indicated that the constant-weight, increasing-frequency format produced more consistent blood lactate than did the constant-frequency, increasing-weight format. Although Edwards [28] and Karlsson et al. [29] showed that fatigue is not due simply to depletion of energy store and accumulation of lactate, there is evidence that they collectively influence it to a great extent [30].

3.2.6 Work-Rest Cycles

Human beings cannot maintain an activity level that is physically demanding for long periods. They need rest periodically to recover from the effects of the task. Rest allowances due to physical activity may be evaluated through extensive time studies or physiological methods. Work measurement through time study leads to approximate results. An example is 15% allowance applied to the normal time in order to calculate the standard time on a stock-handling task. On the other hand, physiological methods aim at determining rest allowances based on changes in bioresponses due to work. These methods may be *strain based* or *metabolic energy expenditure based* Rohmert [31,32] used the working heart rate (heart rate above the resting level) in order to propose rest allowances for static and dynamic muscular work.

An accepted norm in work-rest cycles due to physical activity, using the metabolic energy expenditure method, is that no work-related rest allowance is necessary for jobs that demand energy expenditure of less than the standard (4 or 5 kcal/min). However, it is a common practice to allow for rest when energy demands exceed accepted standards. Throughout the years, various researchers offered suggestions as to allowable rest periods [33,34]. Table 3.4 gives a composite suggestion. Here R_T is the allowed rest time (min), K the energy cost of work (kcal/min), S the accepted standard (4 kcal/min for females, 5 kcal/min for males), T the total expected duration of task (min), and BM is basal metabolism (kcal/min).

An example would be a shoveling task that requires 7.8 kcal/min performed over 100 minutes. Assume that a male of 40 years of age is performing the task. From line 2 in Table 3.4, R_T can be calculated as 51 minutes with $S = 5$ and $BM = 1.7$. This means that 51 minutes of rest must be allowed for each 100 minutes of work time. However, we are not through yet. An age allowance is necessary [35]. Table 3.5 gives the relevant data. With this allowance, the total rest period increases to 53 minutes.

Our experience suggests that in industrial environments, the most effective relief from muscular fatigue is through small periods of rest taken several times during the duration of work as opposed to all taken at one time. This can be reflected in the stan-

TABLE 3.4 REST-TIME REQUIREMENTS

$$R_T = 0 \qquad\qquad \text{for } K < S$$

$$R_T = \dfrac{\left(\dfrac{K}{S} - 1\right) \times 100 + \dfrac{T(K - S)}{K - BM}}{2} \qquad\qquad \text{for } S \leq K < 2S$$

$$R_T = \dfrac{T(K - S)}{K - BM} \times 1.11 \qquad\qquad \text{for } K \geq 2S$$

$$BM_F = 1.4 \qquad BM_M = 1.7$$

Source: Adapted from Refs. [33] and [34].

TABLE 3.5 AGE MULTIPLIER FOR REST PERIODS

Age	Multiplier
20–30	1.0
40	1.04
50	1.1
60	1.2
65	1.25

Source: Ref. [35].

dard time for an operation, based on less than 60 minutes in an hour. For example, a 10-minute fatigue allowance every hour means that one expects only 50 minutes of effective work each hour. Thus the operator can take 10 minutes of break every hour or 5 minutes of rest every 30 minutes.

To refine the physiological methodology and develop a procedure unique to the person performing a physical task, Mital and Shell [36] conducted an experiment using 10 industrial workers from the greater Cincinnati area. The subjects stacked containers in laboratory conditions simulating a palletizing/stacking task. The subjects were allowed to change the weight of the container to set up the task such that the final work rate would be the maximum rate that can be sustained continually for 8 or 12 hours, with breaks spaced throughout the period. Based on the results of the study, the authors developed a computerized algorithm to calculate a rest percentage based on the total energy requirement for the job and the adjusted total energy available for work. Ayoub and Mital [37] provide the listing of the computer code developed.

It is customary in industry to allow for a 15- to 20-minute coffee break 2 hours after start in the morning and after lunch. These periods may be counted in the total rest duration.

3.2.7 Strength and Endurance

Two topics that are related to physical activity are strength and endurance. *Strength is* the maximum force that one can exert voluntarily. It can be measured in kilograms using a dynamometer. Two types of strength must be distinguished. Static strength is measured at static postures, such as standing or seated. Dynamic strength is measured during work. Most of the available data are in terms of static values [38,39]. These are valid when a person is expected to apply forces at static postures. An example is that of activating a push-button to start a machine while standing. Figure 3.11 gives a data set for young males at different elbow angles. Only 5th percentile data are given. It is a common practice to deal with 5th percentile data when the context is strength since 95% of the population can exceed those forces. Using these data a designer may build resistance into a control such as a toggle switch. In general,

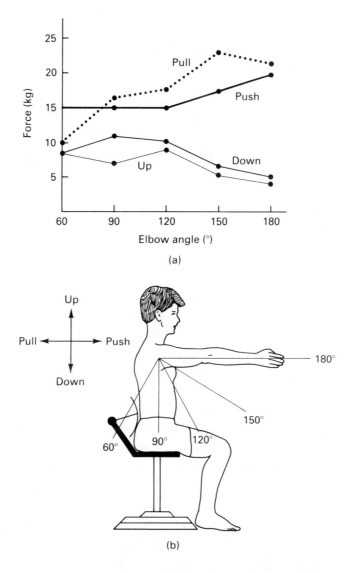

Figure 3.11 Strength in executing various arm movements in seated posture: (a) force exerted (5th percentile) by right hand; (b) elbow angles.

designers use the 5th percentile value as the maximum force to be overcome. It is evident from this figure that push and pull capabilities are the strongest in a seated posture. Furthermore, peak forces can be attained at extended arm posture, as discussed in Section 3.2.1. It is estimated that for the right-handed people, the strength of the left hand is approximately 10% below that of the right hand. Recent research by Kumar et al. [40] showed that young adults are strongest in pulling activity (as opposed to pushing) in the isometric mode in the sagittal plane (see Section 3.3) at medium height (100 cm height tested). Males were stronger (1% to 29%) than females for most tests. With increased postural asymmetry (30° lateral, 60° lateral), the difference between genders declined.

Research by Radwin et al. [41] measured submaximal pinch forces using transducers attached to the distal finger pads. Total resultant and individual finger forces were measured. The index finger exerted the greatest force followed by the middle finger, ring finger, and small finger. Average contribution of individual fingers to the resultant total pinch force was 34% for the index finger, 24% for the middle finger, 21% for the ring finger, and 21% for the small finger. On the other hand, Kinoshita et al. [42] showed that the thumb exerted the largest (>38%) grip force in holding cylindrical objects with circular grips, with the index finger exerting the least (about 11%).

Endurance is the ability to maintain activity over time. In the context of physical activity, endurance refers to the maintenance of effort. If an operator is expected to apply force continuously in order to hold an object, the designer must consider endurance. Research [6,43] suggests that people can maintain their maximum effort only briefly (Figure 3.12). Static effort can be maintained at only 20% of its peak over time, whereas 30% of peak forces can be maintained over extended periods in dynamic work.

Several other characteristics of strength are as follows:

1. Strength is at its peak by the late twenties, and shows continuous decline from then on. At age 65, one has only 75% of the strength of a youth.
2. On the average, women have two-thirds the strength of men.
3. Exercise can increase strength and endurance by as much as 50%.
4. Peak grip strength occurs with shoulder abducted zero degrees, elbow flexed 135 degrees, and the wrist at neutral posture. Deviations in elbow and wrist angles may result in up to 36% decrement in grip strength [44].

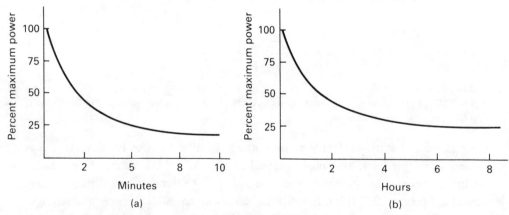

Figure 3.12 Endurance time as a function of maximum power: (a) static work; (b) dynamic work. (Reprinted with permission from *Human Factors*, Vol. 12(3), 1970. After Refs. [43] and [6]. © by The Human Factors Society, Inc. All rights reserved.)

3.3 OCCUPATIONAL BIOMECHANICS

In Sections 3.1 and 3.2 we focused on the various elements of human activity and acceptable standards. They evaluated the effects of physical activity on the body as a whole. On the other hand, engineering analysis of specific stress points in the musculoskeletal system is carried out using the concepts of biomechanics. Ergonomists deal with such concepts very frequently when evaluating physical work.

Biomechanics is concerned with the mechanical elements of living organisms. *Occupational biomechanics* deals with the mechanical and motion characteristics of the human body and its elements. Chaffin and Andersson [30] define occupational biomechanics as a field of science that studies the interrelationships between workers and their tools, workplaces, and so on, in order to enhance performance while minimizing the possibility of musculoskeletal injury. It is an applied science. Since an ergonomist is also concerned with the design of the human-made environment to enhance performance and minimize the risk of injury, there is an excellent parallelism between the two. While an ergonomist's task is more widely defined, occupational biomechanics deals primarily with musculoskeletal cases.

As in other fields of study, occupational biomechanics receives input from various disciplines, and has its methods of analyses and certain outputs. As an interdisciplinary science, the various elements of occupational biomechanics can be displayed using a supplier-customer model. Figure 3.13 gives one such model. The quantitative techniques used are borrowed from the engineering disciplines. The physical sciences contribute information with respect to the physics of equilibrium and motion. The biological sciences provide the basics of anatomy and physiology. As an analysis method, modeling allows us to develop smaller-scale, simplistic representations of the real-life events to solve the problem at hand. *Anthropometry* provides the human body and segment dimensions, including masses and centers of gravity. *Kinesiology* covers the area of human motion. Using kinesiology, body segment motions and the triggering muscular actions can be described. Based on this information, models of motion analysis can be developed. *Bioinstrumentation* deals with the processes of data acquisition and analysis. Data collection via force plates, electromyography, goniometry, and linear-distance measuring devices allow occupational biomechanics to be a very objective field of science.

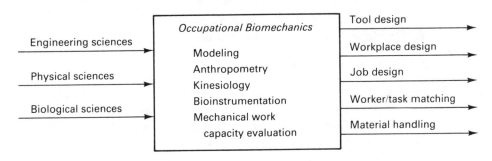

Figure 3.13 Inputs, elements, and areas of impact of occupational biomechanics.

Safe performance of physical activity depends to a large extent on the person's capacity and job demands. These two attributes must be matched. Results of occupational biomechanics analyses can be utilized for hand tool design, workplace design, job design, material-handling task design, and better matching of workers with tasks. Equipped with this information, an ergonomist can also carry out effective evaluations of existing designs and personnel assignments.

The basic ingredient of biomechanical analyses is a set of laws, called Newton's laws. These set the background on which biomechanical modeling is based. The three theories of concern are:

1. A mass will remain at rest or in uniform motion unless an unbalanced external force acts on it.

2. Force is proportional to the acceleration of a mass.

3. Any action will be opposed by reaction of equal magnitude.

These laws are used extensively in biomechanics to describe the state of a body and/or its segments. For instance, if a body is not in motion, the sum of all forces (in whichever direction) and moments acting on it must be equal to zero. This state is also known as the state of *equilibrium*. Force is measured in newtons (N).One kilogram force (also called 1 kilopond, kp) is equal to 9.81 N.

Biomechanical terms that describe motion or action assume a *standard* anatomical posture: standing erect, face forward, holding arms down at the side with palms facing forward (thumb points away from the body). In this posture, the *sagittal plane* splits the body into two equal parts (left and right) in a forward and backward direction. The *coronal plane* passes in a left-to-right direction and splits the body into front and back. The *transverse plane* passes perpendicular to each of these two planes at the abdominal area, splitting the body into the top (headward) and bottom (feetward) halves.

Several other terms used in kinesiology (study of human motion) that describe movement are:

1. *Flexion:* decreasing the angle between body parts
2. *Extension:* increasing the angle between body parts
3. *Adduction:* movement toward the middle of the body
4. *Abduction:* moving away from midbody
5. *Pronation:* face-down, or palm-down position
6. *Supination:* face-up, or palm-up position

3.3.1 Biomechanical Models

As discussed earlier, modeling is used extensively in occupational biomechanics. Since models deal with the evaluation of real situations, valuable information can be derived through them. Furthermore, since a human operator does not have to participate in these analyses, the results can be achieved even faster in office environments. Finally, potentially dangerous conditions (heavy loads, unacceptable postures) can be evaluated without posing any risk to human life. In this section we present several examples of biomechanical models primarily based on Chaffin and

Andersson [30]. They are simple models that one can build on later. However, they get the point across.

Single-segment static model. Assume that an operator is holding a load of 20-kg mass with both hands in the sagittal plane. The load is at about waist height. Furthermore, the operator is anthropometrically a 50th percentile man. Determine the magnitude and direction of forces and rotational moments acting on the man's elbow. Figure 3.14 gives a simple diagram of the situation. Load weight can be calculated as follows:

$$W = mg$$

where: m is the mass of object handled, g the force of gravity and W is weight (measured in newtons, the unit of weight in the metric system). For our case

$$W = (20 \text{ kg})(9.8 \text{ m/s}^2)$$
$$= 196 \text{ N}$$

Under symmetric loading, each hand's load can be calculated as follows:

$$\sum F = 0$$
$$-196 \text{ N} + 2R_H = 0$$
$$R_H = 98 \text{ N}$$

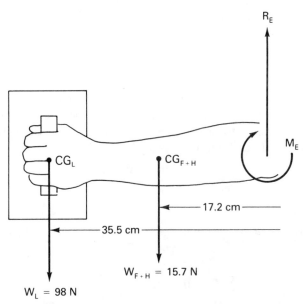

Figure 3.14 Forearm-hand-load model depicting forces and movements. (After Ref. [30]).

It can be assumed that the line of action of the load is passing through the center of gravity of the hand. Furthermore, 50th percentile male anthropometry provides 17.2 and 35.5 cm as the distances from elbow link to the center of gravity of the forearm-hand mass and the hand, respectively. Anthropometric data also provide the weight of forearm-hand mass as 15.7 N for a 50th percentile adult male.

The elbow reactive force R_E can be calculated as follows:

$$\sum F = 0$$

$$-98 \text{ N} - 15.7 \text{ N} + R_E = 0$$

$$R_E = 113.7 \text{ N}$$

This reactive force will be able to resist the translational (downward) motion; however, it cannot stop the rotational (counterclockwise) motion. The elbow moment M_E performs that function and its magnitude is:

$$\sum M = 0$$

$$(-98 \text{ N})(0.355 \text{ m}) + (-15.7 \text{ N})(0.172 \text{ m}) + M_E = 0$$

$$M_E = 37.5 \text{ N} \cdot \text{m}$$

The calculations above made use of two of the Newton's laws concerning the reactive forces and moments, as well as the sum of forces and moments being equal to zero for static bodies, the *equilibrium conditions*.

Two-segment static model. It is possible to treat each body segment as an element in a kinetic chain and work back to all joints affected when handling loads. Depending on the posture, the analyst may have to consider angles between various links when calculating reactive forces. A major goal of the modeler is to determine forces on the most distant joints, such as the L5/S1 disk joint on the vertebral column. In this section we provide an insight into carrying out these calculations in a stepwise manner.

Assume that one wishes to extend the previous analysis one step further. Specifically, shoulder reactive forces and moments are in question. Figure 3.15 gives the free-body diagram. Previously, R_E was determined as 113.7 N and M_E was calculated as 37.5 N· m. Focusing on the upper arm, these forces and moments can be represented by R'_E and M'_E acting in opposite directions but having the same magnitude. Anthropometric data provide values for link distance between the shoulder and elbow, weight of the upper arm, and the distance between the shoulder link and the center of mass of the upper arm.

$$\sum F = 0$$

$$-R'_E - W_{UA} + R_S = 0$$

$$-113.7 - 20.6 + R_S = 0$$

$$R_S = 134.3 \text{ N}$$

$$\sum M = 0$$

$$M_S = (0.132 \text{ m})(20.6 \text{ N}) + (0.329 \text{ m})(113.7 \text{ N}) + 37.5 \text{ N} \cdot \text{m}$$

$$= 77.6 \text{ N} \cdot \text{m}$$

Note that both reactive forces and moments increased, since as the analysis point of reference moves away from the load, additional weights due to body parts are considered along with moment arms. This explains how a 200-N load in the hand can pose a 4000-N compression load on the L5/S1 disk (intersection of the fifth lumbar and first sacral vertebra).

The purpose of these exercises is to develop strain measures corresponding to a given physical loading at various joints. An ergonomist may then compare these with population data on static muscle strength at the corresponding joints to evaluate a given task. Those tasks that pose higher potential stress than voluntary capability will have to be redesigned or automated [45].

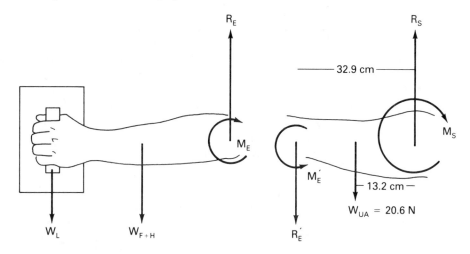

Figure 3.15 Two-segment static model. (After Ref. [30].)

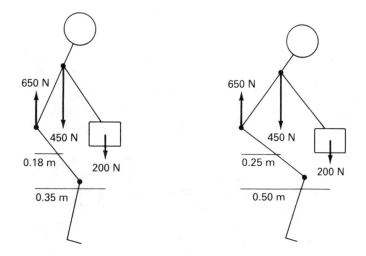

Figure 3.16 A static low-back model.

Back models. Notable effort has gone into modeling the biomechanics of the lower back (see [46–48]). It is not the purpose of this section to review all these (a whole book can be filled by presenting such models). It will suffice to say that the back, especially the lower back, is the most distant to any load handled. Thus notable compression forces, and bending moments may develop there, especially at the L5/S1 joint disk (the most distant vertebral motion segment). In addition to the load, the upper torso will also contribute to reactive forces and moments. Posture is important in evaluating the impact on the lower back. Reviewing Figure 3.16, one can appreciate the 40% additional bending moment created by carrying the load only 15 cm (6 in.) farther from the body. The 450-N load is due to the upper torso.

3.4 CUMULATIVE TRAUMA DISORDERS

A topic closely linked to biomechanics is cumulative trauma disorders (CTDs). In this book we devoted a separate section to CTDs in order to stress their importance since they have become more prevalent (or more extensively reported) in industry. OSHA is now placing specific emphasis on repetitive-trauma cases with citations and standards development. Cumulative trauma disorders were noted in the literature in increasing frequency in the 1980s. Although their existence was noted as early as in 1717 by Ramazzini, such disorders had not been strongly linked to the workplace and the work methods until the early 1980s. Other terms used to describe the same condition are repetitive-motion disorders [49], overuse, cumulative-effect trauma, and cumulative-effect disorders. Putz-Anderson [50] provides an extensive coverage of such ailments.

In short, CTDs are disorders of the softer tissue due primarily to repeated use. Although they range from joint inflammation to muscle soreness, CTDs are commonly reported in the tendons and in the nerves of the upper extremities, including the fingers, the wrist, the forearm and the upper arm, and the shoulder. Joints, muscles, blood vessels, and bursae may also be affected. Workers on assembly operations with short cycle times, such as garment sewing operations, typing, manual packaging, wiring, coil winding, and carpentry, may be particularly susceptible to such injuries. A recent survey by the U.S. Department of Labor, Bureau of Labor Statistics [51] indicates that 23% of all compensable injuries are due to trauma to the upper extremities (Table 3.6). Naturally, this number includes both single-incident (laceration, contusion, fracture, etc.) and chronic injuries, including the cumulative trauma disorders. With lower back pain included (Section 3.5), the CTD-related injury and illness percentage increases to 49.8. "Repeated trauma" cases as reported by BLS display the trend shown in Figure 3.17.

Industries with the highest number of cases are motor vehicles and equipment manufacturers, meat packing, aircraft and parts, men's and boys' clothing makers, and grocery stores. Repetitive motion also resulted in the longest absences from work among the leading events and exposures (median 18 days). In terms of incidence rates, meat packing plants, knit underwear mills, motor vehicles and car bodies, and poultry slaughtering and processing industries lead the way. The 7% drop in CTD cases in 1995 is attributed to heightened interest in and awareness of these injuries and willingness to contain the problem by industry.

The human hand is composed of a complex of tendons, ligaments, nerves, many bones, arteries, veins, and small muscles (Figure 3.18). The extensor and the flexor muscles in the forearm control the posture of the hand. They are connected to the fingers by long tendons. Only the thumb has a notable muscle base. The flexor tendons pass through a tight channel in the wrist called the *carpal tunnel.* Several other anatomical structures

TABLE 3.6 BLS DATA ON INJURIES AND
ILLNESSES BY PART OF BODY—1994

(Nonfatal, Involving Days Away)	
Part of Body	Percent of Injuries and Illnesses
Head, eyes, neck	8.6
Trunk	38.8
Upper Extremities	22.7
Lower Extremities	19.8
General	9.4

Note: Due to rounding, percentages may not add to 100.

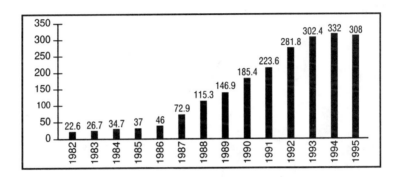

Figure 3.17 Repeated Trauma Cases Reported by Private Industry (thousands) . *Source*: CTD News.

pass through this channel, including the median nerve and the median artery. On the little finger side, the ulnar artery and the ulnar nerve pass next to the pisiform bone. The bones of the wrist connect to the two long bones of the forearm, the radius and the ulna, which in turn connect to the humerus of the upper arm. The biceps muscle connects to the radius, and when the arm is extended, it pulls the radius against the humerus [1].

3.4.1 Causes and Prevention Techniques

Research shows that there are several reasons for CTD:

1. *Unnatural joint posture.* Any time a joint is forced to assume an unnatural position, the risk for CTD is increased. In the case of a hand, maintaining a straight wrist while performing the job will help reduce injury potential. Several unnatural hand postures are shown in Figure 3.19. Research also shows that risk increases notably when joint extremes are involved in motion [52]. Objective data exists for joint mobility ranges.

2. *Forceful application.* Application of forces through hinge joints notably increases injury potential. The wrist is a good example of a hinge joint. To a lesser extent, pivot joints (elbow) are also at risk during force application. Silverstein et al. [53] suggested that a hand force of 39.2 N or more may cause CTDs. There is also some information that suggests exertion exceeding one-third of workers' static muscle strength available for the activity in question may be a causative factor. Using psychophysical technique (see next section), Snook et al. [54] determined maximum acceptable forces for various types and frequencies of repetitive wrist motion for females. These results are consistent with those found earlier by Silverstein et. al.

3. *Repetition of activity.* Highly repetitive tasks of short duration (less than 30 seconds) pose more risk to the employee than do other jobs. When these jobs are performed over months and years, CTD risk increases significantly.

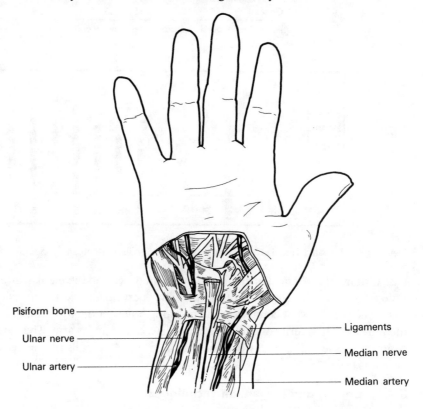

Pisiform bone

Ulnar nerve

Ulnar artery

Ligaments

Median nerve

Median artery

Figure 3.18 Anatomy of the hand.

4. *Individual factors.* Many preexisting conditions seem to aggravate lower-risk cases. These conditions include neuropathy, arthritis, peripheral circulatory disorders, reduced estrogen levels, and small hand/wrist size.

5. *Other.* Localized pressure, and exposure to vibration and cold over time may cause CTDs. A cadaveric study [55] showed that pressure on the base of the palm due to tool use increases carpal tunnel pressure as well as incursion of lumbrical muscles when fingers are flexed as opposed to extended.

Each of the factors above contributes to the risk potential. Furthermore, in combination, they may pose significant risk. For example, a woman textile worker who performs a sewing job involving manipulating heavy cloth hundreds and thousands of times a day in ulnar deviation of the wrist is a candidate for a form of CTD.

Industry studies provide support for the CTD risk factors. Kumar [56] analyzed several high-risk operations in a garment industry with respect to their cycle time, activity elements, repetition and duration of motion, and nonneutral posture of shoulder, elbow, wrist, and hand joints. The severity and duration of deviation of joints from neutral postures were found to be related to the frequency of occurrence of CTD. Imrhan [57] performed a study at a traditional glass manufacturing plant to determine work-related factors that may have contributed to or caused musculoskeletal ailments among workers. Questionnaire responses and video recordings of workers at work were analyzed for risk factors. Results showed that factors related to workplace design, seating, and tool design needed to be improved. Poor posture and extreme joint deviations with forceful applications were to be corrected to combat musculoskeletal stresses.

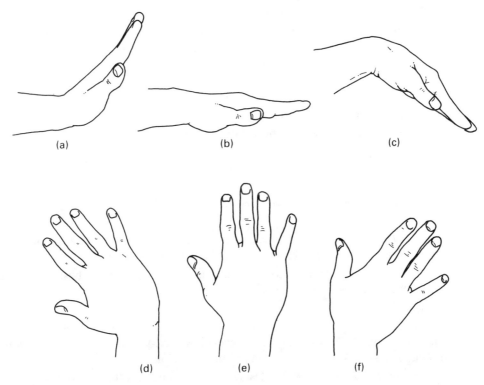

Figure 3.19 Various postures of the hand: (a) dorsiflexion; (b) neutral; (c) palmar flexion, (d) radial deviation; (e) neutral; (f) ulnar deviation.

Prevention techniques. Prevention techniques include administrative and engineering controls. Administrative controls span job rotation, warming exercises, controlling preexisting conditions, and removing time and pace pressures. Engineering controls include automation, job and workplace redesign, tool redesign, and work/rest cycle control. It is important to identify jobs with elevated risk potential and redesign them to achieve a better fit to the worker [58].

Therapeutic approaches. Naturally, prevention is more important than therapy; however, CTDs are treated primarily by rest, immobility, vitamins, anti-inflammatory drugs, and careful heat application. Ultrasound may also be applied. Surgery is the last resort.

3.4.2 Forms of Cumulative Trauma Disorders

This section provides more insight into several common forms of CTDs. Although both the diagnosis and the treatment are done by a physician, the ergonomist must be aware of the various types of such ailments to appreciate the impact of poor design decisions on human well-being and productivity.

1. *Bursitis* is the inflammation of a bursa. It may happen at any extensively used part of the body. Bursae are padlike sacks found in the vicinity of joints. They contain a fluid or synovia which act to reduce friction between tendon and bone, tendon and ligament, or between other structures where friction is likely to occur.

2. *Tendonitis* is inflammation of a tendon due to excessive use, especially when trapped. It is a very common CTD.

3. *Synovitis* is tendon sheath irritation. Tendon sheaths provide lubrication and nutrition to the tendons.

4. *Tenosynovitis* is another common CTD. It is the inflammation of both the tendon and its sheath.

5. *Stenosing-tenosynovitis* is the accompanying swelling and constriction of the tendon sheath.

6. *Stenosing-tenosynovitis-crepitans* is tenosynovitis with swelling that interferes with tendon movements and causes snapping and jerking movements.

7. *Ganglionic cyst* is a tendon disorder that manifests itself as a blister containing synovial fluid that shows up as a bump on the wrist. It is not uncommon to observe several such cysts at the same time on an affected hand.

8. *Trigger finger* develops when a finger is used excessively against sharp edges. The afflicted person can flex the affected finger; however, the finger has to be passively straightened. A click may also accompany the extension.

9. *DeQuervain's disease* manifests itself as tendonitis at the base of the thumb. A diagnostic sign is pain around the thumb when closing the fist or when deviating the thumb toward the little finger.

10. *Raynaud's phenomenon* is also known as *vibration-induced white fingers*. Workers afflicted with this ailment experience vascular attacks (cramps, pain), numbness, or pins and needles when their hands are exposed to cold. It is believed that Raynaud's phenomenon is caused by operating vibrating hand tools over long periods.

11. *Ischemia* is obstruction of blood flow. This could occur, for example, in hands due to a handle consistently being pressed against the palm, such as in paint scraper's operation. Ischemia leads to numbness and tingling of fingers.

12. *Carpal tunnel syndrome* is a common CTD caused by repetitive and forceful application of the wrist at joint extremes. Inflamed tendons, tendon sheaths, and other anatomical structures in the carpal tunnel compress the median nerve, leading to pain, swelling, numbness, tingling, and clumsiness. In advanced cases, wasting of the muscle at the base of the thumb occurs.

13. *Neuritis* of the hand occurs when a nerve is inflamed. For example, repeated use of the wrist in ulnar deviation rubs the ulnar nerve against the pisiform bone, leading to this ailment.

14. *Guyon's canal syndrome* is caused by compression of the ulnar nerve in the Guyon's canal. This canal is formed by the pisiform and hamate bones on the palmar side and the transverse carpal ligament and the ligament between these two bones. Overuse of wrist in palmar flexion and ulnar deviation or constant pressure on the palm may cause this syndrome. Numbness in the little finger and half of the ring finger may indicate Guyon's canal syndrome.

15. *Intersection syndrome* is characterized by pain and redness on the wrist and forearm above the thumb. It is a form of tenosynovitis caused by extensive friction between two sets of extensor and abductor tendon pairs. Overuse of the wrist and hand in wringing, grasping, turning and twisting motions may cause this syndrome.

Other common sites of injury due to cumulative exposure to irritants are the elbow and the shoulder. Several forms of such ailments follow.

1. *Lateral epicondylitis*, also known as "tennis elbow," is irritation of the tendinous attachment of the finger extensor muscles on the outside of the elbow.

2. *Medial epicondylitis*, also known as "golfer's elbow," is irritation of the tendinous attachment of the finger flexor muscles on the inside of the elbow.

3. *Telephone operator's elbow* occurs due to extensive resting of the elbow against a hard or sharp edge. This posture puts pressure against the ulnar nerve and leads to numbness and tingling sensations over and below the little finger, similar to the case when the "funny bone" is hit. Cushions and round edges may help. When the ulnar nerve is irritated at the elbow due to repeated bending and straightening of the elbow, the *cubital tunnel syndrome* may emerge.

4. *Radial tunnel syndrome* is the irritation of the radial nerve in the elbow area. This condition may be caused by a direct blow to the outside (lateral) portion of the elbow or repetitive, forceful pushing and pulling, bending of the wrist, gripping, and pinching. Avoiding repetitive and excessive movement at the elbow and wrist lessens the pain.

Frequently reported shoulder disorders include bicipital tenosynovitis or bursitis, subdeltoid bursitis, rotator cuff irritation, and tear. These are associated with working with the elbow elevated. *Thoracic outlet syndrome* involves compression of the nerves and blood vessels between the neck and the shoulder. Shoulder and arm posture can be controlled by proper positioning of the parts, points of use, and work areas relative to the worker so that the elbow is down and close to the body.

For uncertain reasons, the term "ergonomic injuries/hazards" has for some time been associated with CTDs. For example, OSHA and the State of California have been working on "ergonomic" and "work" standards. In both drafts, the term "ergonomic injury" is associated with musculoskeletal ailments. At the time of preparation of the second edition of this book, only the California standard had made significant progress toward enactment or approval. OSHA has been issuing "ergonomic" citations for at least nine years. However, lately the number of such citations dropped, even though BLS surveys indicate an average 7–8% increase per year in CTD cases. Without a national standard, OSHA has been using the General Duty Clause to issue such citations.

3.5 MANUAL MATERIAL HANDLING

A major portion of human physical activity in industry takes place in manual material-handling activities. There is some manual handling of material in any industry. Even in an office one observes boxes of paper being moved. Most industrial production workers are required to handle parts, subassemblies, containers, and products on their jobs. It is the duty of the ergonomist or engineer to design manual handling jobs so as to minimize risk of injury. Are these jobs posing an inordinate amount of injury risk? Let's review some statistics:

· Approximately 25% of accidents reported in industry each year are related to manual handling of material [53]. Statistics from England place this around 24%.
· 60% of overexertion injuries are due to lifting and lowering [51].
· Overexertion is claimed by over 60% of people suffering from it to be the primary cause of lower back pain.
· Low-back-pain overexertion injuries with significant lost time results in less than one-third of the patients eventually returning to their previous work.
· Overexertion injuries of all types (lift, lower, push, pull, carry, etc.) in the United States occur to about 500,000 workers per year. One in every 200 workers is afflicted by such disorders [60].
· About 50% of self-reported, work-related illnesses are due to back-related problems [60].
· The 25- to 44-year-old population has the highest incidence of back pain.
· Overexertion leads all disabling events cited in a sixth to a third of the cases in every industry division [51].
· With disability costs, total annual cost for back pain reaches $56 billion [61].
· Mechanics and vehicle repair persons; nurses and nursing home attendants; material handlers; operators of extractive, mining, and material-moving equipment; and construction operators are the occupations with the most handling incidents [62].
· Low-back pain related to occupational factors costs $4.6 billion annually [63].
· Overexertion is the most common way in which lost work-time injuries can occur.

Control of manual handling incidents requires a comprehensive program that attacks the problem from several avenues (see Ref. [64]). The four major elements to consider in such a program are given in Table 3.7. Both administrative and engineering controls are necessary in order to contain material-handling-related injuries. Of the four major elements listed in Table 3.7, work organization is probably most amenable to administrative controls. Similarly, material characteristics are probably more controllable by engineering means. The other two, worker characteristics and task characteristics, require a good blend of engineering and administrative controls. Among the worker characteristics are such attributes as age, sex, body weight, anthropometry, strength, endurance, physical fitness, and psychological factors. Although research on some of these variables has led to conflicting results, it is highly recommended that they be treated as notable risk factors. For example, Adams [65] and Sheppard [66] showed that while maximal oxygen uptake (physical work capacity) declines with age, submaximal load oxygen consumption is not affected. Furthermore, Petrofsky and Lind [67] did not observe any changes in isometric strength with age. On the other hand, Montoye and Lamphiear [68] observed some decrease with age.

Evaluation of injury potential due to manual handling jobs follows four major avenues of research:

1. *Epidemiological approach.* This approach characterizes the job, the workplace, and other factors in order to derive significant trends that may lead to incidents/injuries. Historical records are the main input to this procedure.

2. *Biomechanical approach.* Here the researcher is interested in characterizing the forces and moments on various body elements in order to estimate the task and load characteristics that lead to injuries or safe completion of work. The biomechanical approach is applicable to occasional lifting tasks where the frequency of lift is less than 4 lifts per minute. The biomechanical approach has been discussed in Section 3.3.

3. *Psychophysical methods.* In this type of analysis, research subjects adjust the load and task characteristics in order to arrive at a setup that is acceptable to them. This voluntary setting of the task elements is a good estimate of the acceptability of the handling task. The basic assumption of the psychophysical approach is that in any manual handling task, both biomechanical and physiological stresses are present. Karwowski [69]

TABLE 3.7 FACTORS AFFECTING MANUAL HANDLING CAPABILITY

Worker characteristics	Material characteristics
Age	Weight
Sex	Bulkiness
Motivation	Load distribution
Physique, etc.	Handles, etc.
Task characteristics	Work organization
Reach requirement	Work–rest cycles
Frequency of handling	Training, selection
Duration, etc.	Job rotation, etc.

Source: Adapted from Ref. [30].

demonstrated the validity of this assumption. He used the fuzzy sets theory to prove that the acceptability of a lifting task can be determined as a function of biomechanical and physiological stresses.

4. *Physiological approach.* This approach evaluates a given handling task in terms of physiological parameters. In manual handling activities, the physiological stress of interest is the cardiovascular stress. This could be evaluated through physiological responses, such as oxygen consumption, heart rate, blood pressure, and lactic acid accumulation. The physiological approach is applied to repetitive lifting tasks where the weight of the load is within the physical strength of the workers. Many researchers identified measures for determining the physiological limit in lifting tasks, such as the heart rate and the energy cost of work. For example, Petrofsky and Lind [70] used blood lactate accumulation to find the safe limit for lifting tasks. Intaranont [71] used oxygen consumption and minute ventilation to find the aerobic threshold in a lifting task.

The psychophysical approach provides a good review of task characteristics that are acceptable to the general population with almost no changes. If a handling task falls within these guidelines, in general, no administrative or engineering changes/controls are necessary. The physiological and biomechanical guidelines compete and set the upper limit of acceptance. A task whose demands are beyond those limits will have to be redesigned (if limited by physiological approach) or automated (if limited by biomechanics). The epidemiological approach serves to support or test the guidelines developed through other approaches. Furthermore, it may help in developing quick estimations.

Recently, Marras et al. [72] identified five factors which predict well both medium- and high-risk jobs with respect to low back disorder using a tri-axial electrogoniometer. The factors identified were: lifting frequency, load moment, trunk lateral velocity, trunk twisting velocity, and trunk sagittal angle.

Since lifting seems to be the most injurious material-handling activity, more research has been carried out on this type in order to answer the question: How much can a person lift safely? The next section attempts to provide an answer to this question through the results of a government-sponsored research program.

3.5.1 The NIOSH Lifting Guide (1981)

NIOSH's 1981 *A Work Practices Guide for Manual Lifting* [54] has been the most comprehensive approach to controlling the adverse effects of lifting loads that are symmetrically balanced in front of the body. A large variation exists in lifting capability in any normal group of workers. Due to this, the 1981 NIOSH recommendations were based on two levels of hazard. The lower level establishes an action limit (*AL*). This is based on psychophysical studies of acceptable loads in lifting tasks [73]. The higher limit establishes the maximum permissible limit (*MPL*) based primarily on biomechanical studies. If a person was exposed to conditions below the *AL*, no ill effects would be expected. Between the *AL* and the *MPL*, administrative controls are necessary where careful selection and training are required for the lifting task. When the *MPL* is exceeded, the conditions expose most people to injury potential. These jobs will have to be redesigned or automated, in short, controlled through engineering means. Figure 3.20 relates these concepts to weight handled and horizontal location of the load.

The authors of the 1981 NIOSH guide also developed prediction equations for the two levels of hazard that take into consideration the multiplicity of risk factors. At the action limit, the prediction equation is (in metric units):

$$AL = 40(15/H)(1 - 0.004|V - 75|)(0.7 + 7.5/D)(1 - F/F_{MAX})$$

In U.S. customary units:

$$AL = 90(6/H)(1 - 0.01|V - 30|)(0.7 + 3/D)(1 - F/F_{MAX})$$

For the maximum permissible limit,

$$MPL = 3AL$$

where:

· AL and MPL are measured in terms of kilograms (or pounds) for the given job conditions. Then the actual load handled is compared to these values to make the appropriate decisions. MPL is three times the AL.

· H is the horizontal distance (cm or in) from the load center of mass at the origin of the vertical lift to the midpoint between the ankles (lumbar spine). H will assume a minimum value of 15 cm or 6 in (body interference) and a maximum value of 81 cm or 32 in (reach distance).

· V is the vertical distance (cm or in) of the hands to the floor at the origin of the lift. There is no minimum value. The maximum value is 177 cm or 70 in (upward reach).

· D is the vertical travel distance (cm or in) of the object measured by the difference between the final and the initial locations of hands. A minimum value of 25 cm or 10 in and a maximum value of $(203-V)$ cm or $(80-V)$ in are assumed.

· F is the average frequency of lifting (lifts/min) with a minimum value of 0.2 and a maximum value as defined by Table 3.8. Set $F = 0$ for $F < 0.2$.

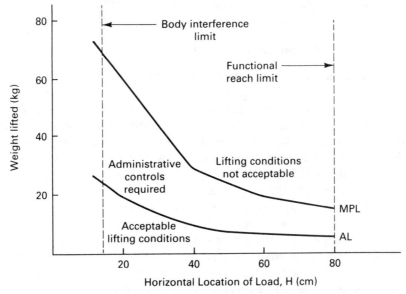

Figure 3.20 NIOSH guide for occasional lifts in the sagittal plane. (Adapted from Ref. [60]).

TABLE 3.8 F_{MAX} FOR DIFFERENT POSTURES

Duration of lifting period	V>75 cm (30 in) (standing)	V≤75 cm (30 in) (stooped)
One hour (occasional)	18	15
Eight hours (continuous)	15	12

Source: Adapted from Ref. [60].

A careful analysis of the equation above indicates that under optimal conditions, one could lift 40 kg with no ill effects expected. This would occur for occasional lifts (less than once every 5 minutes) when the load is held close to the body (*H* is 15 cm) at carrying height (*V* = 75 cm) and it is not lifted far (*D* = 25 cm).

An example will clarify matters. Assume that a lifting task requires that a 15-kg weight be lifted from an initial location of 30 cm above ground through a vertical distance of 20 cm. The load center of mass is 30 cm away from the lower vertebral region. The expected lift frequency is 4 times per minute. Here, *H* = 30 cm, *V* = 30 cm, *D* = 20 cm, *F* = 4 lifts/min, and from Table 3.8, F_{MAX} is 12. Then, *AL* can be calculated as 11.75 kg (25.85 lb) and *MPL* will be 35.25 kg (77.55 lb). Since the load weighs 15 kg (33 lb), this task can be performed under administrative controls only.

The International Occupational Safety and Health Center estimates the maximum safe lifting loads to peak for ages between 20 and 35 at 25 kg (55 lb) for males and at 15 kg (33 lb) for females. In general, females are expected to safely handle 60 to 65% of the load that males can. At age 14 to 16, the suggestions are 15 kg (33 lb) for males and 10 kg (22 lb) for females. The same suggestions are made for ages above 50.

Much less has been defined for one-handed lifting compared to two-handed sagittal-plane lifting. Research [49,74] indicates that at a standing posture, with optimum grip span [6 cm (2.1 in)], a 22-kg (48.4-lb) weight can be lifted occasionally (less than once per 5 minutes). If the operation is done from a seated posture, the object weight should not exceed 4 kg (8.8 lb).

3.5.2 Other Manual Handling Activities

There is much less information available with respect to population capability for lowering, pushing, pulling, and carrying activities than for lifting. Such tasks are prevalent in industry, as can be observed in operating controls and tools, sliding containers and carts, packaging, and clearing jams. Perhaps the most comprehensive study along these lines has been performed by Snook [73] over many years. The tables he presents are based on psychophysical studies. Snook and Ciriello [75,76] presented some modifications to these results in 1991 and 1993. There is even less reliable evidence on combined (lift and push and carry, etc.) manual handling activity. Straker et al. [77] showed that the current recommendations on splitting a combination task into its components for assessment is not acceptable.

In a study of the letter carrier's task, Bloswick et al. [78] redesigned the conventional side-carry (over-the-shoulder) mailbag to include waist support on either one side or both sides of the body. Although the resulting designs did not reduce metabolic load, they did significantly reduce lateral trunk flexor muscle fatigue.

Based on the work by Snook and Ciriello and other research [49], the following summarize this author's recommendations:

1. The lowering activity is just as stressful. Assume only a 5 to 10% increase, compared to lifting, in population capability for such tasks.

2. In a standing posture, the upper force limit for horizontal pushing and pulling tasks is 225 N. In a seated posture, it is 130 N.

3. In a standing posture, an upper limit of 540 N should be used for a "pull-down" activity above head height. At the shoulder level, this value gets down to 315 N. A push-down force of 287 N can be applied safely at elbow height. In a seated posture, these values should be discounted by 15 to 20%.

4. For a transverse or lateral push using only shoulder muscles, design for an upper limit force of 68 N. In general, maximum transverse forces are only 50 to 70% of those that can be developed in horizontal pushes or pulls straight ahead of the body with the same elbow angles.

5. A 12- to 15-kg (26.4- to 33-lb) weight can be carried by both hands through distances of up to 40 to 50 m (132 to 165 ft). As the distance increases, maximum acceptable weight decreases.

Some workers use back belts in manual handling tasks. Many varieties of back belts are available in the market. A NIOSH task force reviewed available research [79] in 1994 on the effectiveness of such belts in preventing injuries. The findings were inconclusive. However, a recent UCLA study [80] at 31 Home Depot stores in California documented injury reduction benefits of back belts.

3.5.3   The Job Severity Index

The job severity index (JSI) was developed in 1978 and validated in the same year and in 1982 under a NIOSH contract research program at Texas Tech University. Its primary use is in the analysis of lifting and lowering activities [81]. The premise of JSI is that the severity of a job in terms of its injury potential is a function of job demands and job capacity. If a person's capacity to perform a job is well below that of the requirement of the job, a fair assumption is that the job can be dangerous for that person. Thus JSI's job evaluation scheme is to measure the degree of mismatch between its requirements and predicted or displayed capacity. The following formula represents the JSI measure:

$$JSI = \frac{job\ demand}{operator\ capacity}$$

This formula indicates that if there is a good match between the two entities, the ratio should come out to be 1.0. Operator's capacity exceeding job demand will produce a JSI value less than 1.0. It is when JSI exceeds unity that we become concerned with the situation. Then the question is: At which value of JSI should we declare the job to be unacceptable? The first answer that comes to mind is: Any time that JSI exceeds 1.0, it should be potentially dangerous. A counter argument is that the operator's real capacity is not necessarily the capacity displayed during testing. It is greater than that. Thus the truly dangerous jobs should display a JSI value greater than 1, with something to spare.

To obtain an acceptable cutoff point, the validation studies focused on industry experience with lifting and lowering tasks. The types of lifting injuries included in the analysis were musculoskeletal injuries to the back and to other body parts, surface tissue injuries due to impact, and other such injuries. A total of 101 jobs involving 385 male and 68 female industrial workers from 28 private companies were used in validation. Other data collected were injury type, injury cause (lifting or nonlifting), number of days lost, and all related expenses. Figure 3.21 provides the results. It displays the frequency of total injuries as a function of JSI. As can be observed, a JSI level of 1.5 seems to be the cutoff point. Beyond this value, there is a substantial increase in injury risk and associated expenses. The developers of JSI are now suggesting a lower cutoff point at around 1.0.

The JSI measure provides much detail for a designer with additional data. It can be used to predict individual lifting capacities. It can answer "what if" questions. The designer can make changes in the job and the operators to achieve a specific ratio that is acceptable.

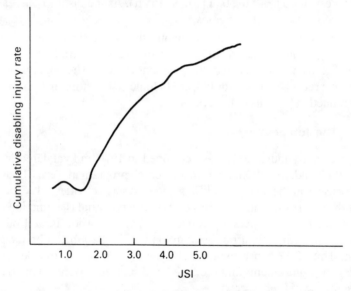

Figure 3.21 JSI versus cumulative disabling injury rate. (Reprinted with permission from *Human Factors*, Vol. 26(6), 1984 by the Human Factors Society, Inc. All rights reserved. Liles et al., "A Job Severity Index.")

The job demand. In the JSI calculation, the task demands are clustered into a job demand via the following formula:

$$JSI = \sum_{i=1}^{n} \left[\frac{\text{hours i}}{\text{hours t}} \times \frac{\text{days i}}{\text{days t}} \right] \sum_{j=1}^{mi} \left[\frac{F_j}{F_i} \times \frac{WT_j}{CAP_j} \right]$$

where

- n is the number of subtask groups
- mi is the number of tasks in group i
- days i is total days per week for group i
- days t is total days per week for the job
- hours i is exposure hours per day for group i
- hours t is number of hours per day that the job is performed
- F_j is lifting frequency for task j
- F_i is total lifting frequency for group i
- WT_j is maximum weight of lift required by task j
- CAP_j is the calculated lift capacity for task j

Population capacity. As discussed before, the lifting capacity models used in JSI have been developed via the psychophysical approach. The primary parameters used are the gender (actual or expected), frequency of lift (lifts/min), range of lift, box size (in the sagittal plane, in inches), population percentage to be accommodated, handles, and twisting motions. The range of lift is measured vertically (in inches) between the initial and the terminal points of lift. The six ranges of lift are:

1. *FK:* floor to knuckle
2. *FS:* floor to shoulder
3. *FR:* floor to reach
4. *KS:* knuckle to shoulder
5. *KR:* knuckle to reach
6. *SR:* shoulder to reach

The range of lift seldom falls within the precise ranges as given above. Table 3.9 provides range assignments for lift conditions that cross the boundaries above.

The predicted lifting capacity calculation proceeds in several steps:

1. An initial capacity is determined using the lifting range, the frequency of lift, and the gender from Table 3.10.
2. The initial capacity is adjusted by the size of the box or load using Table 3.11.
3. A secondary adjustment is made for the percentage of population to be accommodated (Tables 3.12 and 3.13).
4. The final adjustment is due to possible twisting (5% reduction in capacity) and handles (7.2% reduction for absence).

Consider the following example: Containers arrive at a stocking station on a conveyor at the rate of three per minute. Container weight is 12 kg. Conveyor height is 26 in. and the stocking point is 40 in. above ground. The stocking operation involves a twisting motion. Container size is 18 in. The container has well-defined handles. Determine the capacity of 75% of the female population in carrying out this task on a full-time basis for 5 days a week.

TABLE 3.9 LIFT RANGE ADJUSTMENTS

Point of lift initiation[a]	Point of lift termination [a]	Range assignment[b]
0 in. to KL/2	0 in. to KL + 10 in.	1. FK
	KL + 10 in. to KL + 30 in.	2. FS
	KL + 30 in. and above	3. FR
KL/2 to KL	KL/2 to KL	1. FK
	KL to KL + 30 in.	4. KS
	KL + 30 in. and above	5. KR
KL to KL + 10 in.	KL to KL + 30 in.	4. KS
	KL + 30 in. and above	5. KR
KL + 10 in. to KL + 20 in.	KL + 10 in. to KL + 20 in.	4. KS
	KL + 20 in. and above	6. SR
KL + 20 in. and above	KL + 20 in. and above	6. SR

Source: M. M. Ayoub, J. L. Selan, and B. C. Jiang, *A Mini-guide for Lifting* (Lubbock, TX: Texas Tech University, 1983), Table 7, p. 44. Research supported by NIOSH Grants 5R010H00545-02 (1978) and 5R010H00798-04 (1983).

[a]KL, knuckle level.

[b]Range   FK is the floor-to-knuckle range; FS is the floor-to-shoulder range; FR is the floor-to-reach range; KS is the knuckle-to-shoulder range; KR is the knuckle-to-reach range; SR is the shoulder-to-reach range.

TABLE 3.10 INITIAL CAPACITY CALCULATION

	Frequency of lift[a] (lifts/min)	
Range of lift	0.1 < FY < 1.0	$1.0 \leq FY \leq 12.0$
Male		
1. FK	$57.2 \times (FY) ** (-0.184697)$	$57.2 - 2.0 \times (FY - 1)$
2. FS	$51.2 \times (FY) ** (-0.184697)$	$51.2 - 2.0 \times (FY - 1)$
3. FR	$49.1 \times (FY) ** (-0.184697)$	$49.1 - 2.0 \times (FY - 1)$
4. KS	$52.8 \times (FY) ** (-0.138650)$	$52.8 - 2.0 \times (FY - 1)$
5. KR	$50.0 \times (FY) ** (-0.138650)$	$50.0 - 2.0 \times (FY - 1)$
6. SR	$48.4 \times (FY) ** (-0.138650)$	$48.4 - 2.0 \times (FY - 1)$
Female		
1. FK	$37.4 \times (FY) ** (-0.187818)$	$37.4 - 1.1 \times (FY - 1)$
2. FS	$31.1 \times (FY) ** (-0.187818)$	$31.1 - 1.1 \times (FY - 1)$
3. FR	$28.1 \times (FY) ** (-0.187818)$	$28.1 - 1.1 \times (FY - 1)$
4. KS	$30.8 \times (FY) ** (-0.156150)$	$30.8 - 1.1 \times (FY - 1)$
5. KR	$27.3 \times (FY) ** (-0.156150)$	$27.3 - 1.1 \times (FY - 1)$
6. SR	$26.4 \times (FY) ** (-0.156150)$	$26.4 - 1.1 \times (FY - 1)$

Source: M. M. Ayoub, J. L. Selan, and B. C. Jiang, *A Mini-guide for Lifting* (Lubbock, TX: Texas Tech University, 1983), Table 8, p. 45–46. Research supported by NIOSH Grants 5R010H00545-02 (1978) and 5R010H00798-04 (1983).

[a]FY, frequency of lift (lifts/min); **, exponentation (e.g., FY to the power of −0.184697). The initial lifting capacity is 57.2 lb for males and 37.4 lb for females based on the mean capacity for lift based on published data from M. M. Ayoub and S. N. Snook for the various ranges of lift for the 50th percentage and 1.0 lift/min.

TABLE 3.11 BOX SIZE ADJUSTMENT

Range of lift	Box size[a] (in. in the sagittal plane)	
	12 in. $\leq$ BX $\leq$ 18 in.	BX > 18 in.
Male		
1. FK	CAP + 1.65 $\times$ (18 $-$ BX)	CAP + 0.8 $\times$ (18 $-$ BX)
2. FS	CAP + 1.65 $\times$ (18 $-$ BX)	CAP + 0.8 $\times$ (18 $-$ BX)
3. FR	CAP + 1.65 $\times$ (18 $-$ BX)	CAP + 0.8 $\times$ (18 $-$ BX)
4. KS	CAP + 1.10 $\times$ (18 $-$ BX)	CAP + 0.8 $\times$ (18 $-$ BX)
5. KR	CAP + 1.10 $\times$ (18 $-$ BX)	CAP + 0.8 $\times$ (18 $-$ BX)
6. SR	CAP + 1.10 $\times$ (18 $-$ BX)	CAP + 0.8 $\times$ (18 $-$ BX)
Female		
1. FK	CAP + 1.10 $\times$ (18 $-$ BX)	CAP + 0.4 $\times$ (18 $-$ BX)
2. FS	CAP + 1.10 $\times$ (18 $-$ BX)	CAP + 0.4 $\times$ (18 $-$ BX)
3. FR	CAP + 1.10 $\times$ (18 $-$ BX)	CAP + 0.4 $\times$ (18 $-$ BX)
4. KS	CAP + 0.55 $\times$ (18 $-$ BX)	CAP + 0.2 $\times$ (18 $-$ BX)
5. KR	CAP + 0.55 $\times$ (18 $-$ BX)	CAP + 0.2 $\times$ (18 $-$ BX)
6. SR	CAP + 0.55 $\times$ (18 $-$ BX)	CAP + 0.2 $\times$ (18 $-$ BX)

Source: M. M. Ayoub, J. L. Selan, and B. C. Jiang, *A Mini-guide for Lifting* (Lubbock, TX: Texas Tech University, 1983), Table 9, p. 47–48. Research supported by NIOSH Grants 5R010H00545-02 (1978) and 5R010H00798-04 (1983).

[a]CAP, capacity of the lift as determined in Table 3.10; BX, box size (in.)

TABLE 3.12 FINAL ADJUSTMENT DUE TO LIFT FREQUENCY AND POPULATION PERCENTILES

Range of lift	Frequency[a]	
	0.1 $\leq$ FY < 1.0	1.0 $\leq$ FY $\leq$ 12.0
Male		
1. FK	CAP + Z $\times$ 16.86 $\times$ (FY) ** (-0.174197)	CAP + Z $\times$ (16.86 $-$ 0.5964 $\times$ (FY $-$ 1))
2. FS	CAP + Z $\times$ 15.09 $\times$ (FY) ** (-0.174197)	CAP + Z $\times$ (15.09 $-$ 0.5338 $\times$ (FY $-$ 1))
3. FR	CAP + Z $\times$ 14.47 $\times$ (FY) ** (-0.174197)	CAP + Z $\times$ (14.47 $-$ 0.5119 $\times$ (FY $-$ 1))
4. KS	CAP + Z $\times$ 14.67 $\times$ (FY) ** (-0.156762)	CAP + Z $\times$ (14.67 $-$ 0.5534 $\times$ (FY $-$ 1))
5. KR	CAP + Z $\times$ 13.89 $\times$ (FY) ** (-0.156762)	CAP + Z $\times$ (13.89 $-$ 0.5240 $\times$ (FY $-$ 1))
6. SR	CAP + Z $\times$ 13.45 $\times$ (FY) ** (-0.156762)	CAP + Z $\times$ (13.45 $-$ 0.5074 $\times$ (FY $-$ 1))
Female		
1. FK	CAP + Z $\times$ 6.87 $\times$ (FY) ** (-0.251605)	CAP + Z $\times$ (6.87 $-$ 0.1564 $\times$ (FY $-$ 1))
2. FS	CAP + Z $\times$ 5.71 $\times$ (FY) ** (-0.251605)	CAP + Z $\times$ (5.71 $-$ 0.1300 $\times$ (FY $-$ 1))
3. FR	CAP + Z $\times$ 5.16 $\times$ (FY) ** (-0.251605)	CAP + Z $\times$ (5.16 $-$ 0.1175 $\times$ (FY $-$ 1))
4. KS	CAP + Z $\times$ 5.66 $\times$ (FY) ** (-0.258700)	CAP + Z $\times$ (5.66 $-$ 0.1289 $\times$ (FY $-$ 1))
5. KR	CAP + Z $\times$ 5.01 $\times$ (FY) ** (-0.258700)	CAP + Z $\times$ (5.01 $-$ 0.1141 $\times$ (FY $-$ 1))
6. SR	CAP + Z $\times$ 4.85 $\times$ (FY) ** (-0.258700)	CAP + Z $\times$ (4.85 $-$ 0.1104 $\times$ (FY $-$ 1))

Source: M. M. Ayoub, J. L. Selan, and B. C. Jiang, *A Mini-guide for Lifting* (Lubbock, TX: Texas Tech University, 1983), Table 10, p. 49–50. Research supported by NIOSH Grants 5R010H00545-02 (1978) and 5R010H00798-04 (1983).

[a]CAP, capacity of lift as determined in Table 3.11; Z, Z score of population percentage (from normal distribution tables and as shown in Table 3.13); FY, frequency of lift (lifts/min); **, exponent.

TABLE 3.13 *Z* SCORES
FOR POPULATION
PERCENTAGES

Population (%)	*Z* Score
95	−1.645
90	−1.282
80	−0.841
70	−0.527
60	−0.255
50	0.0
40	0.255
30	0.527
20	0.841
10	1.282
5	1.645

From Table 3.10,

$$CAP_a = 30.8 - 1.1(FY - 1)$$
$$= 30.8 - 1.1(3 - 1)$$
$$= 28.6 \text{ lb}$$

From Table 3.11,

$$CAP_b = CAP_a + 0.55(18 - BX)$$
$$= CAP_a + 0.55(18 - 18)$$
$$= 28.6 \text{ lb}$$

From Tables 3.12 and 3.13,

$$CAP_c = CAP_b + Z[5.66 - 0.1289(FY - 1)]$$
$$= 28.6 + (-0.6745)[5.66 - 0.1298(2)]$$
$$= 24.95 \text{ lb}$$

Adjusting for twisting motion yields

$$CAP_d = CAP_c \times 0.95 = 23.71 \text{ lb} = 10.74 \text{ kg}$$

This result means that 75% of the female worker population can do the lifting operation above with a 10.7-kg load over an 8-hour shift. Since the ratio of 12 into 10.7 is less than 1.5, the job seems to be acceptable.

Now let's assume that the task above was embedded in another lifting task where a second conveyor brought containers of 15-in. size with handles that weigh 30 kg. Arrival frequency is 2 per minute. There is no twisting here. This task is also carried out for 8 hours a day, 5 days a week. Assuming that the same population and conveyor parameters apply, determine whether or not the job is acceptable. For the second task, using Tables 3.10–12, CAP_2 can be calculated as 12.51 kg. Using the JSI equation yields

$$\text{JSI} = \sum_{i=1}^{1} \left(\frac{8}{8} \times \frac{5}{5} \right) \left[\left(\frac{3}{5} \times \frac{12}{10.74} \right) + \left(\frac{2}{5} \times \frac{30}{12.51} \right) \right]$$
$$= (1)(0.67 + 0.957)$$
$$= 1.627$$

This job, composed of two lifting subtasks, is unacceptable.

3.5.4 The Revised NIOSH Equation (1991)

In 1991 a panel from academia, industry and government developed a revised NIOSH equation. Although epidemiological support for the new equation is still lacking, the 1991 equation was published [82] in 1993 and detailed information on it is available through a government publication [83] available from the National Technical Information Service. As stated by the developers, the revised equation is based on research findings since the 1981 guide and includes additional lifting task parameters such as asymmetrical lifting and quality of hand-container couplings in addition to expanded range of lifting frequencies and work durations. The rationale for the revised equation includes the 1985 suggestion by the Musculoskeletal National Strategy Committee of stressing engineering controls over administrative for control of lifting hazards, and other changes due to more current research results. The authors state that application of the new equation should be limited to the conditions for which it was designed. A complete list of work conditions not covered by it are listed in [83], including one-handed lifting/lowering, lifting/lowering extremely cold or hot objects, with slip or fall potential, for over 8 hours, in restricted work areas, etc.

With stress on engineering controls, the new equation calculates only one hazard level, the *lifting index* (*LI*). *LI* is a ratio of *recommended weight limit* (*RWL*) and the weight actually lifted (*L*) or planned for lifting. Hence $LI = L/RWL$. If *L* varies from lift to lift, consider both the average and the maximum load. The *RWL* is calculated using the following equation:

$$RWL = LC \times HM \times VM \times DM \times AM \times FM \times CM$$

The parameters in the *RWL* equation above are the load constant (*LC*), horizontal multiplier (*HM*), vertical multiplier (*VM*), distance multiplier (*DM*), asymmetric multiplier (*AM*), and coupling multiplier (*CM*). If *LI* comes out to a number above 1.0, the revised equation recommends redesign of the lifting task, including automation. Any task resulting in a calculated *LI* less than or equal to 1.0 may be acceptable to nearly all healthy American workers, male or female, accustomed to physical labor. If significant control is required, it is recommended that *LI* should be calculated at the beginning as well as at the end of lift. Significant control is primarily determined by care or precaution requirements, such as regrasp of load near the destination of lift, momentary holding of load or positioning/guiding requirements at destination. It is interesting to see a striking similarity in the *LI* and the JSI concept discussed in Section 3.5.3.

In the *RWL* equation, the load constant in metric units is 23 kilograms and 51 pounds in the U.S. customary units. As in the previous equation, other multipliers result in deductions from these values unless the lifting task parameters are optimum. There is a relationship between multipliers and task parameters or task variables as denoted in Table 3.14.

The task variables in the new equation are quite similar to those in the 1981 equation:

· *H* is the horizontal location of hands from midpoint between the ankles. Measure at the origin and at the end of lift (cm or in). *H* assumes a minimum value of 25 cm or 10 in and a maximum value of 63 cm or 25 in. Set *HM* = 0 for *H* > 25 in.

· *V* is the vertical height of hands from the floor. Measure from the large middle knuckle at the origin and at destination of lift (cm or in). There is no minimum value. *V* assumes a maximum of 70 in or 175 cm limited by reach. Set *VM* = 0 for *V* > 70 in.

· *D* is the vertical travel distance between the origin and the destination of lift. *D* assumes a minimum value of 25 cm or 10 in and a maximum value of 175 cm or 70 in. Set *DM* = 0 for *D* > 70 in.

· *A* is the angle of asymmetry which refers to a lift that begins or ends outside the sagittal plane. This is the angular displacement of the load from the sagittal plane. Angle *A* is limited to the range from 0° to 135°. Set *AM* = 0 for A > 135°.

· *F*, the lifting frequency, is the average number of lifts made per minute, as measured over a 15-minute period. Duration is defined as short (≤ 1 h), moderate (> 1 h but ≤ 2 h), and long (> 2 h but ≤ 8 h). Set *FM* to appropriate value using Table 3.15.

· The coupling multiplier (*CM*) is an index evaluating the effectiveness of the hand-container interface. The evaluation is along the lines of good, fair, and poor as per Table 3.16.

Finally, Table 3.17 is used for the appropriate *CM* value.

There may be cases where an operator may be involved with multiple lifting or lowering tasks. The 1981 equation dealt with this condition via a lifting/lowering frequency weighted *AL* and *MPL* calculations. The 1991 equation proposes a different approach to assessing the composite stress via a composite lifting index (CLI). This approach cal-

TABLE 3.14 REVISED NIOSH EQUATION PARAMETERS

		Metric	U.S. Customary
Load Constant	LC	23 kg	51 lb
Horizontal Multiplier	HM	$(25/H)$	$(10/H)$
Vertical Multiplier	VM	$1-(.003\|V\text{-}75\|)$	$1-(.0075\|V\text{-}30\|)$
Distance Multiplier	DM	$82+(4.5/D)$	$.82+(1.8/D)$
Asymmetric Multiplier	AM	$1-(.0032A)$	$1-(.0032A)$
Frequency Multiplier	FM	From Table 3.15	From Table 3.15
Coupling Multiplier	CM	From Table 3.17	From Table 3.17

TABLE 3.15 FREQUENCY MULTIPLIER TABLE (FM)

Frequency Lifts/min $(F)^2$	Work Duration					
	≤1 Hour		>1 but ≤2 Hours		>2 but ≤8 Hours	
	$V<30^1$	V≥30	V<30	V≥30	V<30	V≥30
≤0.2	1.00	1.00	.95	.95	.85	.85
.05	.97	.97	.92	.92	.81	.81
1	.94	.94	.88	.88	.75	.75
2	.91	.91	.84	.84	.65	.65
3	.88	.88	.79	.79	.55	.55
4	.84	.84	.72	.72	.45	.45
5	.80	.80	.60	.60	.35	.35
6	.75	.75	.50	.50	.27	.27
7	.70	.70	.42	.42	.22	.22
8	.60	.60	.35	.35	.18	.18
9	.52	.52	.30	.30	.00	.15
10	.45	.45	.26	.26	.00	.13
11	.41	.41	.00	.23	.00	.00
12	.37	.37	.00	.21	.00	.00
13	.00	.34	.00	.00	.00	.00
14	.00	.31	.00	.00	.00	.00
15	.00	.28	.00	.00	.00	.00
>15	.00	.00	.00	.00	.00	.00

[1] Values of V are in inches.
[2] For lifting less frequently than once per 5 minutes, set F = .2 lifts/minute

culates the sum of the largest single task lifting index and the incremental additions due to each subsequent lifting/lowering task. The process is well described in [83] and [84].

An example may clarify the procedure (Figure 3.22). Assume that a worker manually moves trays of clean dishes from a conveyor at the end of a dishwashing machine and then loads them on a cart with a momentary hold. The tray with dishes weighs 20 lbs. The job takes one hour to complete, and during this period, 300 trays are handled at a fairly constant rate. Worker twists to one side to get hold of a tray and then rotates to the other side to lower it. Trays have well designed hand-hold cutouts. Evaluate the handling risk due to this activity.

Since a momentary hold of trays is required, RWL at both origin and destination of the handling activity need to be assessed:

Hand Location: Origin ($H = 20''$, $V = 44''$), Destination ($H = 20''$, $V = 7''$);
$HM = 10/H = 10/20 = 0.5$ for orig/dest
$VM = 1-(0.0075\,|\,V-30\,|)$; $VM = 0.9$ for origin; 0.83 for destination.
Vertical Distance: $D = 37''$; $DM = 0.82+1.8/D$; $DM = 0.87$
Asymmetric Angle: $A = 30$ degrees; $AM = 1-(0.0032A)$; $AM = 0.9$
Frequency: $300/60\ F = 5$ lifts/min
Duration: 1 h —> $FM = 0.8$ from Table 3.15
Coupling: Well designed cutout, good (see Table 3.16; $CM = 1.00$ from Table 3.17)

RWL (origin) = $(51)(0.5)(0.9)(0.87)(0.9)(0.8)(1.0) = 14.4$ lb; $LI = 20/14.4 = 1.4$
RWL (dest.) = $(51)(0.5)(0.83)(0.87)(0.9)(0.8)(1.0) = 13.3$ lb; $LI = 20/13.3 = 1.5$

TABLE 3.16   HAND-TO-CONTAINER COUPLING CLASSIFICATION

Good	Fair	Poor
1. For containers of optimal design, such as some boxes, crates, etc., a "good" hand-to-object coupling would be defined as handles or hand-hold cutouts of optimal design (see notes 1 to 3 below).	1. For containers of optimal design, a "fair" hand-to-object coupling would be defined as handles or hand-hold cutouts of less than optimal design (see notes 1 to 4 below).	1. Containers of less than optimal design or loose parts or irregular objects that are bulky, hard to handle, or have sharp edges (see note 5 below).
2. For loose parts or irregular objects which are not usually containerized, such as castings, stock, and supply materials, a "good" hand-to-object coupling would be defined as a comfortable grip in which the hand can be easily wrapped around the object (see note 6 below).	2. For containers of optimal design with no handles or hand-hold cutouts, or for loose parts or irregular objects, a "fair" hand-to-object coupling is defined as a grip in which the hand can be flexed about 90 degrees (see note 4 below).	2. Lifting non-rigid bags (i.e., bags that sag in the middle).

1. An optimal handle design has .75–1.5 inches (1.9 to 3.8 cm) diameter, $\geq$4.5 inches (11.5 cm) length, 2 inches (5 cm) clearance, cylindrical shape, and a smooth, non-slip surface.
2. An optimal hand-hold cutout has the following approximate characteristics: $\geq$1.5 inch (3.8 cm) height, 4.5 inch (11.5 cm) length, semi-oval shape, $\geq$2 inch (5 cm) clearance, smooth non-slip surface, and $\geq$0.25 inches (0.60 cm) container thickness (e.g., double thickness cardboard).
3. An optimal container design has $\leq$16 inches (40 cm) frontal length, $\leq$12 inches (30 cm) height, and a smooth non-slip surface.
4. A worker should be capable of clamping the fingers at nearly 90° under the container, such as required when lifting a cardboard box from the floor.
5. A container is considered less than optimal if it has a frontal length > 16 inches (40 cm), height > 12 inches (30 cm), rough or slippery surfaces, sharp edges, asymmetric center of mass, unstable contents, or requires the use of gloves. A loose object is considered bulky if the load cannot easily be balanced between the hand-grasps.
6. A worker should be able to comfortably wrap the hand around the object without causing excessive wrist deviations or awkward postures, and the grip should not require excessive force.

TABLE 3.17 COUPLING MULTIPLIER

Coupling Type	Coupling Multiplier	
	$V<30$ inches (75 cm)	$V\geq30$ inches (75 cm)
Good	1.00	1.00
Fair	0.95	1.00
Poor	0.90	0.90

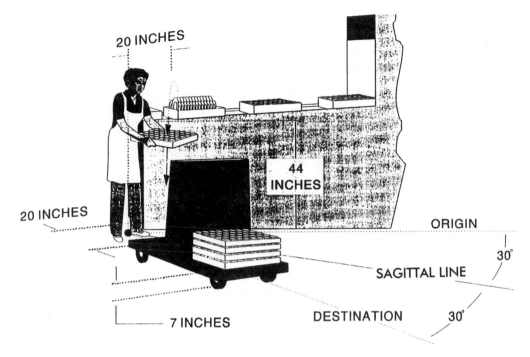

20 INCHES

20 INCHES

44 INCHES

7 INCHES

ORIGIN

30°

SAGITTAL LINE

DESTINATION   30°

Figure 3.22   Dishwashing Machine Unloading

Since *LI* is greater than 1.0, for both cases this task is unacceptable and must be redesigned.

A cross-validation of the NIOSH (1991) limits for manual lifting [85] showed that the assumptions made in the equation for the psychological criterion were valid. However, cross-validation for the biomechanical and physiological criteria did not totally agree with the 1991 equation.

3.6   SUMMARY

Physical ergonomics focuses on the physical elements of the activity. It evaluates human attributes such as strength and endurance and relates them to the task. Furthermore, task elements are also evaluated with respect to their impact on the human being, especially the cardiovascular and musculoskeletal systems. Evaluating the physical capabilities of people relative to task demands is another focus of physical ergonomics.

QUESTIONS

1. Define *stress* and strain.
2. Discuss the relationship between stress and performance.
3. What are the primary sources of job stress?
4. What is metabolism? Distinguish between basal and activity metabolism.
5. Discuss the energy cycle for both anerobic and aerobic work.
6. Discuss the types of muscular effort. Which biological responses (measures) are more appropriate to each?
7. Discuss the behavior of heart rate at rest, work, and recovery.
8. What are the rules of thumb for keeping energy cost of work within acceptable limits?
9. When is it necessary to allow for rest during muscular work?
10. What is the difference between strength and endurance?
11. How are biomechanical models used in ergonomics?
12. What are the reasons for cumulative trauma disorders (CTDs)?
13. How can CTDs be prevented?
14. What is carpal tunnel syndrome?
15. How can material handling injuries be prevented?
16. Discuss the psychophysical approach to evaluating injury potential due to manual material handling.
17. What is the relationship between MPL and AL in the NIOSH (1981) lifting guide?
18. What is the underlying concept in the JSI (job severity index)?
19. What were the reasons for the development of the 1981 NIOSH Revised Lifting Equation?

EXERCISES

1. Select four jobs in your environment. Break each job into a minimum of five elements. Analyze each element with respect to where it falls on the human activity continuum given by Figure 3.1.
2. Calculate the total metabolism for a 70-kg male over a 24-hour period who follows the following regimen:
   · Light assembly: 3 hr
   · Bench soldering: 2 hr
   · Laboratory work: 3 hr
   · Mowing: 2 hr
   · Weeding: 1 hr
   · Resting: 13 hr

3. Calculate the rest period for a mowing task performed over 3 hours. Assume that a female of 40 years of age is performing the task.

4. Modify the first example given in Section 3.3.1 for the case where the forearm makes an angle of 20 degrees with the horizontal. Calculate the elbow reactive force (R_E) and moment (M_E).

5. Assume that a person is pulling on a handle with the forearm in the horizontal direction. Draw a free-body diagram showing all elbow reactive forces and moments. Show the parametric equivalents of R_E and M_E.

6. A job requires a 20-kg box to be lifted from a platform 20 cm above ground to a shelf through a vertical distance of 40 cm. The expected frequency of lift is 200 times per hour. At the origin of lift, the load center of mass is 20 cm away from lumbar spine. Under which conditions can this job be performed?

7. Modify the examples given in Section 3.5.3 for a male operator. Is the combined task composed of two lifting subtasks acceptable? Show your work.

REFERENCES

1. Sanders, M. S., and McCormick, E. J. 1993. *Human Factors in Engineering and Design,* 7th ed. McGraw-Hill, New York.
2. Welford, A. T. 1973. Stress and Performance. *Ergonomics,* Vol. 16(5), pp. 567–580.
3. Corcoran, D. W. J. 1965. Personality and the Inverted U Relation. *British Journal of Psychology.* 52, pp. 267–273.
4. Staff, K. R. 1983. A Comparison of Range of Joint Mobility in College Females and Males. M.S. Thesis. Industrial Engineering, Texas A&M University, College Station, TX.
5. Wolff, J. 1892. *Das Gesetz der Transformation der Knochen.* Hirschwald, Berlin.
6. Astrand, P.O., and Rodahl, K. 1970. *Textbook of Work Physiology.* McGraw-Hill, New York.
7. Green, J. H. 1968. *An Introduction to Human Physiology.* Oxford University Press, London.
8. Grandjean, E. 1980. *Fitting the Task to the Man.* Taylor & Francis, London.
9. Seeley, R. R. 1991. *Essentials of Anatomy and Physiology.* Mosby-Year Book, Inc., St. Louis.
10. Dukes-Dobos, F. N., Wright, G., Carlson, W. S., and Cohen, H. H. 1976. Cardiopulmonary Correlates of Subjective Fatigue. *Proceedings of 20th Annual Meeting of the Human Factors Society,* July 11–16, pp. 24–27.
11. Balanescu, B. F. 1979. Some Aspects Concerning the Dynamic Evaluation of Oxygen Consumption in Exercise Tests. *Ergonomics,* 22(12), pp. 1337–1342.
12. Astrand, P.-O., and Rodahl, K. 1986. *Textbook of Work Physiology,* 3rd ed. McGraw Hill, New York.
13. Rodahl, K. 1989. *The Physiology of Work.* Taylor & Francis, London.
14. Rodahl, K., Vokac, Z., Fugelli, R. Vaage, O., and Maehlum, S. 1974. Circulatory Strain, estimated Energy Output and Catecholamine Excretion in Norwegian Coastal Fishermen. *Ergonom*ics, 17(5), p. 585.
15. Rodahl, K., and Vokac, Z. 1977. The Work Physiology of Fishing. *Nordic Council Arctic Medical Research Report,* 18, p. 22.

16. Datta, S. R., and Ramanathan, N. L. 1971. Ergonomics Comparison of Seven Modes of Carrying Loads on the Horizontal Plane. *Ergonomics*, 14(2), pp. 269–278.

17. Gordon, E. E. 1957. The Use of Energy Costs in Regulating Physical Activity in Chronic Disease. *A.M.A. Archives of Industrial Health*, 16, pp. 437–441.

18. Vos, H. W. 1973. Physical Workload in Different Body Postures, While Working Near to, or Below Ground Level. *Ergonomics*, 16(6), pp. 817–828.

19. Passmore, R., and Durnin, U. V. G. A. 1955. Human Energy Expenditure. *Physiological Reviews*, 35, pp. 83–89.

20. Astrand, I. 1960. Aerobic Work Capacity in Men and Women with Special Reference to Age. *Acta Physiologica Scandinavica*, 49 (Suppl. 169).

21. Bonjer, F. H. 1962. Actual Energy Expenditure in Relation to Physical Working Capacity. *Ergonomics*, 5, pp. 467–470.

22. Brouha, L. 1960. *Physiology in Industry*. Pergamon Press, Elmsford, NY.

23. Christensen, E. H. 1953. Physiological Valuation of Work in the Nykroppa Iron Works. In *Ergonomics Society Symposium on Fatigue*. Floyd, W. F., and Welford, A. T. (Eds.). H. K. Lewis, London.

24. Edholm, O. G. 1967. *The Biology of Work*. World University Library, McGraw-Hill, New York.

25. Volle, M., Brisson, G. R., Perusse, M., Tanaka, M., and Doyan, J. 1979. Compressed Work Week: Psycho-Physiological and Psychological Repercussions. *Ergonomics*, 22(9), p. 1001.

26. Astrand, I. 1960. Aerobic Work Capacity in Men and Women with Special Reference to Age. *Acta Physiologica Scandinavica*, 49 (Suppl. 169).

27. Yates, J. W., Pickering, K., and Karwowski, W. 1990. High Frequency Lifting and Blood Lactate Production. In *Advances in Industrial Ergonomics and Safety II*. Das, B. (Ed.). Taylor & Francis, London.

28. Edwards, R. H. T. 1981. Human Muscle Function and Fatigue. In *Human Muscle Fatigue: Physiological Mechanisms*. Pitman Medical, London, pp. 1–18.

29. Karlsson, J., Funderburk, B., Essen, B., and Lind, A. R. 1975. Constituents of Human Muscle in Isometric Fatigue. *Journal of Applied Physiology*, 38, pp. 208–211.

30. Chaffin, D. F., and Andersson, G. 1984. *Occupational Biomechanics*. Wiley, New York.

31. Rohmert, W. 1973. Problems in Determining Rest Allowances. Part 1: Use of Modern Methods to Evaluate Stress and Strain in Static Muscular Work. *Applied Ergonomics*, 4, pp. 91–95.

32. Rohmert, W. 1973. Problems in Determination of Rest Allowances. Part 2: Determining Rest Allowances in Different Human Tasks. *Applied Ergonomics*, 4, pp. 158–162.

33. Murrell, K. F. H. 1965. *Human Performance in Industry*. Reinhold, New York.

34. Spitzer, H. 1951. Physiologische Grundlagen fur den Erholungszuschlag bei Schwerarbeit. *REFA-Nachrichten*, 2, Darmstadt.

35. Astrand, P. O. 1952. *Experimental Studies of Physical Working Capacity in Relation to Sex and Age*. Munksgaard, Copenhagen.

36. Mital, A., and Shell, R. L. 1984. Determination of Rest Allowances for Repetitive Physical Activities That Continue for Extended Hours. In *Proceedings of the Annual International Conference*. Institute of Industrial Engineers, Norcross, GA, pp. 637–645.

37. Ayoub, M. M., and Mital, A. 1989. *Manual Materials Handling*. Taylor & Francis, London.

38. Diffrient, N., Tilley, A. R., and Harman, D. 1983. *Humanscale 1/2/3, 4/5/6, 7/8/9*. MIT Press, Cambridge, MA.

39. Hunsicker, P. A. 1955. Arm Strength at Selected Degrees of Elbow Flexion. *Technical*

Report 54–548. U.S. Air Force, WADC, Ohio.

40. Kumar, S., Narayan, Y., and Bacchus, C. 1995. Symmetric and Asymmetric Two-Handed Pull-Push Strength of Young Adults. *Human Factors* 37(4), pp. 854–865.

41. Radwin, R. G., Jensen, T. R., Oh, S., and Webster, J. G. 1990. Submaximal Pinch Forces Measured Using Miniature Force Transducers Attached to the Distal Finger Pads. In *Advances in Industrial Ergonomics and Safety II*. Das, B. (Ed.). Taylor & Francis, London, pp. 151–158.

42. Kinoshita, H. Murase, T., and Bandou, T. 1996. Grip Posture and Focus During Holding Cylindrical Objects with Circular Grips. *Ergonomics,* 39(9), pp. 1163–1176.

43. Kroemer, K. H. E. 1970. Human Strength: Terminology, Measurement, and Interpretation of Data. *Human Factors,* 12(3), pp. 297–313.

44. Fredericks, T. K., Kattel, B. P., and Fernandez, J. E. 1995. Is Grip Strength Maximum in the Neutral Posture? In *Advances in Industrial Ergonomics and Safety VII*. Bittner, A. C., and Champney, P. C. (Eds.). Taylor & Francis, London, pp. 561–568.

45. Stobbe, T. J. 1982. The Development of a Practical Strength Testing Program in Industry. Ph.D. dissertation. University of Michigan, Ann Arbor.

46. Chaffin, D. B. 1975. On the Validity of Biomechanical Models of the Low Back for Weight Lifting Analysis. *ASME Proceedings,* 75-WA-Bio-1. ASME, New York.

47. Morris, J. M., Lucas, D. B., and Bressler, B. 1961. Role of the Trunk in the Stability of the Spine. *Journal of Bone and Joint Surgery,* 43A, pp. 327–351.

48. Park, K. S., and Chaffin, D. B. 1974. A Biomechanical Evaluation of Two Methods of Manual Load Lifting. *AIIE Transactions,* 6(2), pp. 105–113.

49. Eastman Kodak Co., Ergonomics Group. 1986. *Ergonomic Design for People at Work,* Vol. II. Van Nostrand Reinhold, New York.

50. Putz-Anderson, Vern (Ed.) 1988. *Cumulative Trauma Disorders: A Manual for the Musculoskeletal Diseases of the Upper Limbs.* Taylor & Francis, London.

51. U.S. Department of Labor, Bureau of Labor Statistics. 1996. *Characteristics of Injuries and Illnesses Resulting in Absences from Work—1996.* USDL–96–163.

52. Silverstein, B. 1985. The Prevalence of Upper Extremity Cumulative Trauma Disorders in Industry. Ph.D. dissertation. University of Michigan, Ann Arbor.

53. Silverstein, B. A., Fine, L. J., and Armstrong, T. J. 1986. Hand Wrist Cumulative Trauma Disorders in Industry. *British Journal of Industrial Medicine*, 43, pp. 779–784.

54. Snook, S. H., Vaillancourt, D. R., and Ciriello, V. M. 1995. Psychophysical Studies of Repetitive Wrist Flexion and Extention. *Ergonomics*, 38(7), pp. 1488–1507.

55. Cobb, T. K., Cooney, W. P., and An, Kai-Nan. 1996. Aetiology of Work-Related Carpal Tunnel Syndrome: The Role of Lumbrical Muscles and Tool Size on Carpal Tunnel Pressures. *Ergonomics*, 39(1), pp. 103–107.

56. Kumar, S. 1990. Analysis of Selected High Risk Operations in a Garment Industry. In *Advances in Industrial Ergonomics and Safety II*. Das, B. (Ed.). Taylor & Francis, London, pp. 227–236.

57. Imrhan, S. 1990. Cumulative Trauma Disorders in Glass Manufacturing. In *Advances in Industrial Ergonomics and Safety II*. Das, B. (Ed.). Taylor & Francis, London, pp. 237–243.

58. Alexander, D. C. 1986. *The Practice and Management of Industrial Ergonomics.* Prentice Hall, Englewood Cliffs, NJ.

59. Rowe, M. L. 1983. *Backache at Work.* Perinton Press, Fairport, NY.

60. National Institute for Occupational Safety and Health. 1981. A Work Practices Guide for Manual Lifting. *Technical Report 81–122.* U.S. Department of Health and Human Services, Cincinnati, OH.

61. Andeson, G. B. J., et al. 1991. Epidemiology and Cost. In *Occupational Low Back Pain.* Pope, M. H., et al. (Eds.). Mosby-Year Book, pp. 95–113.
62. Behrens, V., et al. 1994. The Prevalence of Back Pain, Hand Discomfort, and Dermatitis in the U.S. Working Population. *American Journal of Public Health*, 84, pp. 1780–1785.
63. Snook, S. H., and Jensen, R. C. 1984. Cost. In *Occupational Low Back Pain.* Pope, I. R., Rymoyer, S. M., and Andersson, C. H. (Eds.). Praeger, New York.
64. Chaffin, D. B.1985. Manual Materials Handling Limits. In *Industrial Ergonomics: A Practitioner's Guide.* Alexander, D. C., and Pulat, B. M. (Eds.). Industrial Engineering and Management Press, Atlanta, GA.
65. Adams, W. C. 1967. Influence of Age, Sex, and Body Weight on the Energy Expenditure of Bicycle Riding. *Journal of Applied Physiology,* 22, pp. 539–545.
66. Sheppard, R. J. 1974. *Men at Work.* Charles C. Thomas, Springfield, IL.
67. Petrofsky, J. S., and Lind, A. R. 1975. Aging, Isometric Strength and Endurance, and Cardiovascular Responses to Static Effort. *Journal of Applied Physiology*, 38, pp. 91–95.
68. Montoye, H. J., and Lamphiear, D. E. 1977. Grip and Arm Strength in Males and Females, Age 10 to 69. *The Research Quarterly*, 48, pp. 109–120.
69. Karwowski, W. 1982. A Fuzzy Set Based Model on Interaction Between Stresses Involved in Manual Lifting Tasks. Ph.D. dissertation. Texas Technical University, Lubbock.
70. Petrofsky, J. S., and Lind, A. R. 1978. Metabolic, Cardiovascular and Respiratory Factors in the Development of Fatigue in Lifting Tasks. *Journal of Applied Physiology: Respiratory, Environmental and Exercise Physiology*, 45, pp. 64–68.
71. Intaranont, K. 1983. A Study of Anaerobic Threshold for Lifting Tasks. Ph.D. dissertation. Texas Technical University, Lubbock.
72. Massas, W. S., Lavender, S. A., Leurgans, S. E., Fathallah, F. A., Ferguson, S. A., Allread, W. G., and Rajulu, S. L. 1995. Biomechanical Risk Factors for Occupationally Related Low Back Disorders. *Ergonomics*, 38(2), pp. 377–410.
73. Snook, S. H. 1978. The Design of Manual Handling Tasks. *Ergonomics,* 21, pp. 963–985.
74. Rehnlund, S. 1973. *Ergonomi.* Translated by C. Soderstrom. Volvo Bildungskonconcern, Stockholm.
75. Snook, S. H., and Ciriello, V. M. 1991. The Design of Manual Handling Tasks: Revised Tables of Maximum Acceptable Weights and Forces. *Ergonomics*, 34(9), pp. 1197–1213.
76. Ciriello, V. M., Snook, S. H., and Hughes, G. 1993. Further Studies of Psychophysically Determined Maximum Acceptable Weights and Forces. *Human Factors*, 35(1), pp. 175–186.
77. Straker, L. M., Stevenson, M. G., and Twomey, L. T. 1996. A Comparison of Risk Assessment of Single and Combination Manual Handling Tasks: 1. Maximum Acceptable Weight Measures. *Ergonomics*, 39(1), pp. 128–140.
78. Bloswick, D. S., Gerber, A., Sebesta, D., Johnson, S. and Mecham, W. 1994. Effect of Mailbag Design on Musculoskeletal Fatigue and Metabolic Load. *Human Factors* 36(2), pp. 210–218.
79. National Institute for Occupational Safety and Health. 1994. Back Belts: Do They Prevent Injury? *DHHS (NIOSH) Publication 94*–127. Cincinnati, OH.
80. Kraus, J. 1996. Reduction of Acute Low Back Injuries by Use of Back Supports. *International Journal of Occupational and Environmental Health*, 2(3), Oct.–Dec.
81. Ayoub, M. M., Selan, J. L., and Jiang, B. C. 1983. *A Mini-guide for Lifting.* Texas Technical University, Lubbock.
82. Waters, T. R., Putz-Anderson, V., and Garg, A. 1993. Revised NIOSH Equation for the Design and Evaluation of Manual Lifting Tasks. *Ergonomics*, 36, pp. 749–776.

83. Waters, T. R., Putz-Anderson, V., and Garg, A. 1994. Applications Manual for the Revised NIOSH Equation. *DHHS (NIOSH) Publication 94–110*. Cincinnati, OH.

84. Garg, A., Waters, T. R., and Putz-Anderson, V. 1994. Multiple Task Analysis Using Revised NIOSH Equation for Manual Lifting. In *Advances in Industrial Ergonomics and Safety VI*. Aghazadeh, F. (Ed.). Taylor and Francis, London, pp. 67–70.

85. Hidalgo, J., Genaidy, A., Karwowsky, W., Christensen, D., Huston, R., and Stambough, J. 1995. A Cross-Validation of the NIOSH Limits for Manual Lifting. *Ergonomics*, 38(12), pp. 2455–2464.

_____CHAPTER 4

Information
Ergonomics

The focus of Chapter 3 was on the physical elements of human work. In this chapter we present the remaining major work elements: information acquisition and information processing.

4.1 INFORMATION INPUT PROCESSES

Human beings receive stimuli from many sources. Some of these can be sensed, some cannot. The stimuli are distant or close sources of energy, such as light, thermal energy, mechanical energy, chemical energy, sound, and so on. People can sense those levels of stimuli that are within the spectra to which they are sensitive. Among the human sensory channels are the five classic senses (vision, audition, smell, taste, and touch), known as the *exteroceptors*. These are so called since they deal with stimuli external to the body. On the other hand, *proprioceptors* are stimulated by the actions of the body itself, such as a reach or a sudden turn. Proprioceptors are embedded within the subcutaneous tissue, such as in the muscles and tendons, around the joints, and in the inner ear.

The external stimuli that are within the sensitivity ranges excite the appropriate exteroceptors on a selective basis. Sound excites the auditory channels, light excites the visual channels. Collectively, the external stimuli provide the "data" that are sensed by the human. Perception differs from sensation in that perception involves attaching meaning to the stimuli. Just as there is a difference between "data" and "information," there is a difference between what is sensed and what is perceived. We deal with this topic in more detail later in this chapter.

A special category of the proprioceptors is the *kinesthetic* sense. Receptors of this sense are clustered primarily around the joints. Kinesthetic sense is used to obtain feedback as to the status of our limbs at any given point in time during motion. We will not deal with kinesthesis too much in this chapter, but, it should be remembered that this sense is extensively used after skill acquisition on the job. One of the primary prerequisites of skilled performance is the notable absence of visually guided motions and more reliance on the kinesthetic feedback.

In the remainder of this chapter we dwell on the five exteroceptors with special emphasis on human capability and limitation information on each, followed by information processing and decision functions, and finally, a special section on human–computer interface, a related area that is popular in information ergonomics.

4.2 THE EXTEROCEPTORS

As presented above, these are the five senses in the Aristotelean [1] tradition. In this section they are presented in the order of importance as well as frequency of use during work. Most of the discussion centers around vision, the most important sense.

4.2.1 The Eye

Light is visually evaluated radiant energy. The human organ that is sensitive to light is the eye. As shown in Figure 4.1, the visible spectrum out of the electromagnetic radiant energy continuum is very small [2]. It is estimated that people acquire 80% of the information

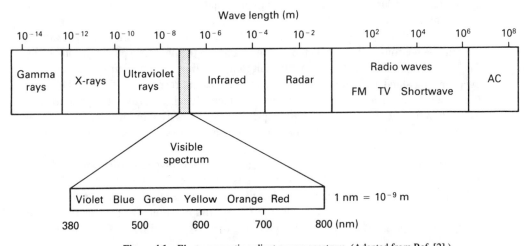

Figure 4.1 Electromagnetic radiant energy spectrum. (Adapted from Ref. [2].)

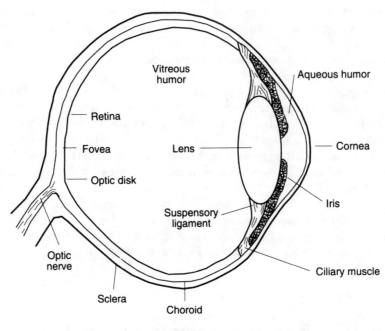

Figure 4.2 The eye.

through the visual channels. The eye is a globe about 2.5 cm (1 in.) in diameter (Figure 4.2). Six muscles attach to the eyeball by means of ligaments. They serve to move the eyeball up–down, right–left, and to corkscrew (roll out or in). Two chambers of the eyeball are separated by the iris and the lens. The anterior and the vitreal chambers are filled with fluids with optical properties that help bend the light rays.

The transparent covering in front of the eye, the cornea, admits light through the anterior chamber and then the pupil, the aperture within the iris. The pupil is a variable aperture that gets larger in dark surroundings and smaller in bright environments. The light is brought to a focus on the retina by the adjustable lens and the vitreous humor. The retina is an outgrowth of the brain that is composed of a complex structure of nervous tissue. Among the neurons that make up most of the retina are the photoreceptors *rods* and *cones*. The cones respond to light levels equivalent to light falling on a white paper 30 cm (1 ft) away from a standard candle, and above. Thus cones are responsible for daylight vision (photopic vision). The rods respond to low illumination levels equivalent to light impinging on earth in full moon, or below. Thus the primary receptors used in low illumination (scotopic vision) are the rods. A combination of rods and cones are used during in-between levels of light (mesopic vision).

The retina contains two distinct regions, called the fovea and the optic disk. The *fovea*, rich in cones, is a shallow pit centered within the yellow spot. The yellow spot contains many cones, increasing in number toward the center of the spot. The fovea is the region of maximum visual acuity under sufficient illumination. On the other hand, rod concentration peaks at around ± 20 degrees off the fovea [3] (Figure 4.3). Thus for best

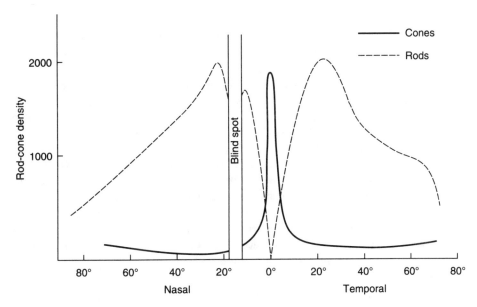

Figure 4.3 Rod-cone distribution on the retina. (Adapted from Ref. [3].)

night vision, one has to look at about 20 degrees off the intended target to get the most of it. The *optic disk* is the exit point from the retina of the nerve fibers that make up the optic nerve. The disk is free of photoreceptors and hence is effectively blind. It is for this reason that the optic disk is sometimes referred to as the blind spot.

The tissue layer between the retina and the external sheathing of the eyeball is called the *choroid membrane*. It is dark in color to absorb the light that is not taken up by the retina. The *sclera*, the outer layer of the eyeball, is a rather tough tissue protecting the structure.

Another important structure in the eye is the *lens*. Its curvature is controlled by the ciliary muscle and the suspensory ligament. The thickening of the lens is accomplished by the relaxation of the muscles, with a shortening of the focal length to bring the nearby objects into proper focus on the retina. A longer focal length is achieved by the flattening of the lens in order to bring the far objects into proper focus. The image of an object on the retina is reversed and inverted. The altering of the power of the lens for proper focusing is called *accommodation*.

Several factors determine the effectiveness of visual functioning. These factors may first be clustered into two groups: the *internal* and the *external*.

Internal factors. The internal factors relate to the human's physiological characteristics. Among them are:

1. *Visual acuity.* Acuity is sharpness of vision. It is the ability to see fine detail. When opthalmologists check visual performance on the Snellen scale, they are checking for visual acuity. A 20/100 vision indicates that the person being checked can see a detail from 20 m whereas a person with perfect vision can see the same detail from a distance of 100 m. Hence, a 20/100 score indicates poor vision. A 20/15 vision is better than perfect since a detail that can normally be seen at a distance of 15 m can be detected by the subject at a distance of 20 m.

A type of visual acuity that is frequently cited is the minimum separable acuity. This is measuring the smallest feature an eye can detect. Such acuity is usually measured in terms of the inverse of the visual angle subtended at the eye by the smallest detail that can be detected (Figure 4.4). The standard is usually 1 minute of arc since its reciprocal is still 1. If a person can discriminate a detail that subtends an arc of 1.5 minutes, the acuity score is 0.67 (1/1.5). A person with better visual acuity has a score greater than 0.67.

Visual acuity depends on several factors, including the shape of the eyeball and the flexibility of the lens. Nearsightedness and farsightedness are anomalies that are caused primarily by the loss of flexibility of the lens. Normal near vision requires that the lens assume a bulged condition to refract the divergent stimulus lines with greater power for the image to fall on the retina. In nearsightedness, the lens remains in a bulged condition even for far objects. Thus the image of far objects forms in front of the retina, blurring vision for such objects. Furthermore, normal far vision calls for the lens to assume a flat condition. In farsightedness, the lens remains in that flat state even for near objects. Lacking the refraction power of the lens, the image forms at the back of the retina. Both of these conditions can easily be corrected by appropriate lenses; however, many people with uncorrected conditions still attempt to function normally, often not being able to detect important signals. In an industrial environment, it is recommended that the workers be checked for visual performance at least once in 2 years. Some jobs demand excellent acuity, such as inspection of completed products for rejects. Other jobs require adequate functioning of the eyes. An example is an analog display on a machine indicating angular speed of a rotating arm. The following is an excerpt from a company's records: "One particular success story is worth mentioning. The quality checker who made the lowest score on both tests became the best 'defect finder/evaluator' in the section after the checker had an eye examination and purchased a pair of eye glasses with bifocal lenses."

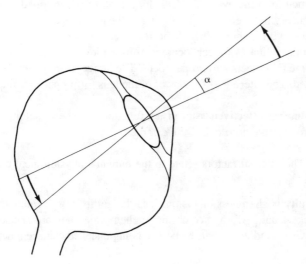

Figure 4.4   The visual angle α.

2. *Depth perception.* A special form of visual acuity, stereoscopic acuity, is the ability to see depth. The fact that the person sees an object with two eyes, thus from different angles, allows one to see depth. The ability to see depth is much less for far objects compared to near objects. Kaufman [4] describes several classes of cues (pictorial, kinetic, and physiological) that an observer uses in perceiving depth. Perspective and interposition are strong pictorial cues. Motion parallax and kinetic depth are examples of kinetic cues. Among the physiological cues are accommodation, convergence, and retinal disparity. Since there are many depth cues, even if some are lost due to reasons such as personal injury or night conditions, others may be sufficient for proper depth perception. Crane operators are normally expected to display good depth perception. There are rather inexpensive means of testing workers for depth perception. If a job requires the ability to see depth, it is a normal expectation that the workers assigned to that job have adequate depth perception.

3. *Near point of accommodation.* As an object is brought close to the eyes, after some point, the object will be seen as two objects. This point is known as the near point of accommodation. It is the nearest point of accommodation without losing details of the object. This measure has been used as an indicator of visual fatigue by various researchers. The hypothesis is that as visual fatigue develops, the near point of accommodation will move farther away from the eyes.

4. *Phorias.* When one looks at an object, the two eyes converge on the object so as to form the two images on the corresponding positions on the two retinas. When this occurs, the two images are fused and one sees only one object. The muscles that control the motion of the eyeball also control convergence. Some people's eyes converge too much, some tend not to converge enough. Such conditions are called phorias. Industrial tests are available to test for phorias both in horizontal and vertical axes. People with phorias do not see double images. They still finally converge, however, with notable stress to the eye muscles. They are vulnerable to rapidly developing visual fatigue.

5. *Color discrimination.* Out of the two photoreceptors in the eye, cones are the ones sensitive to color. This is why one does not see color at low illumination. According to the Young–Helmholz theory, there are three types of pigments in the cones that are sensitive to color: the red, green, and the blue catching pigments. Various combinations of these gives the sensation of different colors in the environment. CIE [5], the *Commission Internationale de l'Eclaire,* also recommends the use of the *calorimetric system* in describing color. This system assumes that any color can be matched by a combination of three spectral wavelengths of light in the red, green and blue regions.

Research shows that as high as 15 to 20% of the population has some degree of color blindness. Males are more apt to have color blindness. *Monochromats,* truly colorblind people, cannot see any color. They see the environment in various shades of gray just like watching a black-and-white TV. Monochromaticity is a rare condition. The most common color discrimination anomaly is deuteranopia. *Deuteranopes* lack the green catching pigment in their cones, *protanopes* lack the red catching pigment, and *tritanopes* lack the blue catching pigment. A number of industrial tasks depend on adequate color perception, such as inspection, remote control operations, machine operation with color displays, and so on. Other cues, such as position and orientation, may enhance human performance in tasks

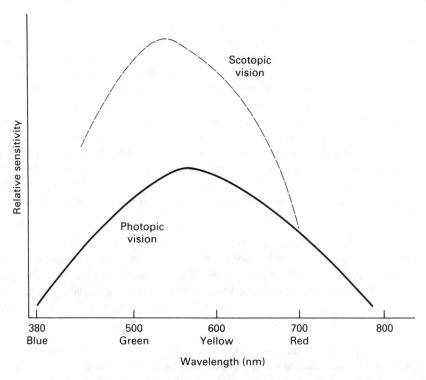

Figure 4.5 Color sensitivity of the eye. (Adapted from Ref. [4].)

requiring color perception. For example, the red light consistently being above other lights helps the monochromat make correct decisions when facing traffic lights. He or she judges the activation of red light when the top shape is brighter than the others. Sensitivity of the eye to different colors also shows variability, as shown in Figure 4.5. In general, the eye is less sensitive to colors that are at the extremes of the visible spectrum than to those in the middle.

6. *Dark adaptation.* When moving from a dark environment to a well-lighted environment, the human eye adapts to the new condition in a few seconds (if not temporarily blinded). However, this is not so in moving from light to dark. It takes up to 40 minutes to adapt to dark, as shown in Figure 4.6. This is due to moving from the use of cones to the use of rods, as well as biological changes within rods for maximum sensitivity to light. This means that in those industrial tasks where light conditions change from light to dark, sufficient time must be allowed for dark adaptation [6].

7. *Age.* It is a fact that visual performance declines with age. Acuity gets worse, cataracts may develop on the lens, and there is a general weakening in the muscles that control the eyeball. A young adult normally reads from a distance of about 30 cm (1 ft). With age, this distance increases. Such a condition is called the recession of near point, or *presbyopia*. The near point of accommodation at age 60 averages around 83 cm (33 in.) [7]. In industry, reading requirements, including character size, may be based on the limiting age of the expected user population, among other factors.

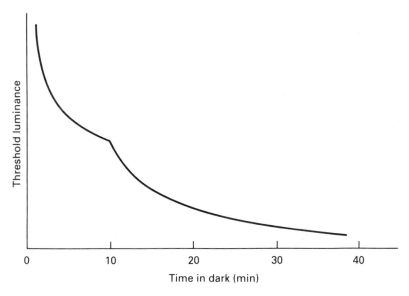

Figure 4.6 Dark adaptation curve. The breakpoint denotes shift from use of cones to use of rods.

8. *Purkinje shift.* Figure 4.5 showed how sensitivity varies as a function of wavelength for scotopic and photopic vision. Less light energy is required to elicit a response to yellow light than to red or blue lights. The light-adapted eye is maximally sensitive to light, with a wavelength of about 555 nm (nanometers), the dark-adapted eye, to 505 nm. This shift from 555 to 505, known as the *Purkinje shift*, is attributed to the fact that the pigments in the rods and cones differ in their selective absorption of wavelengths of light. The industrial implication of this phenomenon is that if an operator is expected to function in low illumination with colored displays, the best colors to use are in the bluish-green region. During daylight conditions, the best colors to use are in the greenish-yellow region. Naturally, other factors must also be taken into consideration, such as population stereotypes, which makes the use of red color for stoplights on cars preferable to other colors, since red is almost always associated with danger [3].

9. *Blind spot.* As discussed earlier, if the image of an object falls only on the optic nerve, the object cannot be seen. However, with binocular vision (both eyes utilized), it is highly improbable that this will be the cause of not being able to see an object. On the other hand, disabled workers with only one eye may experience this problem.

10. *Peripheral vision.* With no eye or head movements, it is estimated that human beings have a vertical visual field of about 160 degrees and a horizontal visual field of 208 degrees. Assuming daylight conditions, acuity is sharpest at center. Color is perceived well within ±62 degrees of the NLS (normal line of sight) for a horizontal visual field, and ±35 degrees of NLS for a vertical visual field. These data have significant implications concerning the location of items within the workspace, especially those that are color coded.

11. *Night blindness.* Some people cannot function visually at night as effectively as they can during daytime. There is some evidence that this may be due to insufficient vitamin A.

12. *Normal line of sight (NLS).* This is the direction of gaze when one is looking at an object. When looking straight ahead at a standing posture, NLS is in general 10 to 15 degrees below the true horizontal. At a seated posture, NLS is 20 to 25 degrees below the horizontal. These data may be useful in designing the height of labels, warning messages, or other displays.

External factors. Among the external factors that affect visual performance are brightness contrast, color contrast, amount of light, duration and movement of target, glare, surrounding brightness, color, size, and position of target in the visual field. These are discussed in more detail in Chapter 9. Visual performance makes significant contributions to psychomotor performance. Figure 3.1 presented an EOG (electrooculogram) as a local measure of psychomotor performance. The EOG measures the activity of the eyeballs. The more intense the activity, the greater the demand for information input. In general, a combination of mental (Section 4.7) and physical bioresponses are used for estimating demands due to psychomotor tasks.

4.2.2 The Ear

The ear is sensitive to variations in air pressure caused by vibrating objects. It is estimated that 15 to 19% of information is acquired by the auditory channel. Two primary attributes of sound that are of interest to an ergonomist are frequency and intensity. Frequency is the number of alternations in sound pressure per unit time caused by changes in density of air molecules. Measure of frequency in cycles per second is expressed in hertz (Hz). The human ear is sensitive to frequencies between 20 and 20,000 Hz. This is the audible range. The ear is not equally sensitive to all frequencies. As displayed in Figure 4.7, it is more sensitive to frequencies between 1000 and 5000 Hz. This has important connotations in industrial deafness, as discussed in Chapter 9. Frequency is also associated with the human sensation of pitch. In general, the higher the frequency is, the higher the pitch is.

The second attribute of sound that interests ergonomists is intensity. Sound intensity is associated with the human sensation of loudness. It can be measured in bels (B) or decibels (dB). OSHA (Occupational Safety and Health Administration) requires manufacturers to expose workers to no more than 90 dB measured on the A-scale, on a time-weighted average basis across an 8-hour shift. Instantaneous sound intensity can be measured by a sound-level meter. A sound dosimeter measures the 8-hour time-weighted average of sound exposure. The A-scale measures sound pressure level as if the human ear is hearing it. In analytic terms, sound pressure level (SPL) measured in dB can be calculated as follows:

$$SPL = 20 \log \frac{P_1}{P_0}$$

where, P_1 is the pressure of the sound to be measured, and P_0 is the standard reference sound pressure. The common sound pressure reference value is 0.00002 N per square

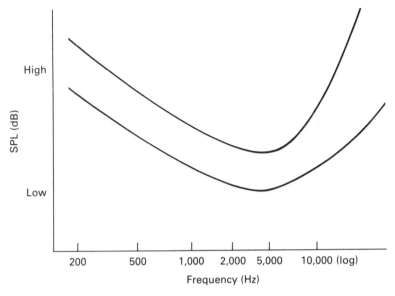

Figure 4.7   Absolute threshold curves. (Adapted from Ref. [12].)

meter, equivalent to the lowest-intensity 1000-Hz pure tone that a healthy adult can barely hear under ideal conditions.

1. *Loudness.* Sound pressure level and frequency combine to determine the perceived loudness of sound. Loudness is measured in *phons* or *sones.* The phon measures the subjective equality of various sounds. The true relative loudness of sound is measured by sones. A 40-dB tone at 1000 Hz is the reference point and is equal to 1 sone. Fletcher and Munson [8] developed equal loudness curves, which combine intensity and frequency of tones that are equal in loudness to the referenced 1000-Hz tone [5]. In general, an increase of 10 dB in sound intensity corresponds to a doubling effect in perceived loudness. Loudness creates annoying effects on people. The louder the sound is, the more annoyance it creates.

2. *Difference threshold.* The smallest detectable change in sound level is called the difference threshold. Generally speaking, difference thresholds increase with stimulus level. The more intense the sound is, the more difficult it is to detect small changes in it. This is an important consideration in auditory coding of information. The smallest difference threshold level for pure tone sounds is on the order of 0.5 dB. For information-carrying changes, the designer must consider intensity differences many times greater than 0.5 dB. Another effect of sound is *masking.* This is a condition where a signal reduces the sensitivity of ear to another signal. In general, the maximum masking effect occurs near the frequency of the masking tone and its harmonic overtones. Wegel and Lane [9] showed that the masking effect spreads to higher frequencies with higher-intensity masking tones. Furthermore, masking effect is more pronounced with intense masking tone.

3. *Ear anatomy.* The ear has three anatomical divisions: the outer ear, the middle ear, and the inner ear (Figure 4.8). The outer ear collects variations in air pressure. They are

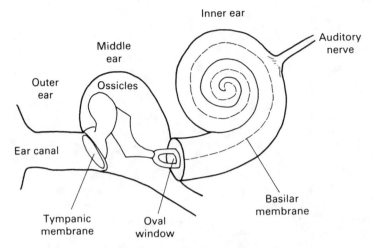

Figure 4.8 Schematic diagram of the ear.

then funneled to the eardrum (tympanic membrane) via the ear canal. In the middle ear are three bones in succession: the malleus, the incus, and the stapes, collectively called the *ossicles*. Their joint purpose is to transmit vibrations from the eardrum to the oval window of the inner ear. Since the area of the eardrum is about 22 times the area of the oval window, by the time it is transmitted to the inner ear, the sound energy is amplified 22 times. When tightened, muscles attached to the ossicles act to protect the inner ear against very intense sounds.

The structure of the inner ear is most interesting. It is a spiral-shaped organ that resembles a snail. It is filled with fluid. With pressure on the oval window, the fluid inside vibrates the basilar membrane, which in turn transmits vibrations to the organ of Corti containing hair cells. Signals picked up by these nerve endings are then transmitted to the brain by the auditory nerve.

4.2.3 Tactual and Olfactory Senses

Relatively speaking, the tactual and the olfactory senses are not as important as vision and audition in efficient human functioning in industrial environments. Skin senses can be used as warning media and are used extensively for blind persons' perceptual needs. Skin is sensitive to pressure, pain, cold and warmth. The olfactory (smell) sense is simple in anatomical structure but complex in its functioning. The smell sense organ is a patch of cells located in the upper part of each nostril. The olfactory hairs in these cells actually carry out the function of detecting different odors. Foul odors in industry (chemical, metallurgical, leather, etc.) may indicate unusual events.

4.3. THE NERVOUS SYSTEM

The nervous system coordinates and regulates body activities. It is responsible for the initiation and control of muscular activity. It brings about responses by which the body adjusts

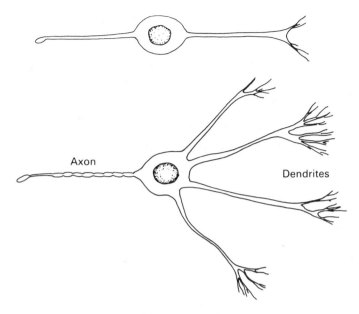

Figure 4.9 Two types of neurons.

to changes in the environment. The basic element of the nervous system is the *neuron*. A neuron is composed of a nerve cell along with a minimum of two nerve fibers (see Figure 4.9). A nerve fiber that brings messages to a nerve cell is known as a *dendrite*. There may be more than one dendrite associated with each nerve cell. Thus multiple paths may send signals to a nerve cell. A nerve fiber that takes messages to another cell is known as the *axon*. A neuron may not have more than one axon. The nerve fibers are coated with a substance called *myelin*, which serves to protect one nerve fiber's messages from another. Hence short circuiting of messages is prevented. Each neuron is complete in itself. It is never connected directly to another neuron. The gap between the axon of a neuron and a dendrite of another neuron is called a *synapse*. With the aid of chemical/electrical processes, the impulse of the preceding neuron is passed to the succeeding one unaltered in one thousandth of a second. Each synapse has a refractory period in order not to overload a neuron.

The nervous system is divided into three major parts: the peripheral nervous system, the autonomic nervous system, and the central nervous system. The central nervous system includes the brain and the spinal cord. The peripheral nervous system controls voluntary activity. The autonomic system covers the balance of nervous control.

4.3.1 The Autonomic Nervous System

The autonomic nervous system controls involuntary body activity, including the functions of glands, smooth muscle tissue, and the heart. It is commonly defined so as to include the sympathetic or thoracolumbar division and the parasympathetic or craniosacral division. Stimulating sympathetic fibers usually produce vasoconstriction in the part supplied, secretion of small amounts of thick saliva, depression of gastrointestinal activity, and acceleration of the heart. In general, these activities occur under emergencies such as fright and are associated with energy expenditure. For example, the sympathetic system takes over when a worker is under an instantaneous health threat. Stimulating parasympathetic nerves generally produces vasodilation of the part supplied, general fall in the blood pressure, contraction of pupil, secretion of plentiful thin saliva, increased gastrointestinal activity, and slowing of the heart. This system takes over the body functions after a safety threat is over.

4.3.2 The Peripheral Nervous System

The peripheral nervous system controls the voluntary activity of the body. Its two major elements are the sensory and the motor systems. The sensory system is responsible for conveying information from the sense organs to the central mechanisms. Sensory neurons have one axon and one dendrite each. Once a message is received in the central nervous system and decisions are made, signals are then sent down through the motor pathways to the appropriate muscles. Motor neuron cells lie in the central mechanisms and their axons travel together in groups. These groups of axons are called *motor nerves*. Upon reaching a muscle, a motor nerve breaks into a number of individual nerve fibers that connect to muscle fibers. The muscle control sensitivity largely lies on the number of muscle fibers controlled by each axon. In the hands and fingers, an axon may control few muscle fibers (three to six fibers). In the legs, an axon controls many more muscle fibers (100 to 200). This leads to a much higher degree of movement-control accuracy with the fingers and arms than with the legs [10].

The speed of transmission of impulses varies with the type of nerve. Motor fibers transmit at the rate of 70 to 120 m/s (230 to 400 ft/sec), and other types transmit at the rate of 12 to 70 m/s (40 to 230 ft/sec). Considering the distances between sensory organs and motor organs (hand) to the central mechanisms, it is not uncommon to have reaction times on the order of 300 to 500 milliseconds. The resting potential of the membrane (myelin) of a nerve fiber is -70 mV, and it is normally polarized (i.e., positive charges predominate the outer surface). Depolarization of the membrane produces the impulse, continuing until a reverse peak of $+35$ mV is reached. Repolarization follows after the impulse passes along the fiber, where the resting potential of -70 mV is restored [11].

4.3.3 The Central Nervous System

The central nervous system is composed of two parts: the spinal cord and the brain. The two are connected by the brain stem reticular formation. The spinal cord, an H-shaped structure made of gray matter, is encased in a long flexible bony column called the *spinal column*. The gray matter contains the cell bodies of many motor neurons. It also makes up the shunting mechanism, where sensory neurons have direct connections to motor neurons forming the reflex arc. These connections have no reference to the brain. One can say that the gray matter is a lower-order information-processing mechanism.

The spinal cord also contains the neurons of the sensory system. The sensory neurons run only short distances with interconnections along the spinal cord. On the other

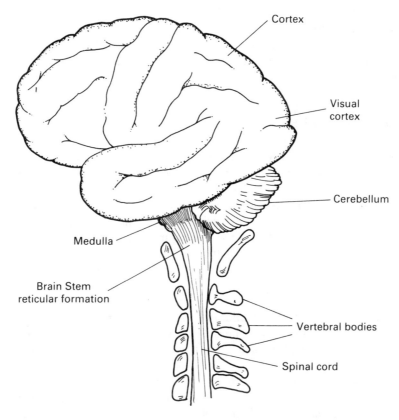

Cortex

Visual
cortex

Cerebellum

Medulla

Brain Stem
reticular formation

Vertebral bodies

Spinal cord

Figure 4.10 The brain.

hand, there are no such connections of the motor system. The cell bodies of motor nerves lie in the gray matter of the brain and the spinal cord.

Several parts of the brain are of more significance in describing human work (see Figure 4.10). The *medulla* connects the spinal cord with the higher centers of the brain. A majority of the cranial nerves also pass through this region. It contains the nuclei of the motor neurons of the autonomic nervous system, which control some vital functions, such as respiration, heart beat, blood pressure, and vasoactivity. The *cerebellum* lies at the base of the skull. It coordinates the messages to be passed to the voluntary muscles. Hence it can be thought of as the center of motor coordination. The cerebellum plays an important role in smoothing activity across many repetitions; thus it is a major center for skilled performance. Kinesthetic receptors have direct connections to the cerebellum and to the gray matter of the spinal cord. Thus it seems that when a repetitive activity is learned, control of its execution is taken from the cortex and is monitored by the cerebellum, and even in some cases, down at the spinal cord.

The *cerebral cortex* is the part of the brain where information is processed and decisions are made. It is also the area of the brain in which information is stored in the memory. Cortex is the most important part of the brain with respect to perception and processing of information.

The *thalamus* is another important part that acts as the relay station in the brain. Its main function is to obtain messages from other parts of the brain, sort them out, and direct them to the correct motor channels to produce the desired effect. In the absence of further instructions from the cerebral cortex, the cerebellum controls muscular activity that is repetitive in nature.

4.4. INFORMATION TRANSFER

Efficient functioning of a system is often possible only when information can be effectively conveyed to the system elements, including the human being. Information can be transferred to a person from multiple sources. There could be person-to-person transfer (verbal instructions), product-to-person transfer (aesthetics, inspection), person-to-machine transfer (controls), environment- or equipment-to-person transfer (displays, codes), among others. In Chapters 10 and 11 we will spend more time on some of the most important design parameters for displays, controls, labels, signs, and forms. Here we would like to cover some issues concerning selection of sense modality for information transfer to human and coding dimensions.

Selection of sense modality. In the beginning of this chapter we discussed several sense modalities. It is estimated that around 95–99% of the information that people gather is channeled through either the eyes or the ears. In human-made systems, designers often question which sense modality to choose to present information to. Table 4.1 gives some suggestions [12]. In general, warning types of messages are presented to both modalities. The next best choice is audition. If an operator is expected to move around, it is best to present the desired information to audition. However, hearing can be trusted for short messages. In any case, making a record of long messages is a good strategy. Such recorded messages can be reviewed if desired.

Coding dimensions. In many cases human beings are presented with coded information rather than direct presentation of the energy source. Hence a code represents a piece or a chunk of data. An example of coded data is a traffic light representing several possible actions for pedestrians and vehicles. Codes are also utilized extensively in industrial environments. Several characteristics of codes are discussed later in this section. Previously, we claimed that people are excellent in terms of relative discriminations

TABLE 4.1 GUIDELINES FOR SENSE MODALITY SELECTION

Select vision if:	Select audition if:
1. The message is long.	Otherwise.
2. The message will be referred to later.	Otherwise.
3. There already is noise in the environment.	The receiving location is too bright or dark.
4. The message does not call for immediate attention.	Otherwise.
5. The user is expected to be stationary.	Otherwise.

Source: Ref. [12].

TABLE 4.2 ABSOLUTE DISCRIMINATIONS AND INFORMATION TRANSFER

Stimulus Dimension	Number of levels that can be discriminated	Information transferred (bits)
Color, surfaces		
Hues	8–9	3–3.2
Hue, saturation, brightness	> 24	> 4.6
Color, lights	10 (3 preferred)	3.4
Shapes	15 (5 preferred)	3.9
Size	5–6 (3 preferred)	2.3–2.6
Brightness	3–5 (2 preferred)	1.7–2.3
Flash rate	2	1
Sound		
Intensity	4–5	2–2.3
Frequency	4–7	2–2.8
Intensity, frequency	8–9	3–3.2
Duration	2	1

Source: Ref. [1].

and poor with respect to absolute discriminations. For example, Mowbray and Gebhard [13] observed that people could differentiate 1800 pitch differences in tone pairs with high accuracy on a relative basis. However, on an absolute basis, they can reliably differentiate about five tones of different pitch. Since codes are presented to human beings one at a time, in perceiving them, people are making absolute decisions. Thus the designer must be aware of human capabilities along coding dimensions while designing codes. The expectation is that the task will not demand more than the person can handle. Otherwise, the person will make errors.

Table 4.2 gives ranges as to the number of levels of each coding dimension that a person can discriminate on an absolute basis. In addition, information transferred in terms of bits is also given assuming equal probabilities (most uncertainty). These then determine human capability along several coding dimensions with respect to the amount of information that can be processed. Under less-than-optimum conditions, lower numbers must be used. Thus if "flash rate" is used to mean different things, the designer should not use more than two different levels.

Whichever code is used to present information to a person, it should possess certain characteristics. Several attributes of codes that are of interest to an ergonomist are:

1. *Detectability.* The code obviously must be detectable by the sensory mechanism to which it is referring.

2. *Discriminability.* Code levels must be discriminable from each other. The greater the separation, the more discriminable code levels are.

3. *Compatibility.* A code that is compatible with its referent is more effective than one that is not. An example is that of increasing frequency of signal representing increasing level of coolant in a cooling system.

4. *Meaningfulness.* The code must be meaningful. The user must understand its meaning very rapidly.

5. *Standardization.* Standard codes are less frequently confused with others. A wailing signal may always denote danger. If this is the case, a user will always correctly interpret the meaning of a wailing signal no matter which environment or factory he or she is working in.

The importance of training for accurate code recognition cannot be overstressed. Confusion about the identity of "friendly" versus "foe" armored vehicles during the Persian Gulf crisis resulted in American forces accidentally destroying 80% of U.S. M1 Abrams tanks and Bradley Fighting Vehicles lost in combat [14].

4.5 SIGNAL DETECTION

A related concept that needs to be covered at this time is *signal detection.* This has been treated formally in the literature via the *signal detection theory.* Its application in industry may take the form of human operators watching for signals in background noise. An example is an inspection operation where an operator is watching for defects in cloth as it is automatically rolled in front of him. The operator is trying to catch and record defects such as tears, color abnormality, and dirt in visual noise (other attributes of the cloth). The discussion below summarizes the major elements of signal detection theory and its implications in designing relevant jobs. Sanders and McCormick [1] and Kaufman [3] provide a more detailed treatment of the subject.

Signal detection theory postulates that a human being always functions within some background noise. The level of noise may change from instant to instant; however, it has an average effect on any sense and is distributed normally over time. The theory further assumes that when a target (signal) is present, the distribution shifts up on a scale of sensation magnitude; however, the variance of the distribution stays constant. This is illustrated by Figure 4.11.

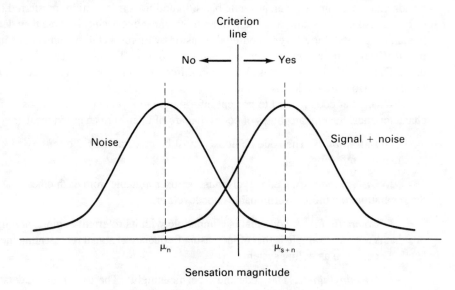

Figure 4.11   Distributions of the signal detection theory.

TABLE 4.3 STIMULUS–RESPONSE ALTERNATIVE
COMBINATIONS

Response Alternatives	Stimulus	
	Noise (n)	Signal and noise (s + n)
Yes (Y)	(Y/n)	(Y/s + n)
No (N)	(N/n)	(N/s + n)

A strong signal produces a large shift, and a weak signal produces a small shift. The difference between the means of the distributions (μ_n and μ_{s+n}) is a measure of the magnitude of the shift and also an index of the observer's sensitivity (k) to the signal. k may also be expressed in standard deviation (SD) terms as:

$$k = \frac{\mu_n - \mu_{s+n}}{\text{SD}}$$

A criterion line is also present in Figure 4.11, which displays the observer's cutoff point for making target presence/absence decisions. If the observer's sensation is excited by a condition that falls to the right of the criterion line, he or she calls for a signal [*yes (Y)* response to the condition]. If not, the response is *no (N)*. In making such decisions, the observer may make mistakes such as calling a signal when no signal is present. The possible conditions are given by Table 4.3.

Assuming that the distributions are normal probability distributions, the four possible conditions indicated in Table 4.3 give the probability of each alternative. The correct decisions are Y/s + n and N/n. The former is the case where the observer correctly identified a signal (*hit*), and the latter is the case where the observer correctly identified a no signal condition (*correct rejection*). The erroneous decisions are Y/n and N/s + n. The former is the case where the observer calls a signal when there is none (*false alarm*), and the latter is a no call when a signal is present (*miss*). In probability terms and with reference to Figure 4.11, Pr(N/n) is the area under the noise distribution to the left of the criterion line. Pr(N/s + n) is the area under the signal + noise distribution to the left of the criterion line. Pr(Y/s + n) is the area under the same distribution to the right of the criterion line. Finally, Pr(Y/n) is the area under the noise distribution to the right of the criterion line. It is obvious that Pr(Y/n) and Pr(N/s + n), the errors, will be minimal when the distributions are well separated from each other. This also helps to maximize the probabilities of making correct decisions.

An interesting exercise is the plotting of the percent of hits versus the percent of false alarms. As discussed previously, the criterion line may be placed anywhere on the sensation magnitude. It can also be slid from left to right to generate pairs of percentages for hit rate and false alarm rate. The resulting curve is known as the *receiver operating characteristic* (ROC) curve. An example is shown in Figure 4.12. If the two distributions coincide (no sensitivity, very weak signal), moving the criterion line from left to right will generate the 45-degree line connecting (0,0) and (100,100). This represents a no-sensitivity condition, designated by $k = 0$. On the other hand, with some deviation between the distributions, a different curve will be generated, as indicated by ROC. In Figure 4.12, point 2 designates

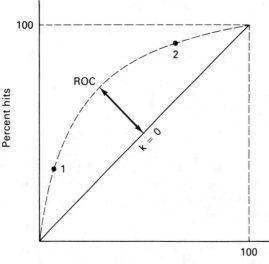

Figure 4.12   The ROC curve.

the case where the criterion line is set low (loose), and point 1 designates the case where the criterion line is set high (tight). If the criterion line is set too low, the observer will be likely to say that a target is present on most trials; however, many false alarms will also be reported. On the other hand, if the criterion line is set too high, many false alarms will be avoided but genuine signals will be missed, also.

The ROC curve will become bulged toward the (0,100) point as the distance between μ_n and μ_{s+n} increases. Hence the distance between the $k = 0$ line and the ROC curve will increase as the magnitude of shift, and hence the observer's sensitivity to the target gets large.

It is easy to see a parallelism between the concepts of signal detection theory and the inverted U curve as described in Chapter 3 and by Figure 4.13. In a relatively simple inspection task, the degree of alertness or arousal depends on the frequency of the stimulus. The inspector is more alert if there are many faults to be detected. Inspection performance increases with increased alertness. This holds up to the peak of the inverted U relationship. If this level is passed, performance will suffer due to overarousal.

The arousal theory was first postulated by Hebb [15]. Broadbent [16] complemented it with information on the variation of performance with level of arousal. For example, in a monotonous job with a low level of arousal, secondary stimulation (such as music) may help maintain alertness.

An interesting application area of the signal detection theory is the medical diagnosis field. Lusted [17] claims that, in general, physicians' detections are less responsive to disease prevalence rate than optimal. This may be due to difficulties in quantifying the consequences of hits, misses, and false alarms. Swets and Picket [18] go into considerable detail in describing the appropriate methodology for applications of signal detection theory in physician performance.

It is obvious from the discussion above that in designing jobs which require human signal detection performance, one must consider the following:

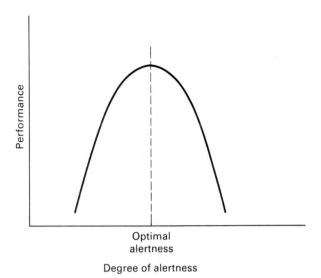

Optimal
alertness

Degree of alertness

Figure 4.13 Relationship between alertness and performance.

1. To the extent possible, separate the signal + noise distribution from the noise distribution. In other words, increase the signal-to-noise ratio.

2. If an overlap is inevitable, the criterion line can be set (loose or tight) based on the expected magnitude of damage due to each type of wrong decision. If the failure to detect signals ($N/s + n$) is much more costly than false alarms (Y/n), the criterion line may be set low. If the reverse is true, the criterion line may be set high. An intermediate setting is indicated for a small difference between the costs.

4.6. INFORMATION PROCESSING

Our sensory mechanisms can pick tremendous amounts of information of all variety, such as sounds (from the typewriter, keyboard, conversation), visual messages (computer screen, newspaper, secretary waving, a fly passing by), and ambient air parameters (temperature, humidity, breeze). However, not all sources of energy sensed can be processed. Only a very small fraction of what is sensed can be processed [19]. Table 4.4 gives an estimate of the

TABLE 4.4 INFORMATION PROCESSING AT VARIOUS STAGES

Process	Maximum flow of information (bits/sec)
Sensory registration	1,000,000,000
At nerve junctions	3,000,000
Conscious awareness	16
Long-term store	0.7

Source: Ref. [19].

amount of reduction that occurs from initial reception of stimuli to flow into permanent memory.

After sensing, processing of information has to proceed smoothly for optimum performance. In some types of human activity, such as reading instructions and operating a simple hand tool, the correlation between input (what is read) and output (action) is quite obvious. In some other tasks, such as remote control operations in a control room, such relationships are not clear. In these cases, many factors affect human performance.

Before discussing various information-processing theories, let's first turn our attention to a process that attempts to quantify information. The information theory proposed by Shannon and Weaver [20] made an important contribution along these lines. The authors devised the term *bit* to designate an item of information. A simple definition of a bit is that quantity of information conveyed by two equally likely alternatives. It is derived from the following formula:

$$H = \log_2 n$$

where n is the number of equally likely alternatives. In the case of two equally likely choices, the formula declares one bit of information since $\log_2 2$ is unity. In the case of an operator attending an instrument panel where one of four lights may light up with equal probability, the information conveyed is 2 bits since $\log_2 4$ is 2. With events having different probabilities of occurrence (p_i), the average information conveyed is given by the formula.

$$H_{av} = \sum_{i=1}^{n} p_i \log_2 \frac{1}{p_i}$$

Although this approach seems to provide an objective means of estimating information processing requirements and then comparing those with human capabilities, it has limitations when applied to human beings. The full significance of stimulus-conveying information cannot be interpreted by information theory. It is valid for comparatively simple situations. Nevertheless, it provides a basis for further understanding of the general area of information processing.

4.6.1 Theories of Information Processing

Throughout the past several decades, many scientists attempted to describe the phases of information processing. It is not the purpose of this book to detail each. The interested reader may refer to [21–24] for more information. Several are described briefly in this section.

1. *Broadbent's filter theory.* In 1958, Broadbent [25] proposed that the entire nervous sytem can be considered as a single-channel system with a limit on the amount of information that can be processed per unit time. He augmented this view by a selective filter that focuses only on a portion of the external stimuli to be processed. Broadbent modified this theory in 1971 to include other aspects of information processing; however, the main outline of his original theory stayed the same.

2. *Bills' blocking theory.* Bills [26] proposed that human beings are limited in terms of continuous information processing. Such work must be interspersed with intervals of interruption. Bills called these "blocks."

3. *Donders' stage model.* As early as in 1868 [27], Donders observed that one could measure the time required to process information based on several mental processes, including sensing, recognition, and action. Until the 1960s, this theory remained dormant. Since then, much attention has been given to this concept.

4.6.2 The Stage Processing Model

Modern analyses of information processing concentrate on identifying and further describing stages in cognitive processing. These can be called the *stages* of information processing. Figure 4.14 gives one such model on which our discussion will be based [28]. Although there are some difficulties in using this model, such as the limitation of serial processing of information, it still gives an insight to a designer as to what takes place in cognitive processing. Engineers/designers may then take appropriate actions to facilitate the process.

Data sensed by the sensory channels are carried to the central mechanisms to be processed after a very brief period of sensory memory (1/4 to 5 seconds). It is at this level that perception takes place. Perception is attaching meanings to sensed stimuli (Figure 4.15). Once perception takes place, data become true information. As an example, take a machine operator who suddenly sees a red light go on. The red light is what is sensed. From previous training and experience, the operator knows that this represents a danger condition (overheat, overload, etc.) and hence may shut down the machine. In this case what is perceived is the need to shut the machine down.

Perception is forming a relationship between new and old experiences. Expectations direct perception. The term *perceptual schema* has been used in the literature to designate the form of mental representation that people use to assign stimuli to certain ill-defined categories. Since the characteristics that define this representation are fuzzy at best, the

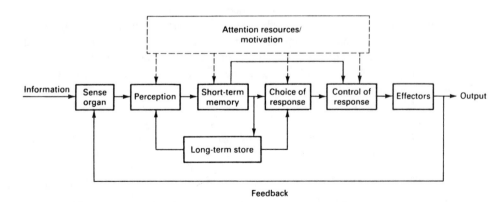

Figure 4.14 Stage processing model. (Adapted from Ref. [28].)

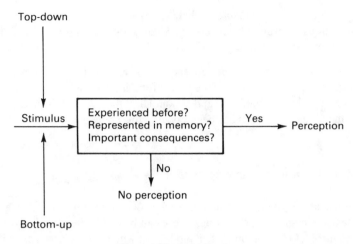

Figure 4.15 Perception.

schema is acquired as a result of perceptual experience with examples (track record) rather than learning a set of rules. Using a set of four random nine-dot patterns, Posner [29] showed that once human beings develop a prototype typical and variability representations around the typical, they are extremely accurate in classifying novel stimuli into one of multiple typical categories.

One's perception of a given situation may not be the same as another's, due to different experiences, background, and conditioning. However, research shows that the perceptual process is flexible. People adapt to perceptual experiences. One's interpretation of a situation may change if different but consistent outcomes are observed several times from the original situation. Research [33,34] also shows that people want to perceive total situations. There is a tendency to link isolated sensory inputs in a fashion to lead to wholeness. An example is a supervisor who develops perceptions of laziness on the part of a subordinate who on several but different occasions displays that behavior. The term *unitization* has been used by researchers to describe verbal perception that starts with analysis of letter features and proceeds toward a perceived word. The work of LaBerge [30] is an example of the existence of such *bottom-up* processing. The subjects in this experiment were required to make rapid judgments about whether a pair of stimuli were identical or different. The stimuli consisted of familiar letters or symbols that were as complex as letters but did not form letters. The experiment required the subjects either to attend or not to attend the stimuli presentations. LaBerge found that although in the attended condition there was no difference in the time to make identity decisions, in the unattended condition nonletter stimuli took considerably longer to process. LaBerge interpreted these results as the letter stimuli being processed automatically, as a unit, with basic features "glued together."

In addition to bottom-up processing, there is also strong evidence that perception follows a *top-down* direction. Such processing allows certain stimulus features to be predicted by context (surrounding units). Tulvig et al. [31] showed that when information redundancy is reduced, the contribution of top-down processing relative to bottom-up processing is also reduced. The perception of objects in the environment is another example of holistic processing. There is also evidence [32] that with familiar symbols, pictorial symbols are comprehended at least as fast as verbal messages.

Designers must take steps not to violate population stereotypes, for in those instances people's perceptions will not necessarily be the same as what is meant. If an operator is used to perceiving yellow signs as always meaning caution, and in one occasion is presented with a yellow sign that now means danger, there is a notable probability that the operator will not take the appropriate measures.

After perception, there is a brief period of retention of the perceived information. Its duration varies with a number of factors, including the sense modality involved. Usually within 5 to 25 seconds, if it is not attended to or rehearsed, the information fades away. Research shows that this period is composed of at least two smaller periods linked in time. Wickens [28] summarized the current state of affairs in this area using Figure 4.16.

The iconic and echoic codes, also known as the *sensory codes*, are raw representations (at sensory level) of visual and auditory stimuli. Sensory memory lasts for about one second for the iconic code and a few seconds for the echoic. The sensory codes do not require the operator's limited attention resources. The visual and phonetic primary codes (working memory or short-term memory) are natural extensions of the iconic and echoic codes, respectively. However, these require the use of limited attention resources. Unlike the sensory codes, they can also be generated from stimuli of the opposite modality. We can form a visual image of something that we hear (such as a beach) or an auditory image of something we see. The working memory or short-term memory usually lasts around 5 to 25 seconds.

Following the arrows from "short-term memory" to "control of response," Figure 4.14 indicates that the human being is sometimes information transmitter, sometimes information processor. A typist who is looking at a source document and copying the document exactly is transmitting information, bypassing decision making. One can claim that in these occasions the cerebellum is probably controlling the action. If the information is attended to (a seemingly novel condition has been detected), decisions are made as to what to do with it. Simultaneously, the long-term memory (semantic code in Figure 4.16) is updated from either visual or phonetic codes. This is more or less a permanent memory store. The long-term memory is unlimited in its information-holding capacity, which is gradually lost by age. However, the rate of information flow into the long-term

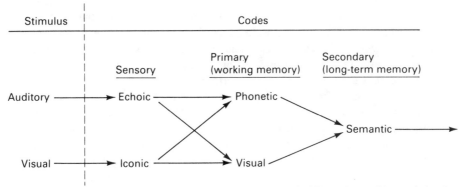

Figure 4.16 Relation between the five codes of memory. (From Ref. [28], reprinted with permission from Charles E. Merrill.)

memory is limited. Information that is not recalled may be forgotten. Therefore, designers use help commands on a computer screen for recall.

The choice of response is also affected by what is already in the long-term store. One's actions will be directed by one's experiences. Again, designers must keep this in mind and not violate population stereotypes. If a person is used to activating a red push-button in case of emergencies, he or she will expect the same until repeated experiences alter this expectation. In the meantime, many errors can be expected in situations where these expectations are violated.

As a consequence of all the previous information-processing steps, people may decide to take an action or may decide that an action is not necessary. If an action is necessary, it may take various forms, including manipulative responses, conversation, and gross body movements (walking, running, etc.). As can be observed in Figure 4.14, the model is augmented by another block, labeled "attention resources/motivation." Focused attention resources and motivation enhance performance at all levels of information processing.

An important element of this model is *feedback*. Das [35] claims that information feedback is generally, but not invariably, effective in improving operator productivity. It is extremely effective in learning situations and simple psychomotor tasks. Although Chapanis [36] demonstrated that not all jobs will benefit from feedback, there are many other studies that focus on its positive effects. In a recent study, Das and Shikdar [37] applied specific work standards and participative standards with feedback to improve performance over a control group (receiving no feedback and no standards) in a large fish processing industry.

4.6.3 The Two Levels of Information Processing

Many researchers agree that people process information on at least two levels [38–40]. These are called the *conscious* and the *automatic* levels. As skill is acquired, one moves from conscious to automatic performance. Conscious processing relies heavily on peripheral sensory feedback. At this level, one heavily attends the task at hand. Performance related to listening, speaking, and thinking are examples of conscious processing. Activities that are consciously controlled add to memory and experiences.

Research suggests that during learning one first acquires accuracy skill. People seem to reach an accuracy level as the basic skills are obtained and for the most part stay with that accuracy level until an outside factor forces a modification. Then, speed skill is acquired. Keele [41] suggests that as one moves toward automatic control, degree of reliance on conscious attention drops, appropriate movements are well planned since successive events are expected, and speed of performance increases.

Automatic performance is considered skilled performance. It takes place only after skill has been developed for a particular activity. At this level, people perform with little or no conscious control. Performance that takes place automatically is not remembered. Automatic performance uses (extracts from) memory but does not seem to feed new information into memory. People may make many mistakes if they are in automatic performance mode and do not respond to novel conditions in time. This is the primary reason why many professional occupations require membership to be retrained periodically for licence maintenance. Pilots are a good example.

4.7 COGNITIVE WORK

Cognitive work exists whenever information must be processed on a job. More and more cognitive work is being created with increasing automation on the shop floor and in the office. As manufacturing jobs decline, more service jobs are created, which mostly use an office environment. In general, office work requires a fair amount of mental work. Engineers, particularly industrial engineers, are interested in productivity improvement. In doing so, their focus should not be limited to blue-collar work but must also include white-collar jobs. Hence an ergonomist is by default concerned with cognitive work and its effects on the physiology and psychology of the worker.

Unlike physical work, cognitive work does not lend itself to detailed analyses with respect to decisions concerning acceptability/unacceptability. More will be said on this later. Decades of research failed to produce scientific tips as to when a person must be pulled from a job due to its potential bodily harm in the short or long term. However, some practical guidelines have been developed which are discussed below under different subtopics.

Vigilance. Vigilance refers to the case where a person is asked to watch for a signal of interest (e.g., a defect on a part, a submarine in the sea, an airplane in the sky) and make correct calls in all conditions [i.e., catch the signal (target) and let nonsignals go by]. Such work exists in all phases of industrial life, including inspection work on the shop floor and proofreading a document in the office. These are cognitive tasks, since the person involved must process incoming information and make decisions as to whether or not a target is present.

For decades, scientists have been studying the effect of such tasks on human performance. The classic studies of Mackworth [42] are good examples. After World War II, Mackworth carried out clock-watching experiments. In one such study, subjects were asked to observe a clock hand that made a full revolution each minute on a scale divided into 100 graduations. Normally, the clock hand jumped one graduation at a time. Occasionally, it would jump two graduations, and these were labeled "critical" signals. The study lasted 2 hours and there were 12 critical signals each half-hour at random intervals. The subjects had to identify each critical signal correctly. Figure 4.17 gives the results of the study. As can be observed, after the first half-hour there were significantly more signals missed. This result is consistent with other similar studies, including search-and-rescue missions at sea [43]. The human attention span seems to be limited to 30 minutes on a routine job.

Many parameters affect performance on vigilance tasks. Research results in this area can be summarized as follows [11]:

· Relative improvements in vigilance performance can be achieved if:
  1. The frequency of signals increases.
  2. The signals are stronger.
  3. The signals are more discriminable from nonsignals.
  4. Periodic feedback is provided to the subject on his or her performance.

· Performance gets worse if:
1. The watch period increases.
2. The between-signal interval shows high variability.
3. The subject is already under stress.

The factors above are very applicable to industry to enhance performance on vigilance tasks, such as the task of a process checker or a control room operator. On those tasks that do not lend themselves to task parameter altering, the best strategy seems to be to control exposure to the task. This can be accomplished by employee rotation every 30 minutes or so.

Operator and surrounding characteristics have also been shown to affect vigilance performance. There is evidence that introverts are better than extroverts in vigilance performance [44]. Noise has been linked to beneficial performance effects. McGrath [45] found that the effects of varying noise consisting of combined music, speech, and so on, had a more powerful effect in improving vigilance performance than did white noise alone. Broadbent [46] showed that high-intensity noise tends to reduce performance. Other environmental factors that have been linked to vigilance performance include ambient temperature and lighting. In a series of experiments, Blackwell [47] found that performance on a vigilance task was significantly improved when the light level was increased from about 100 lux to 10,000 lux. There is also evidence that an ambient temperature over 26° C increases the response time to signals, hence decreasing performance.

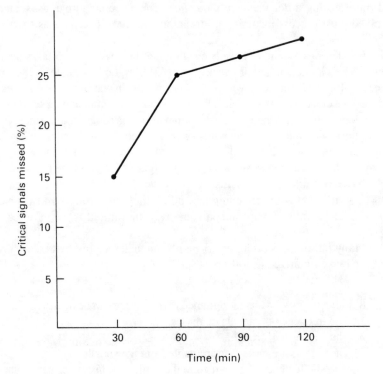

Figure 4.17 Human performance on a vigilance task (After Ref. [42].)

Attention.   Similar to vigilance, attention requires that certain aspects of the environment must be given high priority during performance. Attention also requires processing of incoming information. Two types of attention need to be mentioned [48]. *Selective attention* allows one to focus on one particular aspect of the environment despite other distractions. *Divided attention* allows one to focus on several tasks simultaneously while ignoring others. Research shows that selective attention can be improved with motivation and other aspects of task and environment design, such as coding. A flashing-red indicator light will attract the attention of a control room operator. Divided attention performance can be improved by proper split between input and output requirements. People can process incoming signals in parallel, while output processing seems to be serial in nature. Therefore, if both cannot be controlled in an industrial task, overall performance will improve if demands for output (manipulative actions, conversation, etc.) can be reduced.

Boredom.   Boredom is almost the opposite of attention. In a boring task, an operator is engaged in a monotonous task. The effects of such tasks are weariness, lethargy, and diminished alertness. If a task demands the same requirements from a person over and over again, it becomes boring. An uninteresting job is boring [49–51]. Factors that contribute to boredom are:

1. Short cycle times
2. Few opportunities for bodily movement
3. Warm conditions
4. No contacts with fellow workers
5. People with low motivation
6. People with high abilities, eager to excel
7. Dimly lighted work environments

The remedies for boredom are job expansion, job enrichment, and job rotation. As stated previously, some stress will also help in elevating alertness. Optimal environmental conditions and personal attention to one's job, together with social contacts, will also help.

Mental Workload.   As indicated by Figure 3.1, there are several measures of cognitive work load, although none seems to be sufficient by itself. Among them are psychophysiological indices, such as heart rate variability and electroencephalogram; subjective ratings; and behavioral time-sharing methods, such as the dual-task method.

1. *Psychophysiological indexes.* Heart rate variability (HRV) has been proposed by Kalsbeck [52] as an indicator of mental work load. Heart rate is not regular from one instant to the next. It normally shows variability (up and down) around a mean. The physiological term for this variability is sinus arrhythmia. Several scientists confirmed Kalsbeck's finding of reduction in heart rate variability during mental load. It has been claimed that the reduction in variability is a function of mental stress and increasing concentration to process information.

In addition, EEG (electroencephalogram) has been proposed as a means of measuring

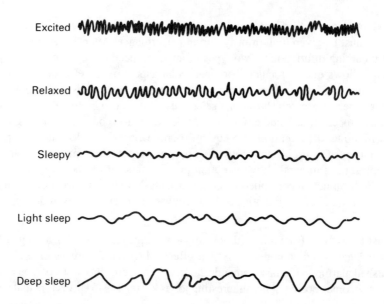

Excited

Relaxed

Sleepy

Light sleep

Deep sleep

Figure 4.18 EEG tracings for different functional states. (Reprinted with permission from *Fitting the Task to the Man,* E. Grandjean, 1980. Taylor & Francis, Ltd., London.)

the functional state of the mind [53], as displayed by Figure 4.18. EEG measures the brain waves. Such tracings may give fairly good clues with respect to the amount of mental work load. However, EEG is a very obtrusive measure. It cannot be collected easily in an industrial environment while the worker is performing a task. Another psychophysiological technique that has been used for this purpose is CFF (critical flicker fusion) of the eye [54,55]. A flickering light will be seen as continuous light at a sufficiently high rate of flicker. At resting state, fusing occurs at around 35 to 40 Hz. Reductions of 0.5 to 7 Hz occur after work under cognitive load. CFF [also known as FFF (flicker fusion frequency)] is also an obtrusive measure.

Grandjean and Etienne [56] used CFF in studies of work load and mental fatigue in different types of work. In general, the observation was that the differences in mental loading must be large for CFF measures to become significantly (statistically) different. Grandjean et al. [57] studied the air traffic controllers at Zurich airport. During the first 6 hours of work, the CFF showed a drop of 0.5 Hz and afterward fell faster until 10 hours into work. The total drop in this period was 2.3 Hz (Figure 4.19).

2. *Behavioral timesharing methods.* These require the simultaneous performance of a side (or secondary) task, along with the primary task of interest. Both tasks are mental tasks. With attention on the primary task, the degree of performance decrement on the side task is then considered an index of mental load. These measures have been shown to be valid in space activities also [58].

3. *Subjective ratings.* Even though their interpretation may not be clear at times, subjective ratings of mental load may be easier to use than all the techniques discussed so far. This is simply asking an operator to estimate the state of his or her mental con-

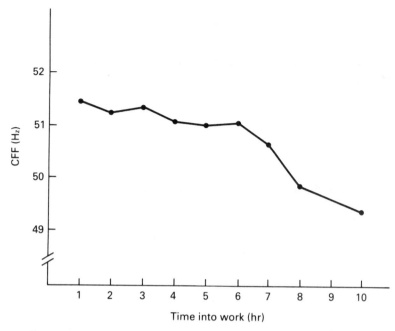

Figure 4.19 CFF drop versus work-hours. (Adapted and reprinted with permission from *Fitting the Task to the Man*, E. Grandjean, 1980. Taylor & Francis, Ltd., London.)

dition before, during, and after a mental task by marking a number or point on a scale. Differences in markings then indicate the degree of mental load. A sample scale along several dimensions is given by Table 4.5. Several subjective assessment techniques that have displayed global sensitivity to operator workload are MCH (Modified Cooper Harper Scale), SWAT (Subjective Workload Assessment Technique) and the NASA-TLX (Task Load Index) procedures. [59]

Mental fatigue. The overstressed muscle will develop a painful phenomenon called the muscular fatigue. This condition is acute and localized. On the other hand, with overburdened cognitive functions, human beings develop mental fatigue, accompanied by a general sensation of weariness. We feel impaired, heavy, and overworked. Both monotony and excessive use of mental functions may cause these symptoms. Mental fatigue requires a general state of rest for recovery. Physical exercise also helps. Several of the mental load measurement techniques may also be applied to measure mental fatigue.

TABLE 4.5 SUBJECTIVE RATING SCALE FOR MENTAL WORK LOAD ASSESSMENT

Task difficulty	Low ——————— High
Time pressure	None ——————— Rushed
Effort	None ——————— Impossible
Frustration level	Fulfilled ——————— Exasperated
Stress level	Relaxed ——————— Tense
Fatigue	Alert ——————— Exhausted

Source: Ref. [48].

In a simulated industrial security task, Murray and Caldwell [60] displayed significant performance penalties with multiple-item monitoring and control, and penalty accumulation even at low levels of complexity. Finkelman [61] documented a high correlation of self-reported fatigue with lack of job control, low pay, and low information-processing demands of a large, regional temporary employment agency in California. Paley and Tepas [62] reported high incidence of self-reported fatigue and higher negative mood scores on the night shift of a rotating 8-hour shift schedule for firefighters.

4.8 HUMAN COMPUTER INTERFACE

Nowadays, many human tasks are being allocated to machines. Transition to machine tasks from human tasks is called *automation*. Especially in an office environment, automation projects take the form of computerization of routine human functions in terms of information documentation, entry, retrieval, and processing. In addition to the hardware functions, an ergonomist must be concerned with effective human–computer interface for best human performance in such occupations. A user-friendly interface will be much more positively responded to by humans than an interface that does not address user needs and characteristics. Such user-hostile interfaces may even lead to a system's removal from the area.

In this section we dwell on designing effective user interfaces in computer systems. We build on the information ergonomics concepts developed earlier in the chapter that apply specifically to information-processing tasks. The potential benefits of improved user interfaces are reduced human-initiated errors, reduced training requirements, increased efficiency, and increased job acceptance. Human beings are very intelligent, but slow and error prone. A computer is very fast and accurate, but dumb. The combination may be explosive. In a subsection later in this section we report on the health effects of VDT (video display terminal) work. Although these effects do not present themselves after a short work spell, prolonged VDT use may lead to certain signs and symptoms, as reported in the literature.

4.8.1 Methodology for Effective Interface Design

Several steps for effective user interface design in computer systems are listed below [63,64].

1. *Define and know the users.* Knowing the users is critical to system design. Users have to be involved with the design early in the design process. Ergonomists design *with* the user, not *for* the user. Who the users are, what they do, and how the designer can help them do their tasks more effectively, are the types of questions that must be answered at this stage. The designers should have direct and ongoing contact with the users. Personal interviews, questionnaires, and "users on the design team" are the best ways to accomplish these objectives.

2. *Define system requirements.* Before any development, a detailed definition of system requirements is necessary. This includes the detailing of an operations plan together with a functions definition. What the system objectives are and how they will be achieved are the questions to be answered.

3. *Define task requirements.* For each operational function, a detailed investigation of user tasks is necessary. Task analysis will show sequential and simultaneous manual and cognitive activities of users. Task analyses are also valuable in allocating functions to the computer and to the user. Skill requirements and hence training needs are other results of task analysis.

4. *Utilize existing design guidelines/develop new ones.* Many suggestions exist in literature as to the effective design of the user interface. In special circumstances, research may be carried out to solve unknowns. Design guidelines describe conventions and practices for developing effective user interfaces. These have been generated after research with human subjects in the laboratory or real-life environments. It is a good practice to review these and use those that apply to the specific design at hand. Sometimes additional research or modifications are necessary before use.

5. *Design the user interface.* This step includes dialogue design, display screen design, equipment selection and design, and training process design, including the training documents. This is the step where all design guidelines and work area design suggestions (see Chapter 7) are put into use.

6. *Develop prototypes.* A prototype is an early version of the system being designed. Their use is primarily in the early design cycle, to reveal flaws in the design and let the user evaluate the requirements firsthand. More often than not, the user may make modifications after a good review of the prototype.

7. *Conduct user acceptance testing.* User acceptance testing (UAT) is the full-scale testing of the software, along with all functionality, screens, and dialogues. The user conducts this testing with help from the developer. Often, the user either makes minor modifications in the requirements or the developer makes modifications on several small items, such as different screen labels and formats. A follow-up to the formal UAT after all modifications are completed will iron out all questions concerning the interface. Virzi [65] showed that 80% of usability problems are detected with four or five users, with the first few detecting most severe problems.

8. *Train the users.* This is mass training of all the users on the system structure and each feature, including the dialogue screens. The training documents developed earlier help in this process. Both in-class and field training are recommended. If parts of the interface are access controlled, users and the functionality that they can access must be matched by selective training.

9. *Conduct follow-up evaluations and enhancements.* Once a system has been designed and put into use, follow-up evaluations reveal mismatches between assumptions and the functionality built into the system. These could be noted and corrected in later releases. Furthermore, as the environment and processes change, user interface requirements change. Hence software capability must be changed for compatibility with the requirements.

4.8.2 Interface Design Guidelines

This section follows up on the general methodology of user interface design given in the preceding section. It provides many focused guidelines for the design of human–computer interface [63]. More specific suggestions are given in Supplement A at the end of this chapter. Furthermore, the reader may refer to Chapters 10 and 11 for additional information on data I/O device design.

General principles

1. *Reduce mental processing requirements.* The computer must aid the user in performing tasks rather than complicate the responsibilities. Designers must build procedures into computer systems to reduce the frequency of mental processing. Requirements to learn complex commands and syntax, to memorize encrypted codes and abbreviations, and to translate data into units before they can be applied to the problem at hand are examples of tasks that demand mental processing.

2. *Allocate functions to the user and the computer based on their relative strengths* [66]. The implication of all research done along these lines is that the human being is best in controlling, decision making, and responding to unexpected events. The computer is best for storing and retrieving data, processing information using prespecified procedures, and presenting options and supporting data to users.

3. *Allow the user to develop effective mental models of system operation.* Extensive rules and syntax, no underlying overall framework, or internally inconsistent conventions can lead to frustration.

4. *Build in as much consistency as possible.* If a consistent set of conventions is not decided upon, documented, and incorporated into all phases of the system, the resulting interface will appear to have a different set of interaction rules for each affected transaction. This will significantly increase mental stress.

5. *Use as many physical analogies as logical.* In this respect, icons are very effective. Direct manipulation of pictorial representations of objects of interest seems more friendly than otherwise.

6. *Build interfaces that capitalize on expectations and stereotypes.* These will minimize requirements to learn new and unfamiliar associations. If the product is to be used only by a specialized group of users, their expectations and stereotypes must be considered. For example, a computer system designed for the Arabic-speaking countries needs to present information in right-to-left order.

7. *Consider stimulus–response compatibility.* More will be presented on the compatibility issue in Chapter 10. However, at this point it is sufficient to say that specific features of control-display compatibility in a computer system must be observed. For example, to move the cursor to the right, the right arrow key should be pressed.

8. *Provide an appropriate balance of ease of learning, ease of use, and functionality.* This could be accomplished by designing for novices, experts, and intermittent users, avoiding excess functionality, providing multiple paths for option selection, and minimizing the consequences of errors through reversible actions.

Dialogue design

1. *Minimize multiscreen transactions.* If possible, carry out all related functionality on the same screen. If not, place frequently used steps first or follow a standardized sequence.

2. *Design for reversible actions.* Provide "undo" functions to allow the operator to reverse an action if he or she desires to do so.

3. *Design for minimum interactive delay.* If possible, respond to interactive requests within two (maximum three) seconds. When the operator has to wait for the system to perform a time-consuming action, use a cumulative dynamic display to create an impression of shorter time passage [67].

4. *Design for minimum file delay.* Users are willing to wait longer for file transactions. To the extent possible, load or save files within 10 seconds.

Data entry and retrieval

1. *Standardize input/output procedures.* This process will help in minimizing training time and errors on the job. Furthermore, users may be able to predict procedures in novel situations.

2. *Reduce keying requirements.* Keying takes time and is error prone. Automatic data transfer between processors to screens is preferred to manual data entry for data already collected.

3. *Inputs should be alterable.* User must have the option of altering input before submitting it for further processing.

4. *Provide help when needed.* Upon request, show correct input formats, valid options, and field lengths.

5. *Standardize prompt locations.* Prompts appearing at same locations on different screens are responded to faster than otherwise.

Control and display devices

1. May use touch devices for selecting. Provide arm support to tasks that require the use of touch devices.

2. Voice input is the preferred input technique when eyes and hands are busy.

3. Use light pens to select, move cursor, and draw.

4. Avoid prolonged light-pen use. Make light-pen-selectable fields large.

5. On keyboards, use fixed function keys for common functions.

6. Avoid frequent switches between keyboard and mouse.

7. Use mouse for cursor control and selecting.

4.8.3 Health Effects of VDT Use

Research with VDT work has led to considerable controversy with respect to its health effects. In two NIOSH-supported field studies, Smith et al. [68,69] reported more health problems (irritability, stomach ache, nervousness) among clerical VDT operators as compared to control groups (no VDT exposure). On the other hand, in two other NIOSH studies, Sauter et al. [70,71] reported that none of the well-being indices related to job stressors and mood states disclosed a strong indication of increased stress among the VDT workers. Dainoff et al. [72] and Evans [73] reported visual discomfort and fatigue, musculoskeletal discomfort, and various subjective physiological and emotional problems. As a note we need to indicate that musculoskeletal and related problems may be due to poor workstation design that requires the operator to adopt awkward postures during work. In a recent study, Morrissey and Bittner [74] had eight subjects experience four conditions (two VDT and two control). In the control conditions, subjects either sat and read quietly or were allowed to perform normal activities during the experimental period across several days. Subjects in the VDT groups displayed motion sickness symptoms on a battery of tests that are sensitive to such symptoms.

From the above and other research, we arrive at some general tentative conclusions on the performance and health effects of VDT use:

1. Clerical VDT operators do not show significant symptoms of excessive stress unless these are caused by poor workplace design or the operators are engaged in repetitive, monotonous, and prolonged VDT use.

2. Computer system breakdowns are very annoying to the operators.

3. According to the U.S. National Academy of Sciences [75], "the level of X-ray radiation emitted by VDTs is far less than the ambient background level of ionizing radiation from natural sources to which the general population is exposed."

4. Grandjean [76] reports that there may be a relationship between facial skin irritations and electrostatic fields around VDTs.

5. There is neither epidemiological nor experimental data to support the idea that VDTs produce cataracts.

6. Similarly, there is no evidence that VDT work causes miscarriages or birth defects.

4.9 CASE ILLUSTRATION 2

The primary objective of this section is to present specific ergonomic applications in an information automation project. The example deals with the external buffer zones at an electronics assembly factory.

4.9.1 The External Zones

The "external zone" terminology was coined at this factory to describe the material buffer locations outside the physical limits of the miniload AS/RS (automated storage and retrieval system). These zones include the external warehouses, high rack and shelving zones, and the unit-load AS/RS. Material buffered in the external zones are mostly large parts, such as frames, fans, fan housings, brackets, and the like. These are purchased and handled in bulk form primarily on pallets (Figure 4.20).

Decisions concerning the routing of material to different areas of the external zone are made at central receiving. An incoming part to be consumed in manufacturing may be routed to several destinations, including the assembly shops, miniload AS/RS, and any physical location within the external zones. If the item is to be buffered prior to manufacturing, it is stored in a bin or rack location. It is picked and dispatched to the point of use after receiving a pull signal or an MRP (material requirements planning—manufacturing planning software) request. Some items may be routed directly to the point of use bypassing the buffer function.

4.9.2 Software Support

In addition to the hardware (forklifts, conveyors, cranes, racks, etc.), material-handling activities require software support. Basic storeroom functions such as receiving, stocking, inventorying (counting), withdrawing, and accumulating require the initial support. Then, interfacing of these functions with other systems, such as factory receiving and the MRP, follow. Finally, all systems are integrated for a closed-loop information flow (Figure 4.21). The primary reason for tracking material in the buffer zones is that the production planning systems need an accurate view of on-hand balances. Production activities for existing orders and purchasing decisions for planned orders rely on such vital information.

Figure 4.20

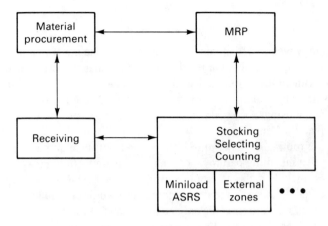

Figure 4.21 Information flow in buffer zones.

4.9.3 Ergonomic Interventions

The factory contracted the task of developing the necessary software support of external zones, including the MRP interfaces, to an outside group. The software was delivered and installed. Two weeks of trials led to the conclusion that it needed a major revision. Use of the system was discontinued for 5 months, during which it was revised extensively. Storeroom operating and engineering personnel worked closely with the developers during this time.

Initial Release. The major problem with the first release of the software was that it did not integrate the physical movement of material with the information flow it supported. Many operations that are normally carried out on the shop floor were not being recognized. This meant that material could move without an accurate record of the move's being captured in the software. Naturally, this would have caused many accountability problems. Data records would be inaccurate, leading to erroneous MRP-initiated transactions. Other problems with the software were as follows:

- Many functions were supported partially. For example, select canceling was possible only by MRP transaction codes. However, operators were required to cancel selects also by AS/RS (the execution software) ID.
- There were notable use problems. A window withdrawal required signing-up to two different systems with 11 screens to be processed. On the average, five different screens had to be processed to track a single material move. The external zones processed thousands of transactions (receive, stock, count, etc.) per day. Thus operators were overwhelmed by the system's requirements (Figure 4.22).
- At times, the software required greater control than was necessary. Many in-transit queues were introduced, with religious tracking requirements accompanied by unnecessary paperwork.
- System response times of 30 seconds or more were quite common. This led to much employee frustration. Daily work could not be completed in time. Unnecessary overtime work introduced significant cost to the company.

Figure 4.22

Improved Release. Both the material-handling engineering and the material storage and distribution (operating shop) departments designated personnel to work with the developers for much-needed improvements in the software. Ergonomic screening of the system functions was carried out for closing the operation/support compatibility gap as well as to solve use problems. Frequent visits to each other's facilities, in addition to electronic communications and intermediate trials, resulted in an improved release of the software. This release was later installed successfully. Specific improvements can be presented in four areas, outlined below.

1. *New function support.* In this category, improvements focused on providing support to those functions that were not supported initially. A detailed task analysis of the external zone operator functions, and comparing the results with those being supported by the first release of the software, revealed the differences. Several examples are:

- Developing of AS/RS ID. This is a five-digit serial number representing the MRP logical store, part identifier, description, assembly shop destination, and the required material quantity. The AS/RS ID allowed the operators to make only one string of data entry to identify the attributes it represented.
- Supporting MRP-generated material count requests.
- Using a new screen for periodic review of the status of any AS/RS ID.
- Supporting return-to-stock receival type.
- Allowing transfer of material from internal to external zones.
- Allowing alien (not reported to the planning systems) parts to be stored in external zones.

Many of the new functions were realized by adding data fields to existing screens.

2. *Improved support for existing functions.* The improvements in this group were aimed at enhancing coverage and effectiveness of functions already being supported, as well as eliminating those that are not necessary. Some examples are:

- Eliminating most in-transit requirements.
- Supporting on-demand printing of external pick request forms.
- Eliminating comcode (part number) check for alien parts.
- Enhancing material select options.
- Allowing material select canceling by AS/RS ID.
- Allowing queues of selects to be canceled.
- Permitting overdelivery for bulk material.
- Allowing growth in external database in terms of number of records.
- Eliminating travel document screens.
- Stopping MRP reporting for alien parts.
- Tracking of in-transit warehouse material. This allows for location updates after stocking. The in-transit record is automatically canceled after a proper location update transaction.
- Updating the MRP with new count versus change in count.

3. *Improving user friendliness.* Here we were interested in allowing a more speedy and error-free data-entry operation with associated reductions in operator frustration. A secondary objective was to make the task of picking more efficient. Examples:

- Displaying sufficient number of material locations on the external pick request form sorted by date of stocking for the total quantity to be picked. This made the material-picking task much more efficient, also allowing for FIFO (first-in, first-out) processing.
- Allowing for multiple transactions such as stock, pick, and count for the same part on the same screen.
- Structuring external location codes to represent unique physical zones. For example, all W-coded locations represent addresses in the warehouse zone.
- Allowing for efficient transaction processing by either combining screens or allowing for fast access between screens. In this respect, function keys were extensively utilized. Using these keys, one can move from one screen to the next without a need to go back to the main menu. Furthermore, a log-off transaction can be performed by one keystroke (the log-off function key).
- Other uses of function keys were exercised. A help function key displays help messages at the bottom of the screen concerning the field pointed to by the cursor. Function keys are also used for moving one field at a time on the screen, redrawing screens, clearing a field, and displaying material locations one page at a time (six locations per page).
- All function key images appear at the bottom of the screen at consistent locations. Such cues minimize error potential.
- Many other internal checks, such as part number, location, and AS/RS ID checks, assure data integrity and negate human-initiated errors.
- One screen allows for several types of receivals to be made, such as "new item"

receival, "return to stock from shortage application" receival, "return to stock from shops" receival, "restock" receivals, and so on.

- Allowing all locations for a part to be displayed across the external zones.
- For quantity updates due to stocking or withdrawal, both the original and the new quantity values are displayed for sanity checks.
- Wherever possible, the space-bar key is used to cycle through allowed options for an entry. The option list is sorted by the expected frequency of selection.
- At the point of receival, an electronic check of the MRP shortage requirement is performed to fill shortages. This allows for one-screen, automated access to two different systems.
- JIT (just-in-time) material handling via MRP is performed the same way. There is no requirement to access other MRP screens for this function.
- Guidance messages are utilized extensively at consistent screen locations, such as NO MORE PARTS AT THIS LOCATION, AT TOP PAGE, and so on.
- Certain screen prompts change automatically depending on the function performed, such as "Quantity ____," with "stocked" or "dispersed" as possible automated selections.

4. *Other.* In this category, effort focused on procedure improvements within and between functions. In addition, certain help functions were considered for operational effectiveness. Examples are:

- Supporting an external transaction report (activity on parts by day, by operator, by location, etc.).
- Mass cleaning of in-transit quantities.
- Printing of external pick request reports on high-speed printers.
- Permitting access to the external main menu from the internal main menu, and vice versa.
- Displaying the "material movement" screen automatically after a receival with non-zero entries in certain fields. Otherwise, the receival screen is redrawn for the next entry. In many cases, an automated move from one screen to the next also retains information to populate relevant fields on the next screen.
- Following an orderly and logical sequence for reporting material move at every stage. For example, on the "External Material Movement" screen, the item to be moved is identified first. Then information concerning all locations accessed to pick the part is entered. Next, confirmation of the pick operation is requested, after which dispersal information is gathered.

4.9.4 Conclusions

The major reason for the failure of the initial release of the external software was ineffective communication between the developers and the users. System requirements were not established clearly, with little understanding of human characteristics and expectations. In the redesign, a two-pronged approach was made to the problem: the value-added approach aimed at clear definition of the necessary functions to be supported, and the user-friendliness approach focused on a human engineering approach to the system interface points for improved use consistent with human expectations[77].

The improved system has been in operation for more than 4 years with minimal complaints. A window withdrawal process now spans two screens, equivalent to an average transaction screen span (Figure 4.23). System response time improved more than 70%, assisted by attached processors. Operators are satisfied with the new release.

Our experience with this case confirmed that systems must be developed *with*, not *for* the user. Close cooperation with the user is necessary for effective system development. Integration of human beings with the development process is a must for any system to succeed, a case for HID (Chapter 1). There is now a formal UAT (user acceptance testing) process at the factory. It is exercised before installing any new system. Those that do not

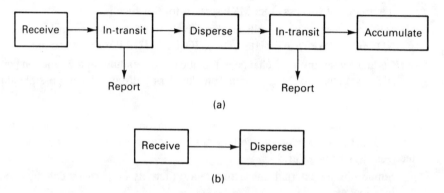

Figure 4.23   Transaction screen spans (a) before, and (b) after modifications.

pass the stringent screening get reworked until both engineering and operating departments are satisfied.

4.10 SUMMARY

Information input and processing are prerequisites of voluntary human activity. Proper acquisition and processing of information helps error-free, fast performance to a great extent. Various sensory, short- and long-term storage and information-processing characteristics determine the efficiency of the total process. Human information input and processing characteristics should define the parameters of information systems designed for human use. The primary bottleneck, number of different items processed per unit time, may be relieved by equipment aids for parallel processing.

QUESTIONS

1. What are exteroceptors? How do they differ from proprioceptors?
2. Discuss the difference between data and information.

3. Discuss the importance of rods and cones in sensing visual stimuli.

4. Why is visual acuity important in visual performance?

5. In what types of jobs would proper depth perception be important?

6. Which photoreceptor is sensitive to color?

7. What is the significance of dark adaptation in human performance?

8. How does age affect human performance?

9. What is EOG? Where can it be used?

10. Discuss the two primary attributes of sound that are of importance to an ergonomist.

11. What is a sound dosimeter? Where is it used?

12. Discuss the masking characteristics of sound.

13. How may difference threshold be of importance to a display system designer?

14. Discuss the major subsystems of the human nervous system.

15. Discuss the importance of the peripheral nervous system in human activity.

16. What is the importance of the cerebellum in skilled performance?

17. Discuss the basic building blocks of human information processing. What input does motivation provide?

18. What is perception?

19. Differentiate between conscious and automatic processing of information. In which level are people more apt to be making errors?

20. Discuss information coding and provide examples.

21. Discuss the ROC curve and its significance in inspection tasks.

22. What is vigilance? How can designers affect vigilance performance?

23. What are some measures of mental work load? Discuss two.

24. Discuss mental fatigue.

EXERCISES

1. A seated operator is expected to monitor a visual display in a dark room. As the designer of the workplace, which ergonomic parameters would you consider with respect to display location, color, and so on?

2. Consider a situation where a signal is detected 500 ms after its display. Decision making consumes 1 second. Hand movement time to counter the display is 300 ms. Calculate the total response time.

3. Select two tasks in your environment; the first must be well known and the other new to the worker. Compare the performance on each. Discuss your observations. What seems to be the major factor that can describe the difference in performance on each task?

4. Consider a display that may present status information on two different system attributes with equal probability. Attribute 1 may assume four possible levels, with probabilities of 0.5, 0.25, 0.125, and 0.125. Attribute 2 may assume three possible levels, with probabilities of 0.3, 0.4, and 0.3. Calculate average information presented by the display.

5. Select a vigilance task in your environment. Discuss ways of improving performance on this task.

6. Select a clerical task that requires data handling (entry/retrieval) using a computer. Evaluate the user interfaces. Discuss your observations while proposing improvements.

SUPPLEMENT A: A SELECTION OF HUMAN–COMPUTER INTERFACE DESIGN SUGGESTIONS

This section provides microlevel design suggestions to optimize the user interface. For further information, refer to Refs. [64,76,78–81].

1. General
   - Make the user feel that he or she has full control over the interaction.
   - Provide adequate feedback to the user.
   - Allow for error detection routines and provide immediate feedback.
   - Minimize training time.
   - Recognize commonly misspelled words.

2. Screen design
   - Information presented must be legible, readable, accurate, and timely.
   - For data entry, the screen design should match a paper form layout.
   - Implicit cues as to data—entry locations and field size are recommended.
   - Messages that appear frequently may be located in standardized locations on the screen.
   - The screen should not have a crowded (cluttered) look.
   - Screen divisions may be separated by blanks or colors or lines.
   - Develop functional category fields. These are fields reserved for particular kinds of data.
   - Use invariant fields on each screen, such as title, page number, date, and so on.
   - Display lists of data in columns.
   - Highlight sparingly.

3. Messages and error messages
   - Provide full messages to novice users and abbreviated messages to experienced users.
   - Detailed information should be available through help options.
   - Short messages are recommended.
   - Positive sentences are best.
   - Active messages are easier to read.
   - Enhance error messages via color, underscore, flash, or reverse video.
   - Error messages should state what the error is and what can be done to correct it.
   - After detecting an error, do not lose data in memory.
   - Allow correction at the point of first error.
   - Display error messages on the entry screen.
   - Highlight fields in error.
   - Show fields requiring missing data.
   - Use auditory signals conservatively.
   - Make error messages polite.
   - Make error messages specific and brief.
   - Keep user errors from destroying valid data.

4. Dialogue design
   - Provide input acknowledgment and progress indicators.

- Do not leave the screen blank when an input is being processed.
- Data-entry sequences should match.
- Indicate to novices how to continue.
- Make menus easily accessible.
- Allow type-ahead entry stacking.
- List menu items in appropriate order.
- Display only active menu options.
- Use a consistent screen location for command entry.
- System response time should not exceed 2 to 3 seconds.
- Consider echoing user entries.

5. Use of color
   - Use color conservatively. Do not overuse.
   - Use color for search tasks, to highlight and to indicate status.
   - Make color codes redundant with other codes.
   - Define each color code and use consistently.
   - Use contrasting color to distinguish data.
   - Display alarms with red, warnings in yellow, and normal conditions in green.
   - Restrict saturated blue to background use.
   - Color may also be used for graphics.

6. Labeling
   - Since labels are descriptive titles, dimensions (lb., in.) should be a part of the label.
   - Labels should be highlighted.
   - Where data-entry fields are distributed across a display, adapt a consistent format for relating labels to entry areas. For example, labels may always be left adjusted and above the corresponding field.
   - Use horizontal labels.

7. Wording
   - Use meaningful words. If abbreviations are used, users must be familiar with them.
   - Words and alphanumeric codes should be left adjusted.
   - Avoid unnecessary punctuation.
   - Both upper- and lowercase characters should be recognized.
   - Use familiar mnemonics, such as F: Female and M: Male.
   - Provide a dictionary of definitions for abbreviations.
   - Use meaningful codes, such as DIR for directory and REW for rewind.
   - Minimize use of jargon.

8. Data handling
   - Present data items in chronological or ascending order.
   - Strings of six or more characters should be presented in groups of two, three, or four with blanks in between.
   - Numeric codes should be right-adjusted.
   - The cursor should be positioned automatically on the next field once data entry in a field is complete.
   - Arrange data to make relationships clear, such as "actual," "predicted," and "difference" values side by side.
   - Place prompts in standard locations.

- Minimize cursor movement.
- Display correct input format for novices.
- Consider showing valid options, such as A: Add and C: Change.
- Show the length of entry fields.
- Do not require leading zeros.
- End input prompts with a colon.
- Input data should be easily alterable.
- Use a standard character for prompt commands.

REFERENCES

1. Sanders, M. S., and McCormick, E. J. 1993. *Human Factors in Engineering and Design*, 7th ed. McGraw-Hill, New York.
2. *Light and Color.* 1968. Large Lamp Dept., General Electric Co., Cleveland, OH. TP- 119.
3. Pirenne, M. H. 1967. *Vision and the Eye,* 2nd ed. Methuen, New York.
4. Kaufman, L. 1974. *Sight and Mind.* Oxford University Press, New York.
5. CIE. 1971. *Calorimetry.* CIE, Paris.
6. Parker, J., and West, V. 1973. *Bioastronautics Data Book.* Superintendent of Documents, Washington, D.C.
7. Illuminating Engineering Society. 1972. *IES Lighting Handbook,* 5th ed. IES, New York.
8. Fletcher, H., and Munson, W. A. 1933. Loudness: Its Definition, Measurement and Calculation. *Journal of the Acoustical Society of America,* 5, p. 91.
9. Wegel, R. L., and Lane, C. E. 1924. The Auditory Masking of One Pure Tone by Another and Its Probable Relation to the Dynamics of the Inner Ear. *Physiological Review,* 23, pp. 266–285.
10. Murrell, K. F. H. 1971. *Ergonomics.* Chapman & Hall, London.
11. Grandjean, E. 1980. *Fitting the Task to the Man.* Taylor & Francis, London.
12. Deatherage, B. H. 1972. Auditory and Other Sensory Forms of Information Presentation. In Van Cott, H. P. and Kinkade, R. G. (Eds.). *Human Engineering Guide to Equipment Design,* rev. ed. U.S. Government Printing Office, Washington, D.C.
13. Mowbray, G., and Gebhart, J. 1961. Man's Senses vs. Information Channels. In *Selected Papers on Human Factors in Design and Use of Control Systems.* Sinaiko, W. (Ed.). Dover, New York.
14. Briggs, R. W., and Goldberg, J. H. 1995. Battlefield Recognition of Armored Vehicles. *Human Factors*, 37(3), pp. 596–610.
15. Hebb, D. O. 1958. *Textbook of Psychology.* W. B. Saunders, Philadelphia.
16. Broadbent, D. E. 1963. Some Effects of Noise on Visual Performance. *MRC Report,* 15.
17. Lusted, L. B. 1976. Clinical Decision Making. In *Decision Making and Medical Care.* Dombal, D., and Grevy, J. (Eds.). North-Holland, Amsterdam.
18. Swets, J. A., and Pickett, R. M. 1982. *The Evaluation of Diagnostic Systems.* Academic Press, New York.
19. Steinbuch, K. 1962. Information Processing in Man. Paper presented at *IRE International Congress on Human Factors in Electronics,* Long Beach, CA.
20. Shannon, C. E., and Weaver, W. 1949. *The Mathematical Theory of Communication.* University of Illinois Press, Urbana, IL.
21. Kahneman, D. 1973. *Attention and Effort.* Prentice Hall, Englewood Cliffs, NJ.

22. Kantowitz, B. H. 1981. Interfacing Human Information Processing and Engineering Psychology. In *Human Performance and Productivity.* Howells, W., and Fleishman, E. A. (Eds.). Lawrence Erlbaum, Hillsdale, NJ.

23. Kantowitz, B. H., and Roediger, H. L., III. 1980. Memory and Information Processing. In *Theories of Learning,* Gazda, G. M., and Corsini, R. J. (Eds.). F.E. Peacock, Itasca, IL.

24. Posner, M. I. 1978. *Chronometric Explorations of Mind* Lawrence Erlbaum, Hillsdale, NJ.

25. Broadbent, D. E. 1958. *Perception and Communication.* Pergamon Press, Elmsford, NY.

26. Bills, A. G. 1931. Blocking: A New Principle of Mental Fatigue. *American Journal of Psychology,* 43, pp. 230–239.

27. Donders, F. E. 1868. Die Schnelligkeit psychischer Processe. *Archiv fuer Anatomie und Physiologie,* pp. 657–681.

28. Wickens, C. D. 1984. *Engineering Psychology and Human Performance.* Charles E. Merrill, Columbus, OH.

29. Posner, M. I. 1973. *Cognition: An Introduction.* Scott, Foresman, Glenview, IL.

30. LaBerge, D. 1973. Attention and the Measurement of Perceptual Learning. *Memory and Cognition, 1,* pp. 268–276.

31. Tulvig, E., Mandler, G., and Baumal, R. 1964. Interaction of Two Sources of Information in Tachistoscopic Word Recognition. *Canadian Journal of Psychology,* 18, pp. 62–71.

32. Whitaker, L. A., and Stacey, S. 1981. Response Times to Left and Right Directional Signals. *Human Factors,* 23, pp. 447–452.

33. Wickelgren, W. 1979. *Cognitive Psychology.* Prentice Hall, Englewood Cliffs, NJ.

34. Lockhead, G. R. 1979. Holistic vs. Analytic Processing Models: A Reply. *Journal of Experimental Psychology: Human Perception and Performance, 5,* pp. 740–755.

35. Das, B. 1990. Information Feedback: The Concept, Application and Research. In *Advances in Industrial Ergonomics and Safety II.* Das, B. (Ed.). Taylor & Francis, London.

36. Chapanis, A. 1964. Knowledge of Performance as an Incentive in Repetitive Monotonous Tasks. *Journal of Applied Psychology*, 48(4), pp. 263–267.

37. Das, B., and Shikdar, A. A. 1990. Applying Assigned and Participative Production Standards with Feedback to Improve Worker Satisfaction in Industry. In *Advances in Industrial Ergonomics and Safety II.* Das, B. (Ed.). Taylor & Francis, London.

38. Fitts, P. M. 1966. Cognitive Aspects of Information Processing: Set for Speed versus Accuracy. *Journal of Experimental Psychology* pp. 849–857.

39. Keele, S. W. 1968. Movement Control in Skilled Motor Performance. *Psychological Bulletin,* 70(6), pp. 387–403.

40. Pew, R. W. 1966. Acquisition of Hierarchical Control over the Temporal Organization of a Skill. *Journal of Experimental Psychology,* 71, pp. 764–771.

41. Keele, S. W. 1973. *Attention and Human Performance.* Goodyear, Pacific Palisades, CA.

42. Mackworth, N. H. 1950. *Researches on the Measurement of Human Performance.* H. M. Stationery Office, London.

43. Donderi, D. C. 1994. Visual Activity, Color Vision, and Visual Search Performance at Sea. *Human Factors*, 36(1), pp. 129–144.

44. Eysenck, H. J. 1959. *The Maudsley Personality Inventory.* London.

45. McGrath, J. J. 1963. *Vigilance.* Buckner and McGrath, New York.

46. Broadbent, D. E. 1957. Effects of Noises at High and Low Frequency on Behaviour. *Ergonomics,* 1(1), pp. 21–29 .

47. Blackwell, R. H. 1959. Specification of Interior Illumination Levels. *Journal of Illumination Engineering,* June.

48. Kantowitz, B. H. 1985. Mental Work. In *Industrial Ergonomics: A Practitioner's Guide.* Alexander, D. C., and Pulat, B. M. (Eds.). Industrial Engineering and Management Press, Atlanta, GA.

49. Wyatt, S. 1944. A Study of Attitudes to Factory Work. *MRC Special Report 292.* MRC, London.
50. Van Beck, H. G. 1964. The Influence of Assembly Line Organization on Output, Quality and Morale. *Occupational Psychology* 38, pp. 161–172.
51. Gubser, A. 1968. *Monotonie im Industriebetrieb.* Hans Huber, Bern.
52. Kalsbeck, J. W. H. 1971. Sinus Arrhythmia and the Dual Task Method in Measuring Mental Load. In *Measurement of Man at Work.* Singleton, W. T., et al. (Eds.). Taylor & Francis, London.
53. Jasper, H. 1974. Quoted from W. F. Ganong, *Lehrbuch der medizinistchen Physiologie.* Deutsche Ausgabe. Springer-Verlag, Berlin.
54. Rey, P., and Rey, J. P. 1965. Effect of an Intermittent Light Stimulation on the Critical Fusion Frequency. *Ergonomics,* 8, pp. 173–180.
55. Hashimoto, K. 1964. Estimation of the Driver's Work Load in High Speed Electric Car Operation of the New Tokaido Line in Japan. *Proceedings of the 2nd International Congress on Ergonom*ics, Dortmund, pp. 463–469.
56. Grandjean. E., and Perrett, E. 1961. Pupil and the Time Effect on the Flicker-Fusion Frequency. *Ergonomics,* 4(1), pp. 17–28.
57. Grandjean, E., Wotzka, G., Schaad, R., and Gilden, A. 1971. Fatigue and Stress in Air Traffic Controllers. *Ergonomics, 14,* pp. 159–165.
58. Manzey, D., Lorenz, B. Schiewe, A., Finell, G., and Thiele, G. 1995. Dual-Task Performance in Space: Results from a Single-Case Study during a Short-term Space Mission. *Human Factors,* 37(4), pp. 667–681.
59. Wierwille, W. W., and Eggemeier, F. T. 1993. Recommendations for Mental Workload Measurement in a Test and Evaluation Environment. *Human Factors,* 35(2), pp. 263–281.
60. Murray, S. A., and Caldwell, B. S. 1996. Human Performance and Control of Multiple Systems. *Human Factors,* 38(2), pp. 323–329.
61. Finkelman, J. 1994. A Large Database Study of the Factors Associated with Work-Induced Fatigue. *Human Factors,* 36(2), pp. 232–243.
62. Paley, M. J., and Tepas, D. I. 1994. Fatigue and the Shiftworker: Firefighters Working on a Rotating Shift Schedule. *Human Factors,* 36(2), pp. 269–284.
63. Brown, C. M. 1989. Methodology for Human Computer Interface Design. In *Industrial Ergonomics: Case Studies.* Pulat, B. M., and Alexander, D. C. (Eds.). Industrial Engineering and Management Press, Atlanta, GA.
64. Brown, C. M. 1988. *Human-Computer Interface Design Guidelines.* Ablex, Norwood, NJ.
65. Virzi, R. A. 1992. Refining the Test Phase of Usability Evaluation: How Many Subjects Is Enough? *Human Factors,* 34(4), pp. 457–468.
66. Price, H. E. 1985. The Allocation of Functions in Systems. *Human Factors,* 27, pp. 33–46.
67. Meyer, J., Shinar, D., Biton, Y., and Leiser, D. 1996. Duration Estimates and Users' Preferences in Human-Computer Interaction. *Ergonomics,* 39(1), pp. 46–60.
68. Smith, A. B., Tanaka, S., Halperin, W., and Richards, R. D. 1982. *Report of a Cross- Sectional Survey of VDT Users at the Baltimore Sun.* National Institute of Occupational Safety and Health, Center for Disease Control, Cincinnati, OH.
69. Smith, A. B., Tanaka, S., and Halperin, W. 1984. Correlates of Ocular and Somatic Symptoms among VDT Users. *Human Factors,* 26, pp. 143–156.
70. Sauter, S. L., Gottlieb, M. S., Jones, K. C., Dodson, V. N., and Rohrer, K. M. 1984. Job and Health Implications of VDT Use: Initial Results of the Wisconsin-NIOSH Study. *Communications of the ACM,* 26, pp. 284–294.
71. Sauter, S. L. 1984. Predictors of Strain in VDT Users and Traditional Office Workers. In *Ergonomics and Heath in Modern Offices.* Grandjean, E. (Ed.). Taylor & Francis, London.

72. Dainoff, M. J., Happ, A., and Crane, P. 1981. Visual Fatigue and Occupational Stress in VDT Operators. *Human Factors,* 23, pp. 421–438.
73. Evans, J. 1987. Women, Men, VDU Work and Health: A Questionnaire Survey of British VDU Operators. *Work and Stress,* 1(3), pp. 271–283.
74. Morrissey, S. J., and Bittner, A. C. 1990. Development of Motion Sickness Symptoms with Prolonged VDU Use: Artifact or Real-Effect. In *Advances in Industrial Ergonomics and Safety II.* Das, B. (Ed.). Taylor & Francis, London.
75. National Academy Press. 1983. *Video Displays, Work and Vision.* Report of the Panel on Impact of Video Viewing on Vision of Workers. NAP, Washington, DC.
76. Grandjean, E. 1987. *Ergonomics in Computerized Offices.* Taylor & Francis, London.
77. Didner, R. S. 1988. A Value-Added Approach to Information Systems Design. *Human Factors Society Bulletin,* 31(5), pp. 1–2.
78. Vassiliou, Y. (Ed.). 1984. *Human Factors and Interactive Computer Systems.* Ablex, Norwood, NJ.
79. Gould, J. D., and Lewis, C. 1985. Designing for Usability: Key Principles and What People Think. *Communication of the ACM,* 28(3), pp. 300–311.
80. Smith, S. L., and Mosier, J. N. 1984. Design Guidelines for User-System Interface Software, *Technical Report ESD-TR-190.* MITRE Corp., Bedford, MA.
81. Bailey, R. W. 1982. *Human Performance Engineering.* Prentice Hall, Englewood Cliffs, NJ.

CHAPTER 5

Engineering Anthropometry

Anthropometry is extensively used by ergonomists in systems design. *Design for use by people* requires that the designer consider human body dimensions in molding workstations and products. This chapter provides the fundamentals of deciding which anthropometric measures are most important for a design case at hand, from where to obtain data, and the proper use of such data. Chapters 7 and 11 build on the foundations established in this chapter.

5.1 DEFINITIONS

Anthropometry is the study of human body dimensions. Humans come in different body sizes and builds. Engineering use of the available information and development of new information for such use is called engineering anthropometry. The ancient Egyptians (1314–1197 B.C.) used a standard unit of measure known as the royal cubit to calculate land area, sculpture, and relief dimensions. The length of the cubit (around 52 cm) and its subdivisions were based on body measurements and ratios. One cubit represented the distance from elbow to tip of the longest finger. It also equaled seven times the width of the palm and 28 finger widths. All these units are marked off on the cubit rod.

Initial surveys of human body dimensions were carried out in the latter part of the fourteenth century. Rather complete anthropometric data existed as early as 1800. Methods of measurement were standardized several times during the early and mid-twentieth century. The most recent standardization occurred in the 1980s by the International Standards Organization (ISO) Technical Committee 159. Standard measurement methods assume repeatable body postures and landmarks [1]. Only during the past few decades were surveys performed specifically for engineering use. Most surveys were carried out for military use. Considerable civilian data also exist for the adult population in the published literature in the Western countries. Unfortunately, young adults and the elderly are not well represented in such data banks.

Large-scale anthropometric surveys are expensive and time consuming. An alternative method is to do specialized surveys to obtain key dimensions. Then, other dimensions are derived from these key dimensions using statistical procedures. Naturally, this method will not lead to accurate data; however, they may be accurate enough for some practical uses.

5.2 DESIGN APPLICATIONS

The primary areas of application of anthropometric data are:

· Clothing design
· Workspace design
· Environment design
· Design of equipment, tools, and machinery
· Consumer product design

Examples of these applications are socks, chairs, helmets, bicycles, kitchen counters, hand tools, beds, desks, tables, car interiors, diving masks, production machinery, tractor seats, and numerous other devices that people use. In short, anthropometric data establish proper sizes of and the dimensional relationships between the *things* people use.

The designer should accommodate the body dimensions of the population that will be using the equipment. Along these lines, playground equipment design problems require the use of child anthropometry, and nursing home design requires the use of anthropometric data from the elderly. In general, universal operability is desired within a population category. That is, at least 90 to 95% of the population within a target user group must be able to use the design. This is desirable since many situations where machinery or equipment are being operated require human interchangeability. Universal operability objectives can be achieved by *adjustable* designs. Such is the case in car seats, where the seat can be adjusted in various directions for driving comfort and capability. A short person may not be able to reach foot controls (gas pedal, brakes, clutch, etc.) to operate a car unless the seat is adjustable in the back-and-forth direction. Adjustability is a prerequisite of good designs, since equipment built according to one set of dimensions seldom accommodates the entire range of body sizes in the user population. Adjustability is also important in products intended for exports, due to diverse human body size around the world [2].

On the other hand, certain aspects of design may call for only one set of dimensions to be used. Such is the case when reach or clearance parameters are being considered. If reach parameter is based on fifth percentile values, 95% of the user population can reach the object in question. Clearances usually call for designs based on 95th percentile values, since most everybody will then be able to fit in the clearance. For example, a maintenance opening in a machine requires the application of this principle (with adjustment for gloves, if necessary) so that 95% of the population will be able to fit their hands through. Other aspects of human anthropometry define diverse environment design parameters such as the eye height determining the maximum height of a front panel, and leg and foot anthropometry defining foot control positioning for a seated production worker. More will be said on this topic in Chapter 7.

5.3 MEASUREMENT DEVICES AND TECHNIQUES

Simple devices exist for measuring body landmark distances (Figure 5.1). These include:

1. Spreading and sliding calipers to measure short distances
2. Anthropometers: straight rods with one fixed and one movable arm with the distance between the two arms indicated on a ruler
3. Tapes to measure circumferences and contours
4. Simple scales for weight measurements
5. Cones and boards with holes for grip circumference and finger-size measurement

The conventional measurement techniques make use of simple devices. Using an anthropometer, one can reach behind corners and tissue folds. Distances are read on a scaled rod from a reference point fixed prior to measurement. The spreading caliper consists of two curved branches hinged at a joint. The distance between the tips of the curved opposing ends is read on a scale attached in a convenient place between the two legs of the caliper. A small sliding caliper measures short distances, such as finger thickness and hand breadth. Skin folds can also be measured by a special caliper. Circular holes of increasing size are used to indicate external finger and limb diameters. There are other special measurement methods, such as shadowing, casting, and multiple probes, as explained by Roebuck et al. [3].

Another conventional measurement technique is the Morant technique. Here a set of grids is used to aid in the measurements. The grids are usually attached on two vertical surfaces that are positioned perpendicular to each other (Figure 5.2). With the subject placed in front of these surfaces, the projections of body landmarks on grids are used for measurements.

The traditional measurement methods use simple devices that are easy to use and inexpensive. They are applied on the body of a subject by a measurer. Their use is time consuming. Data not collected during a measurement session remain unknown until subjects are recalled for additional measurements. Despite their drawbacks, most of the existing measures have been collected through traditional methods [4].

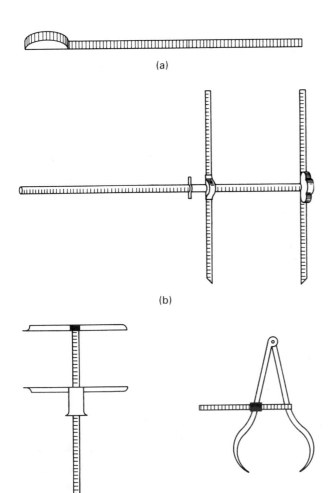

(a)

(b)

(c)

Figure 5.1 Classic anthropometric measurement devices: (a) tape; (b) anthropometer; (c) straight and spreading calipers.

In the 1970s photographic methods emerged as an alternative after the pioneering work of Herron [5]. In the 1980s, extensions in terms of videotaping, use of mirrors, holography, and stereophotometry came into being. These methods overcome many of the disadvantages of conventional methods. Photographic methods allow for storage of many measurements. They can also evaluate the body in three dimensions. Models also exist that develop many data points from recorded (videotape, etc.) pictures of the body and integrate them into entire human body contours using computer programs. Baughman's [6] method is an example of such methods used to generate desired body dimensions from existing data. Photographic methods have several disadvantages, however. The equipment used in such techniques is expensive. It is also difficult to pinpoint body landmarks and parallax distortions must be overcome by some means.

Although they cannot be counted among the measurement techniques, certainly, manikins (two- and three-dimensional) and computerized models [7,8] can be used to gen-

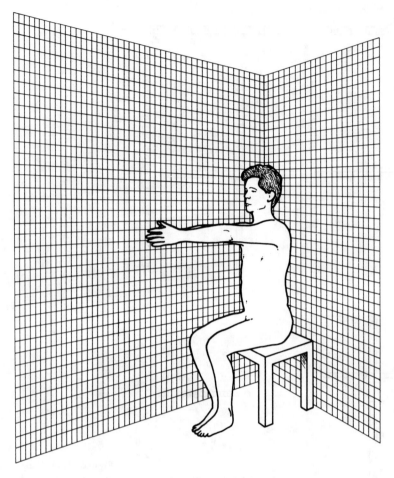

Figure 5.2 Grid system.

erate the anthropometric characteristics of a user population and evaluate proposed designs. Such techniques are especially useful in situations where a real person's presence may not be desired for safety or other reasons.

5.4 HUMAN VARIABILITY

Human beings vary with respect to body measurements. Different ethnic groups, tribes, and nationalities possess varying physiological characteristics that make them different. Differences exist even within the same group, due to gene characteristics. Figure 5.3 gives variability data with respect to one anthropometric measure, stature, between diverse nationalities. One can deduce from these data that there may be as much as 10 to 15% variability between ethnic groups along the same gender. A similar amount of variability may also be expected within a unique group between the 5th and 95th percen-

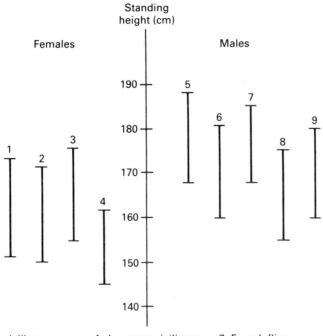

Figure 5.3 Human variability ranges (5th and 95th percentiles). (Reproduced with permission of McGraw-Hill, Inc. From *Human Factors in Engineering and Design, 5/e,* Sanders and McCormick, 1982.)

tile values. One may also conclude that a 25 to 30% difference can be expected in population measures along one anthropometric characteristic when gender, ethnic origin, and percentile ranges are considered simultaneously. This comparison does not even take into account extremes of ethnic groups and percentiles. It is this variability within and between groups that forces designers to consider adjustability [9].

5.5 IS THERE AN AVERAGE HUMAN?

As a follow-up to the discussion above and keeping in mind that there are over 300 anthropometric measurements taken on the body, we can say that there is probably no single person in a given population who is average (50th percentile value) on all accounts. The average design problem must consider several anthropometric dimensions, and the designer should investigate each as to whether an extreme (5th or 95th percentile) or an average value should be used or full adjustability should be integrated into the design. Designing for the 50th percentile person ensures that few people will be satisfied with the design. This is especially true when a given design is expected to be used by many people.

5.6 FACTORS THAT AFFECT ANTHROPOMETRIC DATA

Several factors affect body size. Designers must consider these factors and adjust their designs accordingly. The most important factors are:

1. *Age.* In general, body dimensions increase from birth to the early or late twenties. Roche and Davila [10] studied a sample of Americans and showed that they reached their adult stature at a median age of 21.2 years for males and 17.3 years for females. However, about 10% of males continued growing after 23.5 years and 10% of females after 21.1 years. After that, there is either no change or primarily a decline in stature (see Table 5.1). Trotter and Gleser [11] showed that we begin to shrink in body height at around age 40. The shrinkage accelerates with age, and women shrink more than men. This shrinkage is believed largely to occur in the intervertebral disks of the spine. Borkan et al. [12] suggest that some decrease also occurs in the lower limbs. Other data [13] suggest increases in body breadth by age. Thus it is important for the designer to define the user population as early in the design cycle as possible and take the necessary steps to adjust the design accordingly.

2. *Gender.* As can be observed in Figure 5.3, men are in general larger than women at any given percentile and body dimension except for hip and thigh measurements. Women also exceed men in skinfold thickness. Ethnic differences exist in the magnitude of gender differences. Eveleth [14] showed that such differences are more pronounced in American Indians than in Europeans.

3. *Body position.* Posture affects body size. For this reason, standard positions must be used during surveys. The present standards system owes much to the definitions of Hertzberg et al. [15]. Slumping is one of the major reasons for the variability and a minimum of 2 cm (about 1 in.) deduction in height measures (sitting and standing) must be considered for this factor during design. Restraints also affect data applicability. This is the reason for functional dimensions being larger than static dimensions. Static dimensions must be adjusted somewhat for free body movement during work.

4. *Clothing.* Clothing adds to body size. Another effect is restriction of movement. On both accounts, the extent of the effect is a function of the amount of clothing and the

TABLE 5.1 CANADIAN STATURE VARIATION BY AGE (CM)

	Percentile			
Age	5th	50th	95th	Change from previous value (50th percentile)
18–19	162.8	173.2	183.1	
20–24	162	173.3	184.2	+ 0.1
25–29	159.8	173.5	187.9	+ 0.2
30–34	161	172.9	184.9	− 0.6
35–44	159.2	171.7	184.4	− 1.2
45–54	156.9	169.7	182.9	− 2.0
55–64	153.9	167.6	181.1	− 2.1
>64	153.9	165.4	177.3	− 2.2

Source: Adapted from Ref. [2].

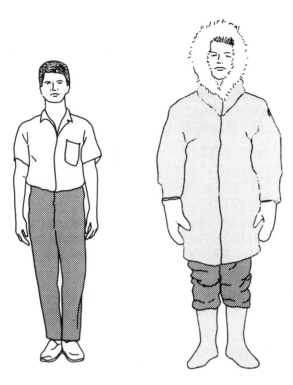

Figure 5.4 Size comparison between 5th-percentile lightly clothed person and 95th-percentile person in Arctic clothing. (Adapted from Ref. [2].)

clothing material [16]. The work space required by a 5th percentile man is much smaller than the working space for a 95th percentile man with heavy clothing (Figure 5.4). Gloves and shoes need special attention. Allowance must be made for gloves if the operator is expected to be wearing them during work. A reach-and-feel clearance around a panel may not be sufficient with heavy work gloves on. Furthermore, all anthropometric measurements are taken in the nude. Hence a work area dimension affected by height (sitting or standing) may have to be adjusted for shoes by almost 3 cm (1.2 in.).

5.7 ANTHROPOMETRIC DATA

Available anthropometric data fall into two categories [17]:

1. *Structural dimensions.* These are taken with the body in various standard and still positions. Another term is *static anthropometry.*
2. *Functional dimensions.* These are obtained with the body in various work postures. Another term for this category is *dynamic anthropometry.*

This section provides specific human body dimensions along both categories for use in design.

5.7.1 Structural Data

The following are several structural dimensions for the adult population. Each dimension is given by a description and a table of corresponding data. For additional data, see [2–4].

1. *Weight.* Nude body weight is measured by a physician scale. Data are given in Table 5.2. Mostly used in the structural support for seats and body restraint systems. For light clothing, add 2 kg (5 lb), and up to 10 kg (22 lb) or more for heavy winter clothing.

2. *Stature.* With the subject standing erect and looking straight ahead, this is the measure taken from the standing surface to the top of the head. Add 2.5 cm (1 in.) for men's shoes, 2 cm (0.8 in.) for caps, up to 8 cm (3.1 in.) for women's shoes, and 3.8 cm (1.5 in.) for steel helmets. Data are given in Table 5.3.

3. *Sitting height.* This is the vertical distance between the seat surface and the top of the head with the subject sitting erect, looking straight ahead, and the knees at right angles. Table 5.4 gives the data. For heavy clothing, add 1.2 cm (0.5 in.) to these values. Furthermore, add 2 cm (0.8 in.) for caps and 3.8 cm (1.5 in.) for steel helmets.

4. *Head length.* This is the maximum length of the head from the most front to the most rear of the head. Table 5.5 gives the data. Adjustments must be made due to the type of head gear worn.

5. *Knee height, sitting.* Vertical distance is measured from the floor to the uppermost point on the knee (see Table 5.6). For light clothing and men's shoes, add 3 cm (1.2 in.); for boots add 4.5 cm (3 in.). For women's shoes and light clothing, add up to 8 cm (3.1 in.).

6. *Popliteal height, sitting.* This is the vertical distance from the floor to the underside of the thigh next to the knee with the subject sitting erect, knees at right angles, and the bottom of the thigh and the back of the knee barely touching the seat pan. Table 5.7 gives the data. Similar clothing adjustments must be made.

TABLE 5.2 WEIGHT CHARACTERISTICS [KG (LB)]

Subjects	Percentile		
	5th	50th	95th
Females			
U.S. civilians	46 (101)	62.4 (137)	89.4 (197)
British civilians	46.6 (102)	60.4 (133)	79.4 (175)
Japanese civilians	39.8 (87)	51.3 (113)	62.8 (138)
Males			
U.S. civilians	58 (128)	75 (165)	98 (216)
Italian military	57.6 (127)	70.25 (155)	85.1 (187)
Japanese civilians	46.1 (101)	60.2 (132)	74.3 (163)
Turkish military	51 (112)	64.6 (142)	78.2 (172)

Source: Adapted from Refs. [2], [4], and [9].

TABLE 5.3   STATURE [CM (IN.)]

Subjects	Percentile		
	5th	50th	95th
Females			
U.S. civilians	150 (59)	160 (63)	170 (66.9)
British civilians	149.5 (58.7)	160.1 (63)	171.2 (67.4)
Japanese civilians	145.3 (57.2)	153.2 (60.3)	161.1 (63.4)
Males			
U.S. civilians	162 (63.7)	173 (68.1)	185 (72.8)
Italian military	160.2 (63.1)	170.8 (67.2)	180.8 (71.2)
Japanese civilians	155.8 (61.4)	165.3 (65.1)	174.8 (68.9)
Turkish military	160.6 (63.2)	169 (66.5)	179.2 (70.5)

Source: Refs. [2], [4], and [9].

TABLE 5.4   SITTING HEIGHT, ERECT [CM (IN.)]

Subjects	Percentile		
	5th	50th	95th
Females			
U.S. civilians	79 (31.1)	85 (33.5)	91 (35.8)
Swedish civilians	82.3 (32.4)	87.3 (34.4)	92.2 (36.3)
Males			
U.S. civilians	84 (33.1)	91 (35.8)	97 (38.2)
Italian military	84.3 (33.2)	89.7 (35.3)	94.8 (37.3)
German Air Force	86.1 (33.9)	91.3 (35.9)	96.5 (38)
Turkish military	84.8 (33.3)	89.7 (35.3)	95.1 (37.5)

Source: Adapted from Refs. [2], [4], and [9].

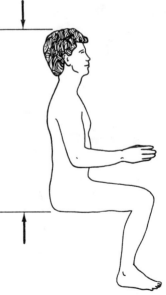

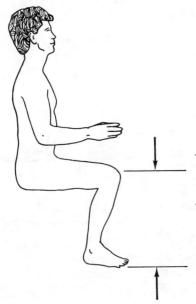

TABLE 5.5 HEAD LENGTH [CM (IN.)]

Subjects	Percentile		
	5th	50th	95th
Females			
U.S. Air Force women	17.3 (6.8)	18.4 (7.3)	19.5 (7.7)
Males			
U.S. Air Force flying	18.8 (7.4)	19.9 (7.8)	21.0 (8.2)
Italian military	18.2 (7.2)	19.3 (7.6)	20.4 (8)
Turkish military	17.5 (6.9)	18.6 (7.3)	19.7 (7.7)

Source: Adapted from Refs. [2], [4], and [9]

7. *Elbow rest height, sitting.* Vertical distance between the sitting surface and the bottom of the elbow is measured with the subject sitting erect, upper arm vertical at the side, and the forearm at a right angle to the upper arm. Table 5.8 gives the data.

8. *Hip breadth, sitting.* This is the maximum horizontal distance across the hips with the subject sitting erect and knees together. Table 5.9 gives the data. Add 1.5 cm (0.6 in.) for light clothing and 4 cm (1.6 in.) for heavy clothing.

9. *Elbow-to-elbow breadth, sitting.* This is the horizontal distance between the lateral surfaces of the two elbows. The subject sits erect, hips and knees together, upper arms vertical and touching the sides of the body lightly. Table 5.10 gives the data. Add 1.5 cm (0.6 in.) for light clothing and 12 cm (4.8 in.) or more for heavy clothing.

TABLE 5.6 KNEE HEIGHT, SITTING [CM (IN.)]

Subjects	Percentile		
	5th	50th	95th
Females			
U.S. civilians	46 (18.1)	50 (19.7)	55 (21.6)
Males			
U.S. civilians	49 (19.2)	54 (21.2)	59 (23.2)
Italian military	49.2 (19.3)	53.4 (21)	57.9 (22.7)

Source: Adapted from Refs. [9] and [4].

TABLE 5.7 POPLITEAL HEIGHT, SITTING [CM (IN.)]

Subjects	Percentile		
	5th	50th	95th
Females			
U.S. civilians	36 (14.1)	40 (15.7)	45 (17.7)
Males			
U.S. civilians	39 (15.3)	44 (17.3)	49 (19.3)
Italian military	36.6 (14.4)	40.3 (15.9)	44.2 (17.4)
German Air Force	40.4 (15.8)	43.8 (17.2)	47.4 (18.7)

Source: Adapted from Refs. [9] and [4]

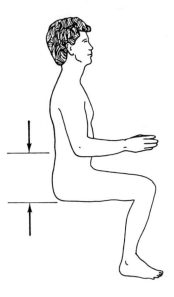

10. *Thigh clearance, sitting.* With the subject sitting erect, this is the vertical distance between the seat surface and the point of intersection of the thigh with the abdomen. Add 0.5 cm (0.2 in.) for light clothing and 3.5 cm (1.4 in.) or more for heavy clothing. See Table 5.11 for the data.

11. *Eye height, sitting.* Vertical distance is measured between the seat surface and the inner corner of the eye. The subject sits erect looking straight ahead. Add 0.7 to 1 cm (0.3 to 0.4 in.) for clothing under the buttocks. Table 5.12 gives the data.

TABLE 5.8 ELBOW REST HEIGHT, SITTING [CM (IN.)]

Subjects	Percentile		
	5th	50th	95th
Females			
U.S. civilians	18 (7.1)	23 (9)	28 (11)
Swedish civilians	19.2 (7.6)	23 (9)	26.7 (10.5)
Males			
U.S. civilians	19 (7.5)	24 (9.4)	30 (11.8)
Italian military	18.8 (7.4)	22.5 (8.9)	26.2 (10.3)
French fliers	22 (8.6)	25.6 (10)	28.8 (11.3)

Source: Adapted from Refs. [9] and [4].

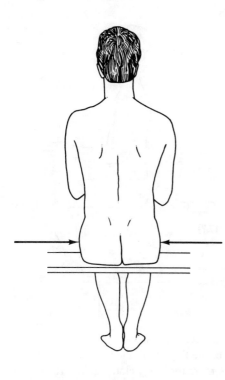

TABLE 5.9   SEAT BREADTH (HIP BREADTH, SITTING) [CM (IN.)]

Subjects	Percentile		
	5th	50th	95th
Females			
U.S. civilians	31 (12.2)	36 (14.1)	43 (16.9)
Males			
U.S. civilians	31 (12.2)	36 (14.1)	40 (15.7)
Italian military	32.7 (12.9)	35.7 (14)	38.7 (15.2)
French fliers	33.9 (13.3)	36.8 (14.5)	39.5 (15.6)

Source: Adapted from Refs. [9] and [4].

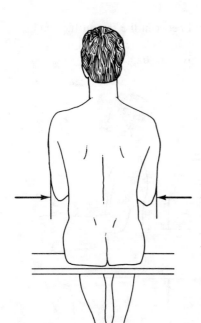

TABLE 5.10   ELBOW-TO-ELBOW BREADTH, SITTING [CM (IN.)]

Subjects	Percentile		
	5th	50th	95th
Females			
U.S. civilians	31 (12.2)	38 (14.9)	49 (19.3)
U.S. Air force fliers	33.8 (13.3)	38.4 (15.1)	43.4 (17.1)
Males			
U.S. civilians	35 (13.8)	42 (16.5)	51 (20.1)
U.S. Air Force cadets	38.3 (15)	42.4 (16.7)	46.7 (18.3)

Source: Adapted from Refs. [9] and [4].

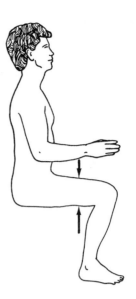

TABLE 5.11 THIGH CLEARANCE, SITTING [CM (IN.)]

	Percentile		
Subjects	5th	50th	95th
Females			
U.S. civilians	10 (3.9)	14 (5.5)	18 (7.1)
Males			
U.S. civilians	11 (4.3)	15 (5.9)	18 (7.1)

Source: Adapted from Refs. [9] and [4].

12. *Arm reach.* Horizontal distance is measured between the tip of the middle finger and the posterior surface of the right shoulder. Subject stands erect with ankles and knees together. The arm is extended to its maximum. Add 1.5 cm (0.6 in.) for light clothing and light gloves. Add 2 cm (0.8 in.) or more for heavy clothing and gloves. For manipulation by fingers, subtract 8 cm (3.2 in.). For whole hand manipulation, subtract 12 cm (4.8 in.). See Table 5.13 for data.

13. *Hand length.* This is the distance from the base of the thumb to the tip of the middle finger of the right hand. The hand is extended straight on the arm. Add 0.4 cm (0.16

TABLE 5.12 EYE HEIGHT, SITTING [CM (IN.)]

	Percentile		
Subjects	5th	50th	95th
Females			
U.S. Air Force women	68.7 (27)	73.7 (29)	78.8 (31)
Males			
U.S. Air Force fliers	76.2 (30)	81.0 (31.8)	86.1 (33.9)
Italian military	73.1 (28.8)	78.0 (30.6)	82.9 (32.6)
French fliers	77.5 (30.5)	83.4 (32.9)	87.7 (34.5)

Source: Adapted from Ref. [4].

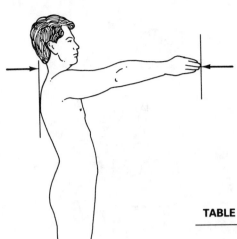

TABLE 5.13   ARM REACH [CM (IN.)]

	Percentile		
Subjects	5th	50th	95th
Females			
U.S. Air Force pilots	75.4 (29.8)	80.8 (31.8)	86.6 (34.1)
Males			
U.S. Air Force cadets	83.1 (32.7)	89.4 (35.2)	96 (37.7)
Truck/bus drivers	83.6 (32.9)	90.7 (35.6)	97.5 (38.4)

Source: Adapted from Ref. [2].

in.) for light gloves and 0.8 cm (0.32 in.) or more for heavy gloves. Table 5.14 gives the data.

14. *Foot length.* This is the horizontal distance from the back of the heel to the tip of the longest toe. Subject stands erect with his weight equally distributed on both feet. Add 3 cm (1.2 in.) for men's shoes, 3.7 cm (1.5 in.) for boots and 2 cm (0.8 in.) for women's shoes. Table 5.15 gives the data.

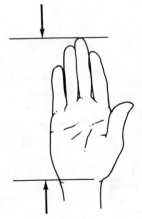

TABLE 5.14   HAND LENGTH SITTING [CM (IN.)]

	Percentile		
Subjects	5th	50th	95th
Females			
U.S. Air Force women	16.9 (6.7)	18.4 (7.3)	20.1 (7.9)
Swedish civilians	16.3 (6.4)	17.9 (7.1)	19.6 (7.7)
Males			
U.S. Air Force fliers	17.8 (7)	19.1 (7.55)	20.5 (8.1)
Italian military	17.6 (6.9)	19.0 (7.5)	20.4 (8)
French fliers	17.7 (6.95)	19.2 (7.6)	20.4 (8)

Source: Adapted from Refs. [4] and [2].

TABLE 5.15 FOOT LENGTH [CM (IN.)]

Subjects	Percentile		
	5th	50th	95th
Females			
U.S. civilians	22.1 (8.7)	24.1 (9.5)	26.2 (10.3)
Swedish civilians	22.8 (8.9)	24.6 (9.7)	26.3 (10.3)
Japanese civilians	21.1 (8.3)	22.6 (8.9)	24.1 (9.5)
Males			
U.S. civilians	23.8 (9.3)	26.2 (10.3)	28.2 (11.1)
Italian military	24.6 (9.7)	26.5 (10.4)	28.4 (11.2)
French fliers	24.7 (9.7)	26.5 (10.4)	28.5 (11.2)
Japanese civilians	22.8 (9)	24.4 (9.6)	26.0 (10.2)

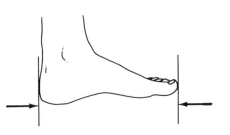

Source: Adapted from Refs. [4] and [2].

5.7.2 Functional Dimensions

This section gives several functional dimensions. For additional data, see [18–20].

1. *Prone length and height.* Subject lies in a prone position with feet together and straight, arms maximally extended and together, with fists clenched. Prone length is measured horizontally from the extreme point of the fist to the tip of the toes (see Table 5.16). Prone height is the vertical distance between the floor and the highest point on the head when the head is raised maximally with the chest still on the floor (see Table 5.17). Appropriate clothing allowances apply here, too.

2. *Squatting height.* This is the vertical distance between the floor and the top of the subject's head when the subject balances on his or her toes with the body erect. See Table 5.18 for the data. Shoe allowances are appropriate.

TABLE 5.16 PRONE LENGTH, *B* [CM (IN.)]

Subjects	Percentile		
	5th	50th	95th
U.S. Air Force male	215.1 (84.7)	228.8 (90)	243.3 (95.8)

Source: Adapted from Ref.[2].

TABLE 5.17 PRONE HEIGHT, A^a [CM (IN.)]

Subjects	Percentile		
	5th	50th	95th
U.S. Air Force male	31.2 (12.3)	36.8 (14.5)	41.6 (16.3)

Source: Adapted from Ref.[2].
^aSee Table 5.16.

TABLE 5.18 SQUATTING HEIGHT [CM (IN.)]

Subjects	Percentile		
	5th	50th	95th
U.S. Air Force personnel	103.6 (40.8)	110.7 (43.6)	119.4 (47)

Source: Adapted from Ref.[2].

3. *Crawling length and height.* The subject rests on his or her knees and flattened palms with the arms and thighs perpendicular to the floor. Crawling length is the horizontal distance between the most forward point on the head and the most rearward point at the toes. Crawling height is the distance between the floor and the highest point on the head. Tables 5.19 and 5.20 give the data.

Appendix B gives other anthropometric data for design use.

5.7.3 Correlations of Body Measurements

Some body dimensions are significantly correlated with each other, while some are not. Table 5.21 gives selected correlation coefficients among U.S. Air Force personnel body dimensions. The higher the correlation coefficient, the more closely related are the relevant body dimensions. Such data can be used to generate one dimension using another in a design case economizing from detailed measurements. [21]

5.7.4 Anthropometric Data of Wheelchair Users

Goldsmith [22] observed that wheelchair users are handicapped three times: first the condition that put the person in the wheelchair; second, the user's eye level in that posture is 40 cm (15.7 in.) below that of standing people; and third, the user is trapped in a cumbersome and space-consuming vehicle. For these reasons the ergonomist must consider mod-

TABLE 5.19 CRAWLING LENGTH, *B* [CM (IN.)]

Subjects	Percentile		
	5th	50th	95th
U.S. Air Force male	125.2 (49.3)	135.1 (53.3)	147.8 (58.2)

Source: Adapted from Ref.[2].

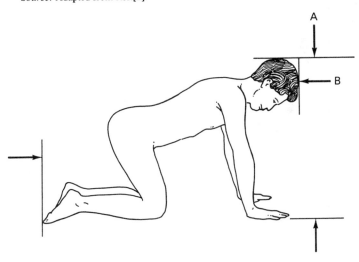

ified reach, access, clearance, and working-level dimensions for wheelchair users. Assuming a random sample of population from any age group, Pheasant [23] provided the data given by Table 5.22 for wheelchair users.

5.7.5 Considerations for Pregnancy

Pregnancy makes certain activities difficult to perform. As pregnancy advances, bending and reaching activities become limited. The body dimensional changes during pregnancy must be considered whenever the situation warrants, such as garment design for pregnant women. Pheasant [23] provides data on two important anthropometric dimensions as given by Table 5.23.

TABLE 5.20 CRAWLING HEIGHT, A^a [CM (IN.)]

Subjects	Percentile		
	5th	50th	95th
U.S. Air Force male	66.8 (26.3)	72.1 (28.4)	77.5 (30.5)

Source: Adapted from Ref.[2].
^aSee Table 5.19

TABLE 5.21 SELECTED CORRELATION COEFFICIENTS FOR ANTHROPOMETRIC DATA, U.S. ADULTS[a]

	1	2	3	4	5	6	7	8	9	10
1. Age										
2. Weight	.113									
3. Stature	-.028	.223								
4. Chest height	-.028	.515	.949							
5. Waist height	-.033	.483	.923	.927						
6. Crotch height	-.093	.422	.856	.866	.897					
7. Sitting height	-.054	.359	.786	.681	.580	.453				
8. Popliteal height	-.102	.457	.841	.843	.883	.880	.485			
9. Shoulder circumference	.091	.299	.318	.300	.261	.212	.291	.182		
10. Chest/bust circumference	.259	.831	.240	.245	.203	.147	.171	.114	.822	
11. Waist circumference	.262	.832	.224	.212	.142	.132	.167	.068	.720	.804
12. Buttock circumference	.105	.856	.362	.334	.278	.217	.347	.149	.744	.766
13. Biacromial breadth	.003	.452	.378	.335	.339	.282	.349	.316	.555	.401
14. Waist breadth	.214	.852	.287	.260	.215	.195	.216	.133	.715	.801
15. Hip breadth	.105	.809	.414	.380	.342	.283	.376	.221	.632	.647
16. Head circumference	.110	.412	.294	.251	.233	.188	.287	.194	.327	.340
17. Head length	.054	.261	.249	.218	.208	.170	.244	.175	.201	.196
18. Head breadth	.122	.305	.133	.097	.089	.066	.132	.075	.245	.271
19. Face length	.119	.228	.275	.220	.226	.199	.253	.193	.162	.172
20. Face breadth	.233	.453	.190	.160	.142	.099	.185	.098	.401	.421

	11	12	13	14	15	16	17	18	19	20
1. Age	.234	.219	.149	.146	.194	.095	.118	.190	.189	.089
2. Weight	.824	.886	.495	.768	.770	.403	.304	.290	.264	.358
3. Stature	.279	.360	.456	.329	.348	.331	.318	.136	.267	.199
4. Chest height	.216	.289	.412	.266	.276	.284	.284	.085	.222	.162
5. Waist height	.238	.336	.409	.293	.318	.306	.297	.123	.225	.200
6. Crotch height	.221	.246	.380	.277	.225	.294	.280	.089	.205	.172
7. Sitting height	.236	.383	.384	.277	.379	.294	.275	.136	.248	.146
8. Popliteal height	.186	.201	.327	.249	.181	.235	.253	.087	.185	.189
9. Shoulder circumference	.775	.717	.581	.719	.606	.330	.248	.252	.217	.313
10. Chest/bust circumference	.796	.674	.370	.706	.551	.273	.204	.255	.176	.273
11. Waist circumference		.722	.382	.886	.600	.281	.149	.267	.174	.310
12. Buttock circumference	.852		.396	.668	.893	.310	.214	.238	.180	.269
13. Biacromial breadth	.288	.355		.401	.361	.311	.239	.178	.266	.211
14. Waist breadth	.936	.849	.327		.576	.292	.168	.263	.182	.296
15. Hip breadth	.724	.895	.340	.760		.265	.183	.188	.155	.215
16. Head circumference	.309	.330	.251	.310	.288		.692	.430	.273	.299
17. Head length	.158	.195	.179	.164	.166	.779		.115	.311	.113
18. Head breadth	.265	.252	.188	.268	.227	.521	.058		.174	.497
19. Face length	.129	.186	.187	.151	.161	.315	.289	.148		.144
20. Face breadth	.412	.394	.278	.410	.364	.464	.131	.660	.206	

Source: Published with permission. Kroemer, et al., 1986. *Engineering Physiology: Physiologic Bases of Human Factors/Ergonomics*, p. 18. Elsevier, Amsterdam.
[a]Women above the diagonal, men below.

TABLE 5.22 ANTHROPOMETRIC ESTIMATES FOR WHEELCHAIR USERS (mm)

Dimension	Men Percentile 5th	Men Percentile 50th	Men Percentile 95th	SD	Women Percentile 5th	Women Percentile 50th	Women Percentile 95th	SD
1. Floor to vertex	1260	1335	1410	45	1180	1265	1355	53
2. Floor to eye	1150	1220	1290	43	1080	1160	1235	50
3. Floor to shoulder	965	1080	1100	40	910	985	1065	47
4. Floor to elbow	625	685	745	37	610	670	730	36
5. Floor to knuckle	370	435	500	41	330	405	480	45
6. Floor to top of thigh	620	650	680	18	565	600	635	21
7. Floor to top of foot	120	150	180	19	165	190	215	16
8. Floor to vertical grip reach	1550	1665	1785	71	1460	1570	1680	67
9. Knee from front of chair	80	140	200	37	55	120	180	37
10. Toes from front of chair	360	435	505	43	305	370	435	40
11. Forward grip reach from abdomen	370	455	540	51	330	410	490	49
12. Forward grip reach from front of chair	250	315	385	41	175	240	305	39
13. Sideways grip reach from side of chair (shoulder—grip length)	580	645	710	38	520	580	640	37
14. Shoulder breadth (bideltoid)	390	445	500	33	330	380	420	30

Dimension	Minimum		Mean	Maximum		
15. Overall length of wheelchair	915		1075	1445		
16. Overall breadth of wheelchair	560		615	645		
17. Height of armrests	705		735	770		

Source: Ref. [23]. Published with permission.

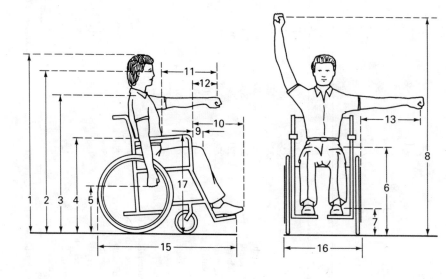

TABLE 5.23 ESTIMATED ANTHROPOMETRIC CHANGES DURING PREGNANCY

Dimension	Month of pregnancy							
	2–3	4	5	6	7	8	9	10
1. Abdominal depth								
5th percentile	195	210	225	250	275	300	315	330
50th percentile	245	260	280	300	320	345	360	375
95th percentile	295	310	335	350	360	385	405	425
SD	29	29	34	29	26	26	27	28
2. Forward grip reach from front of abdomen								
5th percentile	390	375	350	335	320	295	280	260
50th percentile	460	445	425	405	385	360	345	330
95th percentile	530	515	500	475	450	425	410	400
SD	42	42	46	42	40	40	41	42

Source: Ref. [23]. Published with permission.

5.8 APPLICATION METHODOLOGY

Anthropometric data for engineering use are best presented in percentiles. Extreme values represent outliers and may be disregarded in many applications. Given a set of raw data on a specific anthropometric characteristic, the percentiles may be found by the following procedure:

1. Find the mean.
2. Find the standard deviation.
3. Find the factor (see Table 5.24) F_1 corresponding to a specific percentile point. Multiply this factor by standard deviation and add to (50th percentile and above) or subtract from (up to 50th percentile) the mean.

TABLE 5.24
PERCENTILE
CALCULATION
FACTORS

Percentile	F_1
5th	1.645
10th	1.282
25th	0.674
50th	0
75th	0.674
90th	1.282
95th	1.645

The application of the data to a design case is according to the following procedure:

1. Determine the body dimensions important in the design.
2. Define the user population.
3. Select the percentage of the population to be accommodated.
4. For each body dimension identified:
   · Look up the value from data tables.
   · Apply the data.
   · Make adjustments in the data if necessary (slump, clothing, etc).

Alternatively, one may use the "fitting trials" method as described by Jones [24,25]. In this method, subjects are asked to make adjustments on a dimension until the design "feels" best for the task for most people. It is expensive and time consuming to carry out fitting trials for design problems; however, they may be justified in certain situations such as lack of available data and disputed design cases.

5.9 CASE ILLUSTRATION 3

A problem existed at the stocking stations of a mini-load AS/RS (automated storage and retrieval system) of a leading electronics manufacturer. At these stations, operators fill the bin delivered by the crane with material arriving in a tote over a roller conveyor. After the stocking operation, empty totes are circulated to other need points in the system. The bin is taken by the crane to its location in the rack structure.

The problem was concerned with access requirements to an overhead cable trolley conveyor for hooking empty totes after stocking. The conveyor was designed at such a height that it was impossible to reach the hooks comfortably even with the tote extended. This problem did not exist at other interface points in the handling system. Furthermore, cost considerations came into the picture and the conveyor height was not reduced. Instead, a step stool was considered to enable the stocker to reach the moving hooks comfortably. Figure 5.5 gives the geometrical relationships between the worker and the equipment [25]. A side platform prevented the operator from getting close to the moving hooks. This platform was not removed since it helped the operator set loads on temporarily during bin stocking. As measured, the height of the hooks from the floor was 250.2 cm (AD). The problem was to calculate the distance EF. Distance AB was calculated using the Pythagorean theorem. The hypotenuse of the right-angle triangle CBA is equal to the functional reach for 5th percentile females (66.5 cm) plus the remaining tote length (54.9 cm). This anthropometric characteristic was chosen based on the fact that if a 5th percentile female can reach that distance, almost everyone can. Hence

$$CA = 66.5 + 54.9 = 121.4 \text{ cm}$$

The distance CB is equal to 95th percentile male thigh-link to thigh-surface length (7.7 cm) plus the distance from the tip of the side bar to the projection of hooks (58.4 cm). Hence

$$CB = 58.4 + 7.7 = 66.1 \text{ cm}$$

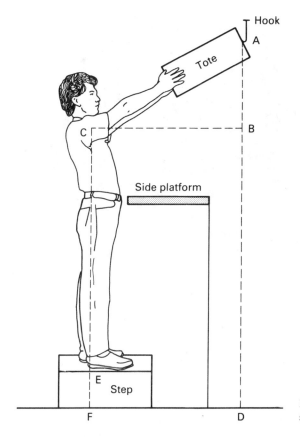

Hook

A

Tote

C ---- B

Side platform

E
Step

F   D

Figure 5.5 Workstation geometry at the stocking station.

Then

$$AB^2 = (121.4)^2 - (66.1)^2$$
$$AB = 101.82 \text{ cm}$$

Then

$$BD = AD - AB = 250.2 - 101.82 = 148.38 \text{ cm}$$

The length CE is equal to the 5th percentile female shoulder pivot height (119.9 cm) plus 2.5 cm for shoe adjustment. Hence, $CE = 122.4$ cm. Finally,

$$EF = BD - CE$$
$$= 148.38 - 122.4 = 26 \text{ cm}$$

EF sets the minimum height of the step stool. The minimum length can also be found easily. This will be based on the 95th percentile male foot length (28.2 cm) plus a shoe adjustment of 3 cm. Hence the minimum length of the step is 31.2 cm. Similarly, the minimum width is

$$2(\text{95th percentile male foot breadth} + \text{shoe adjustment})$$

$$= 2(10.4 \text{ cm} + 0.76 \text{ cm}) = 22.32 \text{ cm}$$

For a comfortable stance, one may more than double the foregoing result, to 50 cm.

Figure 5.6   Step-stool in the foreground.

The stool was designed and installed at the stations (Figure 5.6). The surface is of rough finish to minimize slip potential. It is permanently fixed on the floor under the side bar, minimizing slip and trip hazards due to mispositioning.

5.10 SUMMARY

Engineering anthropometry deals with the development of human body dimensions and use of such data in design. All types of design may benefit from engineering anthropometry: product, workplace, and workspace design. Two basic types of data used are the functional and structural data.

QUESTIONS

1. What is anthropometry? What is engineering anthropometry?
2. Is there evidence that people used anthropometric data early in history to develop human-made objects and structures?
3. What are some of the areas of application of anthropometric data?
4. Discuss the concept of adjustable design.
5. List some of the measurement devices for body landmark distances.
6. Is there such a thing as an average human? Discuss.
7. List and discuss the factors that affect body size.
8. Discuss the two categories of anthropometric data.

9. Given a set of raw data, explain how a person may calculate percentile scores to summarize anthropometric characteristics.

10. List the steps of applying anthropometric data to design.

EXERCISES

1. Should a designer whose focus is the U.S. market only be concerned with ethnic variability? Support your answer with a detailed discussion.

2. Take a tape and measure the standing height of 30 males and 30 females. Plot the results (height versus frequency). What do you observe?

3. Take a tape and measure the standing height of either 40 adult U.S. males or females. Develop a table of stature with 5th, 50th and 95th percentile values. Compare the results with available data in the literature. Explain any difference.

4. Measure the standing height and sitting height of 20 adult males. Calculate the Pearson product-moment correlation coefficient between the two data sets. Compare the results with available data in the literature. Explain any difference.

5. Using available literature, obtain static anthropometric data for three measures on the elderly. Compare results with corresponding data for the younger population. Explain differences.

REFERENCES

1. Kroemer, K. H. E. 1987. Engineering Anthropometry. In *Handbook of Human Factors*. Salvendy, G. (Ed). Wiley-Interscience, New York.

2. Hertzberg, H. T. E. 1972. Engineering Anthropology. In *Human Engineering Guide to Equipment Design*. Van Kott, H. P., and Kinkade, R. G. (Eds.), U.S. Government Printing Office, Washington, DC.

3. Roebuck, J. A., Kroemer, K. H. E., and Thomson, W. G. 1975. *Engineering Anthropometry Methods*. Wiley, New York.

4. NASA. 1978. *Anthropometric Source Book*. Vol. I: *Anthropometry for Designers*. NASA Scientific and Technical Information Office. Houston, TX.

5. Herron, R. E. 1973. Biostereometric Measurement of Body Form. In *Yearbook of Physical Anthropology 1972*. American Association of Physical Anthropologists, Vol. 16, pp. 80–121.

6. Baughman, L. D. 1982. *Segmentation and Analysis of Stereophotometric Body Surface Data Report*. AFAMRL-TR-81–96. Air Force Aerospace Medical Research Laboratory, Wright Patterson Air Force Base, OH.

7. Bonney, M., and Case, K. 1976. The Development of SAMMIE for Computer Aided Work Task Design. *Proceedings of 6th Congress of the International Ergonomics Association and Technical Program for the 20th Annual Meeting of the Human Factors Society*, Santa Monica, CA. Human Factors Society. College Park, MD pp. 340–348.

8. Bittner, A. C., Dannhaus, D. M., and Roth, J. T. 1975. Workplace-Accommodated Percentage Evaluation: Model and Preliminary Results. In *Improved Seat, Console, and Workplace Design*. Ayoub, M. M., and Halcomb, C. G. (Eds.). Pacific Missile Test Center, Point Mugu, CA.

9. Sanders, M. S., and McCormick, E. J. 1987. *Human Factors in Engineering and Design*, 6th ed. McGraw-Hill, New York.

10. Roche, A. F., and Davila, G. H. 1972. Late Adolescent Growth in Stature. *Pediatrics*, 50, pp. 874–880.

11. Trotter, M., and Gleser, G. 1951. The Effect of Aging upon Stature. *American Journal of Physical Anthropology*, 9, pp. 311–324.

12. Borkan, G. A., Hults, D. E., and Glynn, R. J. 1983. Role of Longitudinal Change and Secular Trend in Age Differences in Male Body Dimensions. *Human Biology*, 55, pp. 629–641.

13. Annis, J. F. 1995. Aging Effects on Anthropometric Dimension Important to Workplace Design. In *Advances in Industrial Ergonomics and Safety VII*. Bittner, A. C., and Champney, P. C. (Eds.). Taylor & Francis, London, pp. 27–33.

14. Eveleth, P. B. 1975. Differences between Ethnic Groups in Sex Dimorphism of Adult Height. *Annals of Human Biology*, 2, pp. 35–39.

15. Hertzberg, H. T. E., Churchill, E., Dupertius, C. W., White, R. M., and Damon, A. 1963. *Anthropometric Survey of Greece, Turkey and Italy*. Pergamon Press, Oxford.

16. Sharp, E. D. 1964. A Comparison of Three Full-Pressure Suits in Terms of Control Activation Time. *Report AMRL-TR-64-126*. Aerospace Medical Research Labs, Wright Patterson Air Force Base, OH.

17. Herzberg, H. T. E. 1968. The Conference on Standardization of Anthropometric Techniques and Terminology. *American Journal of Physical Anthropology* 28, pp. 1–16.

18. Hertzberg, H. T. E., Emanuel, I., and Alexander, M. 1956. The Anthropometry of Working Positions. I: A Preliminary Study. *Report WADC-TR-54-520*. Wright Air Development Center, Wright Patterson Air Force Base, OH.

19. Kennedy, K. W. 1964. Reach Capability of the USAF Population. *Report AMRL-TDR-64-59*. Aerospace Medical Research Labs, Wright Patterson Air Force Base, OH.

20. Gifford, E. G., Provost, J. R., and Lazo, J. 1964. Anthropometry of Naval Aviators. *Report NAEC-ACEL-533*. U.S. Naval Air Engineering Center, Philadelphia.

21. Kroemer, K. H. E., Kroemer, H. J., and Kroemer-Elbert, K. E. 1986. *Engineering Physiology: Physiologic Bases of Human Factors/Ergonomics*. Elsevier, Amsterdam.

22. Goldsmith, S. 1976. *Designing for the Disabled*, 3rd ed. RIBA, London.

23. Pheasant, S. 1986. *Bodyspace: Anthropometry, Ergonomics and Design*. Taylor & Francis, London.

24. Jones, J. C. 1963. Fitting Trials: A Method of Fitting Equipment Dimensions to Variation in the Activities, Comfort Requirements and Body Sizes of Users. *The Architects Journal*, February, pp. 321–325.

25. Jones, J. C. 1969. Methods and Results of Seating Research. *Ergonomics*, 12, pp. 171–181.

26. Winship, W. S., and Pulat, B. M. 1987. Ergonomic Applications in Automated Storage and Retrieval Systems. In *Trends in Ergonomics/Human Factors IV*. Asfour, S. S. (Ed.). Elsevier, Amsterdam, pp. 103–109.

_____CHAPTER 6

Human–Machine Systems Design

In this chapter we present the general steps for designing human–machine systems. Following these steps largely ensures that the system will meet its objectives with minimum errors. Equipped with this information, an ergonomist can analyze a human–machine system at a gross level for improvements. This chapter also serves as lead-in to the next three chapters, where we discuss the design of work systems.

6.1 HUMAN–MACHINE SYSTEMS

Ergonomists, designers, and engineers frequently face problems that involve functional planning for combinations of humans and equipment. The basic model that may guide a designer here is a person–object model. An example could be the design of a hand tool. The object is the hand tool, the person is the user. The objective is to design the hand tool such that it is handled most effectively by the person.

The person–object model manifests itself in industrial environments as the human–machine model (Figure 6.1). The human–machine system ordinarily functions as a closed-

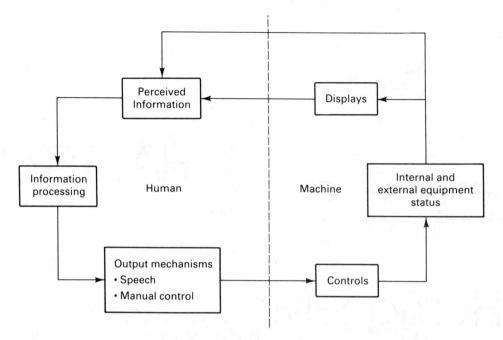

Figure 6.1 A closed-loop human-machine system. (After Ref [1].)

loop system. The equipment status is displayed to the operator via displays. The human being perceives this information and processes it. There may or may not be a following action. If an action follows, it may take the form of speech output and/or a manipulative response mostly using controls. This may, in turn, affect the machine and change its status. The human being observes the change and may or may not take subsequent actions, depending on the status displayed, in order to bring the machine to a desired level or maintain its status. Observe that in the case of a hand tool, the external status may be perceived without any equipment-mounted display [1].

Figure 6.2 gives the major elements of a human–machine system. The system functions within an environment (physical, social). Various inputs are transferred into outputs through a combination of hardware, software, firmware, and human elements based on a task schedule. Outputs can be modified via a feedback loop to the inputs.

Each element of the human–machine system needs particular attention to its subelements:

1. *Environment.* The possible effects of the environment on other system elements must be considered. Vibration, heat, cold, and isolation are examples (see Chapter 9).

2. *Hardware.* Subelements include displays, controls, chairs, equipment dimensions, and their layout. Certain arrangements and designs are better responded to by human beings than by others.

3. *Software,* Menu structures, screen layout, messages, and manuals are sample subelements. Certain designs are more user friendly than others.

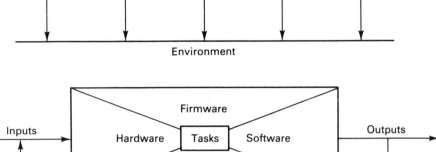

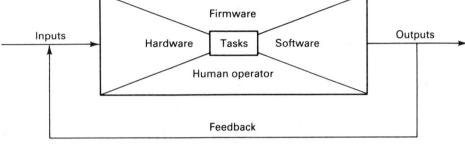

Figure 6.2 Major elements of a human-machine system. (After Ref. [1].)

4. *Firmware.* Software-loaded hardware is known as firmware. Compatibility between the hardware and software, and labeling for proper identification, are important factors.

5. *Humans.* Human beings and equipment complement each other for system success. Human variables to be considered by system designers include age, sex, physique, training, education, and so on.

6. *Tasks.* Both human beings and equipment carry out tasks to achieve system objectives. Assuming that the equipment is functioning properly, human task variables that must be addressed include strength requirements, speed of performance, accuracy requirements, and task sequence.

System classes.   The person's role in the system in terms of coverage, degree of involvement, and importance varies with the degree of automation. In the simplest role, a person acquires information, makes decisions, and takes an action if necessary. Task demands in terms of cognitive, physical, and psychomotor skills are at a minimum. Such jobs require training and job experience for best decision-making performance. An example may be the task of a control room operator in a refinery.

At the opposite extreme, most system functions are assigned to the human being. He or she has to time share between various input stimuli, make frequent decisions, and apply forces routinely. A cyclist is probably operating in this mode. Many other roles for the human being fall between these two extremes.

In terms of degree of manual versus machine control, Sanders and McCormick [2] consider systems in three general classes:

1. *Manual systems.* These are high-flexibility systems where an operator applies forces to a hand tool or other job performance aid and adjusts task parameters (force, pace, etc.) based on the information received.

2. *Mechanical systems.* In these systems, power and several other functions are provided by machines. A human being is responsible for guiding the machine, primarily via controls, to achieve a desired system response. A chip removal activity on a piece part using a lathe is an example.

3. *Automated systems.* Automated systems do not need any operator attention. Direct dialing without telephone operator service is an example. Equipment is capable of performing all functions, including sensing, decision making, and action functions. However, these systems are also configured to accept human input when necessary. Contingency analyses lead to features that allow human intervention.

6.2 REQUIREMENTS AND FUNCTIONS ANALYSIS

Each human–machine system has one or more objectives to fulfill. The system is designed for a purpose. Conversely, for any new system, one can set forth a set of requirements which define that system's goals or objectives. It is extremely important for the designer to analyze those requirements carefully and design system functions to achieve the requirements.

Coupled with system requirements may be *system limitations.* The system may have to function under certain restrictions. Some of these may be stated as hard constraints (must) and some soft (would be nice). System limitations, especially hard constraints, set the boundaries within which the system should work. An example of system needs with requirements and constraints specified is the design of a human-driven bill-of-material development system that achieves minimum 98% accuracy. Here the requirement is that the designed system must attain a minimum of 98% accuracy, and the limitation is that the primary tasks must be assigned to human beings.

Once system requirements and limitations have been specified and detailed, the next step is to carry out a thorough functions analysis to specify which functions are necessary and their feasible combinations to achieve the objectives under the constraints specified. This step includes the listing of all operational functions and developing the parameters of each through task analysis. Each operational function needs to be performed by either machine(s) or human(s) or both. This phase is frequently called *function allocation* [3,4]. The execution of any operational function involves many combinations of several basic functions, including sensing, information processing, and action functions.

Information sensed is matched with the information stored in order to attach meaning to it. Processing of information utilizes what is in the store to make effective decisions. Action functions are also guided by these learned experiences. Each task in any operational function may also be performed by either a human or a machine. A human being's flexibility across the basic functions spectrum is his or her major advantage over machines.

In simple terms, the design process of a merchant marine ship's bridge may illustrate the various steps up to this point. Table 6.1 summarizes these steps. Each major operational function may then be analyzed into its subfunctions prior to a detailed task analysis. For example, the communication function will involve subfunctions of search, talk, and end talk.

It is also important to carry out contingency analysis as part of functions analysis. A contingency is any situation that is nonroutine and may have to be dealt with in performing

TABLE 6.1 INITIAL STEPS OF A SHIP BRIDGE
DESIGN PROCESS

System requirement: Safe transport of goods and passengers
from one port to another

System limitations: Operation restricted to rivers

Operational functions: Control of traffic
Navigation
Communication
Supervision of ship

required tasks and functions. In many cases, contingencies arise from unpredictable system behavior. Brainstorming sessions as well as previous experiences may guide the system designer to plan for what-if analyses. Contingency analysis may indicate new functional requirements that were not thought of during design requirements specification.

Significant research has been conducted with respect to the relative capabilities and limitations of people and machines [5–7]. Some of the relative strengths of human beings are that they:

· Sense unexpected stimuli.
· Apply general principles to solve specific problems.
· Sense low levels of stimuli.
· Adapt responses to changing stimuli.
· Generalize from observations.
· Develop novel solutions to problems.

Machines are generally better at:

· Performing several tasks at the same time
· Making consistent responses to the same stimuli
· Retrieving data details from memory reliably
· Making repetitive preprogrammed responses
· Applying high forces over a period of time
· Storing large amounts of data with much detail

System designers use such relative capability/limitation information in making function and task allocation decisions. Naturally, each project requires case-dependent review since specialized machinery developed in the 1980s and early 1990s possesses capabilities that did not exist at the time the above-mentioned research projects were carried out. Good examples abound in artificial intelligence. There are now commercially available software and machinery that can recognize voice, act as system expert (after programming), perform machine-vision check for defects, and so on. Another area that requires mention here is the use of robotic equipment. Recently, Nof et al. [8] and Nof [9] focused on detailing characteristics of robots versus human beings. A representative evaluation is given by Table 6.2. Users of such data should exercise similar cautions.

TABLE 6.2 HUMAN AND ROBOT CHARACTERISTICS

Characteristic	Human	Robot
1. Energy efficiency	(a) Low (10–25%)	(a) Relatively high (120–135 kg)/ (2.5–30 KVA)
	(b) Improves if work is distributed rather than massed	(b) Relatively constant regardless of work load
2. Fatigue and downtime	(a) Within power ratings, primarily cognitive fatigue (20% in first 2 hr; logarithmic decline)	(1) No fatigue during periods between maintenance
	(b) Needs daily rest, vacation	(b) Preventive maintenance required periodically
	(c) Various personal problems (injuries, health, absenteeism)	(c) No personal requirements
3. Differences in characteristics	100–150% variation expected	Only if designed to be different
4. Reaction speed	$\frac{1}{4} - \frac{1}{3}$ sec	Ranges from long to negligible delay from receipt of signal to start of movement
5. Signal processing	(a) Primarily single channel; can switch between tasks	(a) Up to 24 input/output channels; can be increased; multitasking can be provided
	(b) Refractory period up to 0.3 sec	(b) Limited by refractory period
6. Social/psychological needs	(a) Emotional sensitivity to task structure—simplified/enriched; whole part	None
	(b) Social valve effects	
7. Intelligence	(a) Can use judgment to deal with unpredicted problems	(a) No judgment ability of unanticipated events
	(b) Can anticipate problems	(b) Decision making limited by control program
8. Memory	(a) No capacity limitation	(a) Memory limited by controlling facility
	(b) Not applicable	(b) Memory partitioning possible for enhanced data storage/retrieval efficiency
	(c) Directed forgetting very limited	(c) Can forget fast with a command
	(d) Very limited working register $\simeq$ 5 items	
9. Reasoning	Inductive reasoning	Good deductive reasoning Poor inductive capability
10. Wrist movement	Roll $\simeq 180°$ Pitch $\simeq 180°$ Yaw $\simeq 90°$	Roll $\sim$ 100–575° 35–600°/sec Pitch $\sim$ 40–360° 30–320°/sec Yaw $\sim$ 100–530° 30–300°/sec Right–left traverse (uncommon) 1000 mm, 4800 mm/sec Up–down traverse (uncommon) 150 mm, 400 mm/sec

(continued)

TABLE 6.2 HUMAN AND ROBOT CHARACTERISTICS

Characteristic	Human	Robot
11. Arm	(a) Articulated arm comprised of shoulder and elbow revolute points	(a) One of four primary types: Rectangular Cylindrical Spherical Articulated
	(b) Two arms	(b) One or more arms
	(c) Right–left traverse range: 432–876 mm	(c) Right–left traverse: 100–6000 mm 100–1500 mm/sec
	Up–down traverse range: 1016–1828 mm	Up–down traverse 50–4800 mm 50–5000 mm/sec
	Up–down rotation: avg. $= 249°$	Up–down rotation: 25–330° 10–170°/sec

Source: Adapted from Ref. [9].

6.3 TASK ANALYSIS

The next step in human–machine system design is the analysis of each subfunction for the basic building blocks or tasks, skill and knowledge requirements, workplace assignment, and error potential [10]. This process is known as task analysis [11]. Hence tasks are finer breakdowns of functions. Combinations of tasks in time sequence define functions. Tasks may further be broken into subtasks if need arises. All this information may be tabulated for each operational function broken into tasks and finally, subtasks. It may be beneficial to record information such as task frequency, category (operator, maintenance, etc.) and duty cycle time. These data may then be analyzed for *job allocation,* where tasks and subtasks are assigned to personnel positions based on common skill-level requirements and *workplace assignment.*

As a part of task analysis, equipment malfunctions and human error potential must be identified. This is known as *error analysis.* The objective is to develop a system success probability based on task success probabilities and interrelationships between tasks to make sure that the system will achieve its goals with a degree of confidence. If analyses show otherwise, changes in function and task compositions may be necessary. Swain [12] proposes that in many cases, human tasks may have to be aided by other means in order to improve functional success probabilities. The objective is to design systems that are either fault tolerant or that allow the operator to commit minimal errors.

6.4 TIME PLANNING

Time planning involves a review of system functions based on temporal loading [13]. The purpose is to identify resource (human, machines) allocation problems and optimal use of resources, as well as making sure that the human–machine system can achieve the goals within a planned period. The benefits of time planning may be on-time mission completion, minimization of overtime and delays, and high resource utilization.

Kurke [14] developed OSD (operational sequence diagramming) as the initial stage of time planning. OSD displays the functions and their elements in a sequential logic. It is useful for developing a pictorial description of system operation, indicating the relationships between functions, tracing the flow of information, and identifying the inputs and outputs of functions. OSD uses special symbology to document events and attributes [15].

When time information is added to any flowcharting technique, the result is a *time-line analysis*. Such an analysis will clearly display the overload and underload conditions for the resources together with specific instances of occurrence. The expected result of time-line analysis is better resource utilization. Gantt charts, human–machine process charts, and precedence diagrams are examples [16]. A precedence diagram may also be used for generating task groupings based on task sequences and skill requirements for job identification and workplace assignment [17].

6.5 HUMAN RESOURCES PLANNING

The system design process is not complete without human resources planning and implementation. According to Meister [18], this is necessary for adequate staffing of the operational functions. The four major steps of human resources planning are:

1. *Developing staffing requirements.* This step includes quantification of operator needs by operational function, and then for the total system, as well as identifying skill requirements. The basic data for this step come from the detailed task analysis.

2. *Selection.* The second step is to obtain the human resources to attend the system. These resources may already be available in the organization, or they may have to be obtained from outside. This step may also include testing applicants for the special skills required and assigning them to tasks or job functions on that basis. However, one must be extremely careful here. There must be clear scientific evidence that the skills tested for are actually needed on the job, and lack of those may pose potential injury risk to the performer and/or others in the vicinity. Otherwise, roads will be opened for lawsuits.

3. *Training.* Training is needed when it is desired to upgrade one's skills and background for adequate performance on the job. It may not be administered to all potential personnel. Training may take several forms, including in-class and on-the-job training. Selection and training are discussed in more detail in Chapter 12.

4. *Follow-up.* It is recommended that periodic follow-up be carried out on the personnel for some time after actual job assignment. This is necessary for making personnel changes when there is a demonstrated need.

6.6 DETAILED DESIGN AND ACQUISITION

Detailed design involves reviewing the preliminary plans in more detail and making final decisions on all steps discussed above. The acquisition phase includes detailing, developing, and purchasing of all equipment, software, and other resources.

6.7 TESTING

Although testing is being discussed as the last step in human–machine system design, it is not necessarily conducted only at the end of the development process. Tests can be developed throughout the planning, preliminary and detailed design, development/acquisition, and operational stages. Testing is performed to ensure that:

1. System requirements are being fulfilled.
2. Defects are caught and corrected early.
3. Undesirable features have not been introduced into design.
4. Design decisions are of good quality.
5. There is optimal interaction between people and equipment [19].

Tests and evaluations may concentrate on the hardware, software, firmware, and personnel subsystems as well as on the operational relationships among all. *Mock-ups* are favorite tools of the ergonomist during testing. They may be static or functional. Static mock-ups may be constructed of plywood, cardboard, or plastic material and are inexpensive. However, many dimensional checks with respect to the equipment and the user can be performed using static mock-ups. Functional mock-ups allow for powering of various equipment on the mock-up for quasi-operational testing. However, they are more expensive.

Simulation, modeling, and other quantitative techniques allow the ergonomist to make strategic decisions concerning alternative schedules of system activities, relative positioning of equipment and workstations, and the like. Several simulation packages have been designed for use along these lines [20]. Other test techniques include *two- and three-dimensional templates, drawings, checklists, questionnaires, and so on.* The interested reader may refer to Refs. [19 and 21] for more information.

6.8 DESIGN SYNTHESIS

Up to this point, we concentrated on detailing the more distinct steps of a human–machine system design cycle. This process frequently manifests itself as a cycle, since after completing one step, the designer almost always goes back to the previous steps and makes corrections or revisions. Thus the designer iterates between alternative methods of achieving functional and total system objectives until the best and/or workable combination is obtained. Figure 6.3 gives a summary of the processes discussed for designing human–machine systems. The flow is in general from top to bottom; however, upward iterations are very common. Following this general structure assures good compatibility between requirements and expected system results.

6.9 CASE ILLUSTRATION 4

Several years ago, in a major electronic products manufacturer's factory, a decision was made to provide extra support to the factory's receiving operations. System requirements were set at increasing productivity and accountability at the areas affected. The constraints

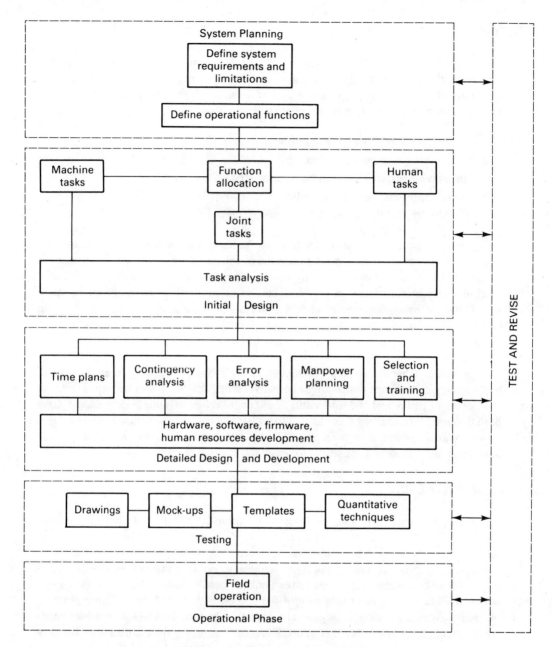

Figure 6.3 Major elements of systems development cycle.

were that only receiving operations will be of concern, no additional personnel should be needed, and state-of-the-art technology must be used (Table 6.3). An ergonomist worked in close cooperation with plant industrial and material handling engineers in structuring and detailing of the problem. As a first step, an operations analysis was carried out to get familiar with the method of material handling in the receiving areas. The information gath-

Requirements: Increase productivity and accountability
Constraints: No new personnel
Receiving dock only
State-of-the-art technology

ered was then matched with the system requirements to develop operational functions to be supported (Table 6.4).

At first, four required functions were identified (OF1–4). Contingency analyses showed that two more were to be added to the list (OF5,6):

OF1. This function is necessary to gather receiving data from vendor shipments for inventory adjustment and subsequent payment to vendors. It spans 16 subfunctions, from recording bill-of-lading number to marking gross weight of the shipment.

OF2. The requirement of productivity increase signaled the need for increasing material velocity in the areas affected. Investigations showed that of 16 data elements collected on each receipt, only four were necessary to check the integrity of shipment, determine inspection requirements, and make routing decisions. Others could be entered during or after the movement of the material. Thus it was decided that the new system should support a two-stage receipt function. The first stage will involve entry of four data elements into the system, after which the material could be moved. During the second-stage entry, the remaining elements would be entered.

OF3. The accountability requirement indicated that each unit (box, bag, etc.) of receipt must be tracked as to its current location within the confines of the dock. Each unit's association with a given receival would also be maintained.

OF4. To check the integrity of shipment (early or overdelivery, delivery of wrong or unneeded material), the receiving system needed to obtain information from the corporate procurement system as to the expected receivals. Similarly, receival information had to be passed to the procurement system in order to update their records.

OF5. As the functions were being developed, it was observed that support had to be extended to various dock operations, including prebagging (bagging of boxed items in

TABLE 6.4 OPERATIONAL FUNCTIONS FOR
RECEIVING

OF1: Support receival data gathering
OF2: Increase material velocity (two-stage receival)
OF3: Track material location
OF4: Communicate with procurement system
[a]OF5: Support dock operations
[a]OF6: Delivery verification

[a]Identified after contingency analysis.

small quantities) and inspection. This was to display work instructions to the operator and collect work results data.

OF6. A reanalysis of the accountability requirement showed that the docks were accountable for material from dock exit to actual delivery to its destination. Thus this operational function was added to the list.

The human resource and technology restrictions led to decisions to develop a computer-based system with product ID (identification) scanning capabilities as opposed to manual data entry. However, some keyboard entry functions were retained. Based on an initial function allocation scheme, OF1, 3, and 6 were assigned to people. Others would be conducted by the computer with human help. OF4 was the only function that needed almost no human involvement.

Task analysis then followed for each function. Operational sequence diagrams were developed (Figure 6.4). The results of task analysis indicated workstation, manning, and skill requirements. A detailed error analysis led to the development of software features for editing and error recovery. Decisions were made as to the machine–human allocation of tasks for each function. Existing dock personnel were then trained based on task assignments. The tracking technology used was bar-code identification and scanning of each box. In addition, remote data input technology using radio frequency was implemented.

During the detailed design and development phase, all necessary equipment, including bar-code printers, scanners, hand-held radio terminals, CRTs (cathode ray tubes), line printers, and labels were purchased. The development of software involved many design and development iterations. Each software feature developed was immediately tested. Purchased equipment were tested at delivery to the plant. All human workstations were carefully designed and developed for seated and standing work considering anthropometric and task requirements (Chapters 5 and 7). Workstation arrangement within the dock space followed operation clusters as well as sequential relationships between operations.

When all software and hardware were ready, a detailed user acceptance testing (UAT) was performed on the total system. Several deficiencies in the software and hardware communications were observed and corrected. Before actual use, a massive load test was performed with human operators and many receival units. This test showed that a larger-capacity central processor (computer) was necessary to improve system response time. One was purchased and installed. The system's maximum receipt capability was determined by a queueing analysis. After all this, production use was rather smooth.

The system has been in operation for several years. During this time, several features have been added to the system, reflecting support to additional requirements. Training courses have been retained for periodic refreshers and to train personnel new to the area. The system developed was accepted by the employees and management. Since all manage-

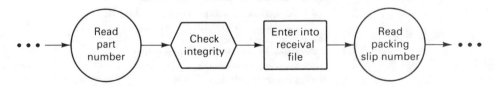

Figure 6.4 Section of OSD for OF1.

ment concerns were integrated into the design, the total solution gained speedy acceptance. Ergonomic aspects of the design (task allocation, workstation design, job design, software ergonomics, training, job aids, etc.) contributed significantly to the employees' satisfaction with the system. As a result of this implementation, productivity in the receiving dock doubled after an initial drop. The drop in productivity was attributed to inefficiencies during the learning period and some resistance to change.

6.10 SUMMARY

The human–machine system is the model that the ergonomist uses in conceptualizing a system to be designed or in evaluating a system that already exists. The input–process–output sequence is important in identifying the elements involved, especially the customers and their needs. The ergonomist uses a step-by-step process in designing new human–machine systems. The process starts with mission, function, and task analysis and ends with design synthesis after resource analyses.

QUESTIONS

1. What is a human–machine system? Why does an ergonomist deal with a human–machine system?
2. What are the major elements of a human–machine system? Discuss each.
3. What are the three system classes with respect to the degree of human versus equipment control? Why is it important that an ergonomist be able to conceptualize systems as such?
4. Discuss system requirements and functions analysis. Why is it important that the ergonomist understand system requirements before system design?
5. What is function allocation? Discuss with an example.
6. Which design step follows function and subfunction analysis? How does it differ from function analysis?
7. Discuss the importance of time planning and human resources planning in system design.
8. Why is testing necessary in human–object system design? Which system elements should testing focus on?
9. What is the significance of system design cycle?
10. Discuss OSD. What is its significance in design?

EXERCISES

1. Select a human–object system in your environment. Analyze its constituting elements in terms of hardware, software, tasks, human(s), and environment. Discuss each element and its relationships with other elements.
2. Select a manual human–object system in your environment. How can it be converted first into a mechanical, and then into an automated system? What may be some roadblocks?

3. Select a human–object system in your environment. Detail its operational functions. Carry out a detailed task analysis for one of these functions and display the results on a flowchart with allocated resources.

REFERENCES

1. Meister, D. 1971. *Human Factors: Theory and Practice.* Wiley-Interscience, New York.
2. Sanders, M. S., and McCormick, E. J. 1993. *Human Factors in Engineering and Design*, 7th ed. McGraw-Hill, New York.
3. Fitts, P. M., and Posner, M. I. 1967. *Human Performance.* Brooks Publishing, Belmont, CA.
4. Bekey, G. A. 1970. The Human Operator in Control Systems. In *Systems Psychology.* DeGreen, K. B. (Ed.). McGraw-Hill, New York.
5. McCormick, E. J. 1970. *Human Factors Engineering.* McGraw-Hill, New York.
6. Fitts, P. M. 1962. Functions of Men in Complex Systems. *Aerospace Engineering,* 21(1), pp. 34–39.
7. Estes, W. K. 1980. Is Human Memory Obsolete? *American Scientist,* 68, pp. 62–69.
8. Nof, S., Knight, J., and Salvendy, G. 1980. Effective Utilization of Industrial Robots: A Job and Skills Analysis Approach. *AIIE Transactions,* 12(3), p. 216.
9. Nof, S. 1985. *Handbook of Industrial Robotics.* Wiley, New York.
10. Miller, R. B. 1953. A Method for Man–Machine Task Analysis. *Report WADC-TR-53-136.* Wright Air Development Center, Wright Patterson Air Force Base, OH.
11. Van Cott, H. P., and Kinkade, R. G. (Eds.). 1972. *Human Engineering Guide to Equipment Design.* U.S. Government Printing Office, Washington, DC.
12. Swain, A. D. 1964. Some Problems in the Measurement of Human Performance in Man–Machine Systems. *Human Factors,* 6, pp. 687–700.
13. ANSI. 1982. *Industrial Engineering Terminology* (ANSI Standard Z94.0). American National Standards Institute, New York.
14. Kurke, M. I. 1961. Operational Sequence Diagrams in Systems Design. *Human Factors,* 3, pp. 66–73.
15. Folley, J. D., Jr., Altman, J. W., Glaser, R., Preston, H. O., and Weislogel, R. L. (Eds.). 1960. Human Factors Methods for System Design. *Report AIR-B90-60-FR-225.* The American Institutes for Research, Pittsburgh, PA.
16. Niebel, B. W. 1988. *Motion and Time Study.* Richard D. Irwin, Homewood, IL.
17. Pulat, B. M., and Pulat, P. S. 1985. A Computer Aided Workstation Assessor for Crew Operations. *International Journal of Man–Machine Studies,* 22, pp. 103–126.
18. Meister, D. 1982. Systems Design, Development, and Testing. In *Handbook of Human Factors.* Salvendy, G. (Ed.). Wiley-Interscience, New York.
19. Chapanis, A., and Van Cott, H. P. 1972. Human Engineering Tests and Evaluations. In *Human Engineering Guide to Equipment Design.* Van Kott, H. P., and Kinkade, R. G. (Eds.) U.S. Government Printing Office, Washington, DC.
20. Pulat, B. M. 1984. A Computer Aided Multi-Man–Machine Work Area Design and Evaluation System. *Technical Report.* North Carolina A&T State University, Greensboro, NC.
21. Meister, D. 1982. System Effectiveness Testing, In *Handbook of Human Factors.* Salvendy, G. (Ed.). Wiley-Interscience, New York.

CHAPTER 7

Work Area Design

7.1 INTRODUCTION

One of the major tasks of an industrial ergonomist is design of work areas. Since such areas must comfortably accommodate workers while offering an arrangement for maximum human effectiveness, the task of work area design is a prime candidate for extensive ergonomic involvement. The work area concept covers a continuum from a simple tool sharpening station to a control room for remote activity monitoring and control. In the first case there is one operator working with one piece of equipment; in the other, there are multiple operators working at multiple stations. The primary objective is to accommodate the workers in the resulting design such that they can work with the equipment, machinery, panels, and other people effectively with no health or safety hazard [1,2]. Naturally, the system's requirements must also be met by the resulting design as discussed in Chapter 6. Furthermore, we expect the workers to feel comfortable with the design. As design ideas form and implementation progresses, it is a good and strongly recommended practice to review the system's performance periodically against the objectives stated above.

In this chapter we dwell primarily on workstation design where one operator works with one or more pieces of equipment. A section is also devoted to workspace design where

167

multiple operators interact with multiple equipment and with each other. The design of the equipment is covered in Chapters 10 and 11.

It should be remembered that design inherently involves trade-offs. Designing for 90 to 95% of the user population is an example. Here 5 to 10% of the population may not be comfortable with the design; however, the majority will. Other trade-offs may include major functionality, such as visibility and available space. More space may be needed to make a label more legible than is available. In such cases the designer will implement the alternative that presents the most gain for the effort. Evidence for the need to make trade-offs comes from many studies. In one, Bendix and Hagberg [3] investigated the relationship between desk slope and use preference in a reading and writing task. The desk slopes tested ranged between 22 and 45 degrees. Their results showed that trunk posture improves with desk slope. For the reading task, subjects clearly preferred the steepest desk. On the other hand, the same subjects preferred the flat desk for writing. A trade-off is obviously necessary here if the same desk surface is to be used for both purposes.

7.2 GENERAL PRINCIPLES

Several generic design principles/conventions apply to all workstation/workspace design cases. These may be called the general principles and are discussed in this section.

1. *Consider functional requirements.* System requirements with respect to needed functionality filtering all the way down to human and equipment tasks must be adhered to. The resulting design must allow the operator to carry out the required tasks in the sequence called for. Specific equipment needs for each task must be considered.

2. *Consider visibility.* Visibility of primary and adjunct displays, controls, other equipment, other workers, task area, and so on, is of primary importance (Figure 7.1). Making sure that everything that the worker must be able to see is visible in the best conditions is one of the primary tasks of the designer. After all, human performance will be degraded significantly if signals and information of interest cannot be seen. A cardinal rule here is to present data at right angles to the line of sight in order to minimize visual parallax. Naturally, if the operator is expected to walk around, this consideration may have to be traded off with others. In an experiment that evaluated the effect of angle of hard copy on VDT (video display terminal) text editing performance, Brand and Judd [4] showed that as the angular discrepancy between the hard copy and VDT screen increased, so did editing times and errors. Furthermore, the researchers showed the desirability of locating hard copy adjacent to the screen.

3. *Consider hearing requirements.* Another information input parameter that the designer must consider is the audibility of signals that the operator must be able to hear. These include buzzers, alarms, and other auditory displays within and outside the work area. Proper and timely acquisition of such information is of primary importance.

4. *Consider clearances.* Clearances for the trunk, the legs, the head, the knees, the feet, and so on, are essential in any design to be used by human beings. Access and egress space must be provided at each work area. Proper clearances will allow for comfort and ease in grasping and operating equipment and controls. They are especially helpful to a

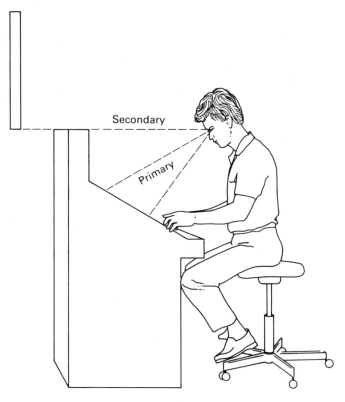

Figure 7.1 Consider visibility requirements.

person in moving around to relieve static loading and other discomfort. Finally, clearances isolate the worker from potential injury. In designing clearances, consider clothing worn by the operator and mobility requirements during work in addition to anthropometric characteristics of the user population.

5. *Consider reach and manipulation requirements.* These determine dimensional factors with respect to the operation of controls, equipment, seat adjustments, and the like. Such requirements can be based on 5th percentile reach characteristics of the user population. Other items that may be reached are parts bins, tools, parts, fixtures, cabinets, and manuals. Normal and maximum work areas, as defined by a sweep of the forearm and the arm, respectively, are some guides with respect to locating work components within the work area (Figure 7.2) [5]. A good rule of thumb is to locate the most frequently used items within the normal work area, and all others within the maximum work area.

The definition of normal and maximum work areas attracted considerable attention of the researchers in the past. The data reported in Figure 7.2 are very close to those proposed by Farley [6], a widely used data set by designers of industrial work areas. Farley assumed fixed elbow position when defining normal work areas in the horizontal plane. Squires [7] pointed out that the elbow does not stay at a fixed point but moves out and away from the body during motions where the forearm pivots. Using this concept, Squires developed two parametric equations to define a modified normal work area in the horizon-

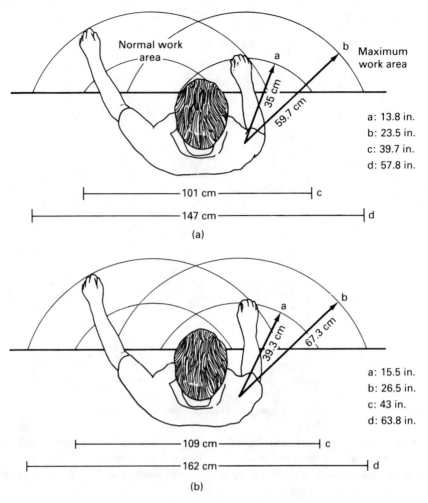

Figure 7.2 Normal and maximum work areas for (a) women and (b) men.

tal plane based on anthropometric data for the 10th percentile male. Squires' curves extend the normal work area away and lateral from the body. However, Squires did not adjust for human variability. Finally, Konz and Goel [8], using 40 men and 40 women selected to be representative of the U.S. population, defined the normal work area for 5th, 50th, and 95th percentile males and females (Figures 7.3 and 7.4) [9].

6. *Consider population stereotypes.* As discussed earlier, stereotypes are defined by operator expectancies. The designer must check with operational personnel to assure that existing stereotypes will not be violated. It is a good practice to maintain the relative placement of equipment, tools, parts, and so on, for similar work areas.

7. *Consider psychosocial factors.* A disorganized, inconvenient, and unattractive workplace will be frustrating to a worker. On the other hand, a well-organized, reliable, simple, safe, and attractive work area will motivate the worker. Human beings prefer compatible display/control and well-planned motion/equipment location relationships. Elias et

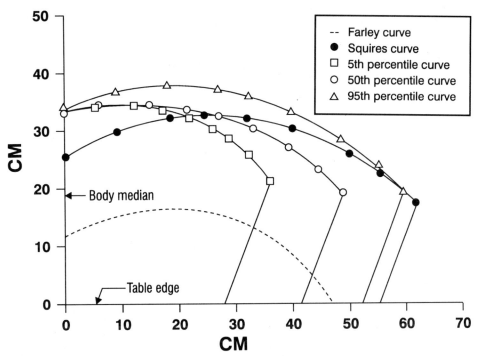

Figure 7.3 Normal working area for males. (From Konz [9], reprinted with permission of Publishing Horizons, Scottsdale, AZ)

al. [10] compared the results of a questionnaire distributed to two groups (81 and 89) of women operators, 20 to 39 years old, engaged in either off-line data acquisitions or dialogue activities. The former group's task was characterized by such attributes as frequent but of short duration (average 1 second) and fragmented with almost no cognitive demands. They found that 70% of the data acquisition operators and 28% of dialogue operators were dissatisfied with their jobs. The chronic neuropsychical disorders were more frequent with the data acquisition operators than in dialogue.

8. *Specify environmental factors.* The work area should protect the worker from undesirable effects of environmental factors such as heat, humidity, noise, extremes of light, glare, vibration, and cold. Independent seat suspension may alleviate stress from vibrating machine platforms. Proper orientation of displays and other equipment may minimize the effects of glare. Colors may help reduce the ill effects of long duty periods and clutter.

9. *Investigate possibility for standardization.* Benefits of standardization are numerous. Savings in hardware development and training time are some. Furthermore, there is lessened chance for operator error. Naturally, a poor design repeated many times is more detrimental than different designs. Hence the prototype of standardized work areas must be evaluated carefully.

10. *Consider the total system.* Each work area must be designed with the total system in mind. Operational relationships in the system define relative locations of the various work areas. Such relationships also define the equipment needs at each work position.

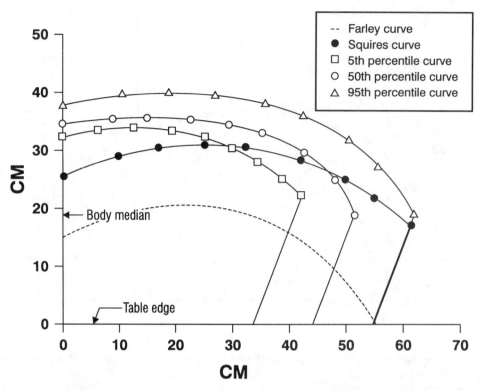

Figure 7.4 Normal working area for females. (From Konz [9], reprinted with permission of Publishing Horizons, Scottsdale, AZ)

11. *Design for maintainability.* Any work area must be maintainable. Work benches, equipment, and tools require periodic maintenance. While human engineering the work area, the designer should not complicate the maintenance person's task. The best method of ensuring maintainability features is to go through the design as if the maintenance activity was being carried out at the station. A review of the tasks involved and the various posture and visibility requirements will provide the necessary data for design for repairability and maintainability [11,12].

12. *Allow for various work postures.* At his or her discretion, the operator should be able to alternate between sitting and standing postures. This may require adjustability in work surface height and other relevant dimensions and angles.

13. *Minimize safety hazards.* Accident potential at the work area must be minimized. For example, sufficient friction between the sole and the floor will avoid slip and fall accidents.

14. *Consider fixed locations for work components.* Fixed locations for tools, materials, and controls will eliminate ineffective motion elements such as search and select.

The general design considerations listed above and in Section 7.3 define the minimum design criteria for work areas. Specific design parameters for each are given in Sections 7.4 through 7.7.

7.3 ADJUSTABLE DESIGN

Adjustability in design is a desired feature, since many factors affect design parameters. Individual, task, and other factors, when taken in combination, dictate that work area design process should visit a check step to make the design adjustable. This is an excellent way to obtain a good fit between the person and the task. Although adjustable features are provided at a work area, they may not be used. Usability depends on how much time and effort are necessary to make the adjustments, as well as the operator's perception of resulting benefits. The designer may rest assured that such features will not be used if notable effort and time are required to make the adjustments with minimal perceived gains.

In a priority rank order, the following adjustability approaches should be investigated in work area design [13]:

1. *Workplace adjustments.* The shape of the workplace may be adjusted by cutouts to minimize reach requirements. This will also prevent protrusion of chairs into aisles. Work surface height and inclination may be adjustable to accommodate people of different sizes. Reach requirements can also be reduced by orientation adjustments in the workplace relative to conveyors and other equipment.

2. *Worker position adjustment relative to work place.* Change in seat height allows vertical adjustability of operator position. Chairs with lockable casters allow for horizontal adjustability. Platforms also help in changing operator position with respect to work surface for standing operators. Footrests help resolve the problem of unsupported legs. However, footrest height must be adjustable relative to the seat height. Armrests help the operator maintain arm position during work without excessive fatigue.

3. *Workpiece and tools adjustment.* The worker can maintain a comfortable and safe working posture if the workpiece can be adjusted via clamps, vises, jigs, and so on. Fixturing is used extensively in circuit board assembly for best work angles for viewing and arm positioning. Multitiered gravity bins help bring parts to within the normal work area. Lift tables, levelators, and lowerators help adjust workpiece height for the next operation. Hand tools could be attached to tracks, spring mechanisms, or tension reels for effective use.

7.4 WORK POSTURE/DESIGN RELATIONSHIPS

There is a relationship between work area design and the expected work posture in that area. In general, an operator will either sit or stand during work, although both postures should be allowed for in the ideal design. In a seated posture, there will be more design considerations in terms of visibility, clearances, and so on. A seated posture allows for reduced static loads to maintain body postures, improved blood circulation, and a feeling of balance along with less fatigue development. A standing operator will experience greater physiological load. Standing for extended periods in the absence of leg movements may lead to blood and body fluid accumulation in the legs. This will result in swelling and possibly varicose veins. If the operator is allowed to move around, the problem potential will be minimized [14].

Long periods of sitting will also be a health disadvantage. Grandjean [15] claims that poorly designed chairs and work postures may lead to back and neck aches, curvature of the spine, and slackening of the abdominal muscles. Field studies also support these views. Hunting et al. [16] observed prevalent neck pains, neck stiffness, and arm/hand pains among 119 accounting machine operators engaged in reading figures from coupons and typing them into a keyboard in constrained seated postures. The working day was 8.5 hours. On the average, 5 to 6 hours was spent on the job with keying speeds between 8000 and 12,000 strokes per hour. Figures 7.5 and 7.6 give some of the complaints from the accounting machine operators. In one case these incidences are also compared with complaints from a group of shop operators of the same age engaged in general shop activities.

Since neither sitting nor standing is preferred for long periods, the best alternative is to provide work areas where the operator can alternate between sitting and standing. We next review several specific design considerations for seated and standing work areas. Then some guidelines will follow with respect to the use of each.

7.4.1 Seated Work Areas

Sitting workplaces are best in the following situations [13]:

- Precise foot control actions are required.
- Fine assembly or writing tasks are predominant.
- No large forces are required, such as handling weights greater than 4.5 kg (10 lb).
- The items being handled do not require the hands to work at an average level greater than 15 cm (6 in.) above the work surface.

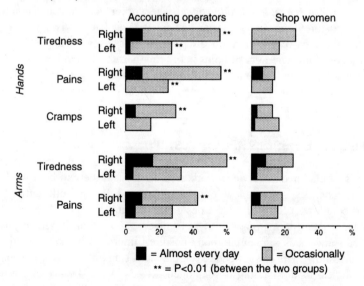

Figure 7.5 Hand and arm impairment in accounting machine operators and shop women. (Reprinted from *Applied Ergonomics* (11), Hunting et al., "Constrained Postures in Acct. Machine Operators," copyright 1980, with kind permission from Elsevier Science Ltd., The Boulevard, Langford Ln., Kidlington 0X5 1GB, UK.)

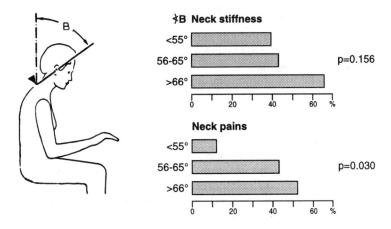

· A high degree of body stability or equilibrium is required.
· Long work periods are expected.
· All items needed within the task cycle can easily be supplied within the seated workspace.

Critical dimensional factors include [1]:

· Correct eye position relative to viewing tasks
· Seat height, depth, back angle, and foot rest position
· Clearance for the legs and the knees
· Height and depth of work surfaces
· Hand and foot reach requirements

Several other design considerations include:

· Supply bins should be placed within 41 cm (16 in.) to the right or to the left of the center of the workplace and not more than 50 cm (20 in.) above (preferably within 25 cm or 10 in.).
· Occasional reach-overs are acceptable. This helps mobility needs.
· Place larger and heavier objects closer to the operator.
· Work surface height depends on elbow height and the nature of the task. Perception of fine detail warrants higher surfaces.
· Force application warrants a lower work surface height (see Section 7.5).
· Lehman and Stier [17] propose that the most comfortable head position for seated

operators is when the angle between the line of sight and horizontal is 32 to 44 degrees.

- Seated workplaces should be provided with adjustable chairs and footrests (see the seating design section).
- Arm pads minimize the impact of possible sharp edges of the work surface on the arms.
- Armrests on the chairs allow for extending comfortable ranges of reach.

Figure 7.7 gives several suggested dimensions for seated work areas. Work surface could be 81 cm (32 in.) high provided that the chair height is adjustable between 51 and 66 cm (20 to 26 in.). Knee clearance for footrests is 66 cm (26 in.). Without a footrest, this could be dropped to 46 cm (18 in.). Thigh clearance is 20 cm (7.9 in.). Adjustable footrest height is between 2 and 23 cm (0.78 to 9 in.). Footrest depth is 30 cm (12 in.). The center of the footrest may be 42 cm (16.5 in.). in front of the edge of the work surface.

Many times VDT work requires a seated posture. Musculoskeletal problems are prevalent among VDT operators. Complaints are mostly focused on the back, neck, arms, shoulders and occasionally in the legs. These complaints are attributable to static muscular loading, biomechanical stress, and repetitive work. Chronic disability due to VDT work may only develop after long exposure. Training, good job design, proper workstation design, adequate rest breaks, reporting ergonomic problems and employing countermeasures are remedies [18].

7.4.2 Standing Work Areas

Standing work areas are best in the following situations [13]:

- No leg room or leg and knee clearance is available.
- Heavy objects are being handled [more than 4.5 kg (10 lb) in weight].
- High, low, or extended reaches are frequently required.
- There are frequent demands for downward force application (wrapping, packing, and so on).
- Mobility is a must.

Some of the critical dimensional factors are [1]:

- Proper eye position with respect to viewing requirements
- Reach distances
- Work surface height and depth

Several other design considerations include:

- A smooth and level standing surface must be provided.
- There should be room for occasional moving around and bracing.

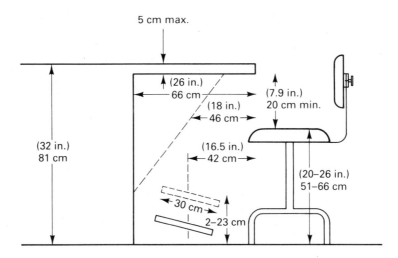

Figure 7.7 Guidelines for seated work area design. (Adapted from Ref. [13]; reprinted courtesy of Eastman Kodak Company.)

· A nonslip and padded standing surface is recommended.
· For standing workers, the most comfortable attitude of the head is when the angle between the line of sight and the horizontal is between 23 and 37 degrees.
· For information display, reading distance and display size should be proportional to readout precision.
· If two hand reaches are required, design for shorter reach distances.
· Forward reach can be extended through a hip bend for an extra 36 cm (14 in.) for occasional standing tasks where sustained activity is not necessary, such as activating a switch or marking a record. A waist bend may allow for 20 cm (7.9 in.) extra reach.
· Work surface height depends on the task.

Figure 7.8 gives several dimensional factors for standing work areas. A nominal 10 cm (4 in.) for knee clearance is recommended. A foot clearance of 13 × 10 cm (5 × 4 in.) is also suggested. For upward reach requirements, maximum reach distance should not exceed 203 cm (80 in.). It is a good practice to make provisions for sitting down even in operations that require a standing posture. To relieve stress and static loading, the operator may sometimes sit down, such as during automatic machine cycle times or possible slack times (due to unavoidable delays). Floor mats and/or shoe inserts with some cushioning have recently been shown to reduce perceived fatigue and discomfort over concrete or extremely stiff mats (usually used to offer slip resistance) [19]. However, mats that are extremely soft may not be desirable.

7.4.3 Sit/Stand Work Areas

Sit/stand operations are best for long primary tasks with frequent standing requirements. A minimum of 51 cm (20 in.) leg clearance is recommended. A fixed work surface height at

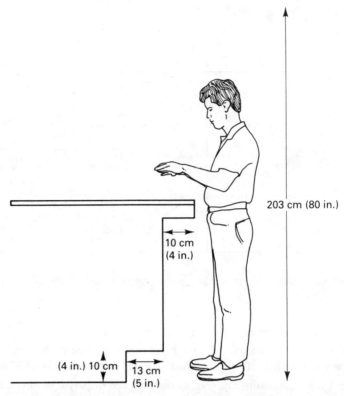

Figure 7.8 Several dimensions for standing work areas. (Adapted from Ref. [13]; reprinted courtesy of Eastman Kodak Company.)

102 cm (40 in.) seems to provide the best trade-off between sitting and standing. For standing posture, check for excessive static loading on the back and shoulder muscles.

7.4.4 Design Trade-offs

If one work area type has to be selected over others, the following suggestions are in order [13]:

- If extended reaches and exertion of forces are required frequently, the work area type that optimizes them should be the choice.
- If critical visual tasks are involved, the type that optimizes them is the choice.
- The duration of each work element is also important. Work area type that optimizes elements with long durations should be selected.

Innovative work areas such as the one shown in Figure 7.9 may also be investigated. This figure gives a portable computer station that allows for best visual performance with respect to different hardware in the system. Green and Pulat [20] developed a computerized procedure to decide on the work area type based on dexterity, mobility, and force exertion requirements. Several critical work area dimensions are also generated as a result of this procedure, depending on the characteristics of the user population.

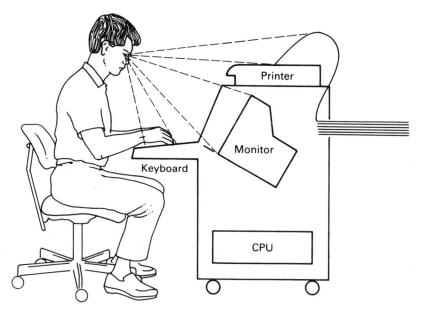

Figure 7.9 Portable work area.

7.5 WORK SURFACE HEIGHT

Working height of the hands and arms should not be overlooked. If the height of the hand is too excessive, the operator will try to compensate for the situation by raising the shoulders and/or abducting the arms. In time this posture will be tiring and may lead to severe pain and cramps in the arms and the shoulder. If the working height of the hands is too low, the operator may bend over or slump to have good access to the work components and improved visibility. This posture is also unhealthy and causes backache. In short, for best human performance the work surface height should be optimal.

Many variables affect the best height of work surfaces. The most important are gender of the worker, type of task, and workplace type. Other variables include such factors as personal preferences and work habits. A good rule of thumb is to *design working heights at 5 to 10 cm (2 to 4 in.) below elbow height* unless the task involves fine manipulation and seeing requirements or force application.

When the work area type is added into the picture along with task type, a set of suggestions emerge. Ayoub [21] offers the recommendations given by Tables 7.1 and 7.2 for seated and standing activities respectively. As can be observed from these tables, fine manipulation requires heights above the elbow level, and coarse work with force application calls for work heights significantly lower than the rule of thumb. Hence it is desirable to have the working height adjustable to suit the user as well as the task type. If this is impractical, height can be designed on the basis of the tallest operators with elevated platforms for smaller workers, although this may create a trip hazard at the workplace. In addition to adjustability in the work surface height, adjustability in the work surface inclination is also desired. This is to provide better visibility of the task elements to the user. The recommended range of inclination is between 0 and 75 degrees.

TABLE 7.1 RECOMMENDED WORK SURFACE HEIGHTS FOR SEATED OPERATIONS

	Work surface height [cm (in.)]	
Type of task	Male	Female
Fine work	99–105 (39–41)	89–95 (35–37)
Precision work	89–94 (35–37)	82–87 (32–34)
Writing, light assembly	74–78 (29–31)	70–75 (27–29)
Coarse or medium work	69–72 (27–28)	66–70 (26–27)

Source: Modified table published by permission. M. M. Ayoub, 1973. "Work Place Design and Posture," *Human Factors,* 15(3), pp. 265–268.

Typing usually requires a lower work surface height since the height of the keyboard defines the working height. Grandjean [15] concludes that the tabletop height for such activities should be 68 cm (27 in.) for men and 65 cm (25 in.) for women. For seated operations, office desks without a typewriter call for a working height of 74 to 78 cm (29 to 30 in.) for men and 70 to 74 cm (27 to 29 in.) for women. The recommendations above have been somewhat confirmed by Cox [22]. In a fitting trial, subjects selected the desktop height range between 66 and 84.5 cm to match the full range of adult users.

7.6 SEATING

Much has been done with respect to comfortable seating in industrial environments. Poor seating can lead to many ailments, including fatigue and poor performance. It is important for a chair to provide correct posture and features that help performance rather than hinder it. In a seated posture, the muscles in the back and the spine are not relieved of stress. Poor posture may make the situation worse. Nachemson [23] and Nachemson and Elfstrom [24] showed that compared to standing erect, sitting imposes more intervertebral disk pressure, an average of 60% more. Sitting erect creates 25% less pressure than sitting with trunk bent forward. Andersson et al. [25] claim that there are other variables that affect disk pressure,

TABLE 7.2 RECOMMENDED WORK SURFACE HEIGHTS FOR STANDING OPERATIONS

	Work surface height [cm (in.)]	
Type of task	Male	Female
Precision work with elbows supported	109–119 (43–47)	103–113 (40–44)
Light assembly	99–109 (39–43)	87–98 (34–39)
Heavy work	85–101 (33–40)	78–94 (31–37)

Source: Modified table published by permission. M. M. Ayoub, 1973. "Work Place Design and Posture," *Human Factors,* 15(3), pp. 265–268.

such as backrest inclination, amount of lumbar support, and hand rests. The authors showed that disk pressure drops with increased lumbar support and backrest angle, and arm support.

Naturally, in an industrial environment, workers do not work with a posture that calls for back inclination of the trunk. The body is more or less inclined forward slightly. However, the back support should allow frequent backward mobility to relieve static loading on the back. Although armrests are not recommended for industrial chairs where mobility of the trunk is desired, the arms can be supported by other means, such as pads on the work surface. Kramer [26] suggests that for proper nourishment of the intervertebral disks, an alternate sitting and standing posture is required. The ensuing pressure differentials trigger the in and out flow of tissue fluid from the disks.

The following list provides ergonomic suggestions with respect to industrial seating design (Figure 7.10 and Table 7.3):

· The seat height should be adjustable between 38 and 51 cm (15 to 20 in.). Greater heights jeopardize stability.
· The height of the back rest must be adjustable between 10 and 25 cm (4 to 10 in.) above the seat.
· Recommended backrest width is 33 cm (13 in.).
· The backrest must provide lumbar support to the user. Occasionally, the user must be able to lean backward.
· The seat must be fitted with multiple legs for additional stability.
· Recommended seat surface depth is 40 cm (16 in.). Allow for 43 cm (17 in.) as width.
· The seat should be cushioned to give way about 2.5 cm (1 in.).
· Fabric should be used as a seat cover.

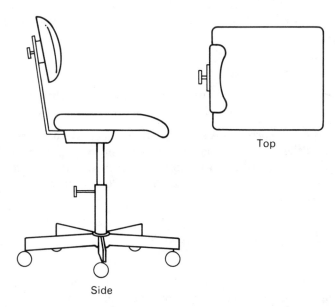

Side

Figure 7.10   Industrial chair.

TABLE 7.3 RECOMMENDED CHAIR DIMENSIONS

Feature	Dimension [cm (in.) or degrees]
Seat height	38–51 (15–20)
Backrest	
Height (from seat)	10–25 (4–10)
Width	33 (13)
Seat surface	
Depth	40 (16)
Width	43 (17)
Slope	3–5°
Armrest height (from seat)	20 (8)

· Armrests are generally not recommended. In certain cases they may be used, such as in tasks that require high hand and arm stability. If armrests are used, recommended distance between the top of the armrest and the compressed seat is 20 cm.

· The seat should slope 3 to 5 degrees backward with a slight dip in the middle for extra body stability. The front edge of the seat should be curving downward.

· Footrests are strongly recommended (see Figure 7.7).

Helander et al. [27] evaluated the adjustability features of chairs and found the following:

· Control discernibility and control feedback significantly improved users' understanding of adjustability.

· Controls with long levers were preferred to controls with short levers or push buttons.

· The chair with the greatest number of adjustability controls took significantly greater time to adjust and was also judged to be the most comfortable.

7.7 OTHER USE AREAS

People use areas other than their workstations during work. These areas also require attention. Several such areas are listed below with specific design suggestions.

1. *Aisles and corridors.* Aisles and corridors should be designed such that minimum clearance guidelines are met when the system is being fully utilized. Minimum aisle width for two-person passage is 137 cm (54 in.); for three persons it is 183 cm (72 in.). Minimum requirement for a two-wheel hand truck is 76 cm (30 in.). Other considerations include:

· Mark traffic guides on floors.

· Keep aisles clear.

· Avoid having doors opening to corridors.

· Avoid blind corners.

2. *Floors and ramps.* Floors and ramps must allow maintenance of stability and not lead to slip and trip accidents. They should be easily maintainable. Other considerations include:

- Building entrances should have mats to minimize tracking in of snow, water, and mud.
- Slip-resistant floor coatings (sand in paint, etc.) are recommended. Swensen et al. [28] determined that people could identify slippery surfaces quite reliably. Their subjects did not slip at a COF (coefficient of friction) of 0.41 but lost their footing at a COF of 0.2.
- Cracks, depressions, and other irregularities on floors must be repaired as soon as possible.
- Rugs or mats on the floors help maintainability where feasible, such as locations where parts may be dropped.
- Ramps should not be lined up directly with doors and stairs.
- Ramps should have a nonskid surface with handrails on each side.

3. *Conveyors.* Conveyors are used to link manufacturing areas on which raw materials, work-in-process goods, and finished products are moved. Some assembly tasks are carried out on conveyors. Their height, width, and other characteristics affect the effectiveness of people working with them. Some considerations are:
- Levelators and lowerators aid loading and unloading activity.
- Conveyors must be accessible from both sides.
- Height and width depend on the type of task. Goods on conveyors must be within reach distances.
- Gates in conveyors must be provided for people and equipment traffic when necessary.

4. *Stairs and ladders.* Attention to the design of stairs and ladders minimizes the risk of injury due to misstepping, slipping, and falling. Normal walking gait must be taken into account when designing stairs.
- Step height should be between 16 and 18 cm (6 to 7 in.).
- Step depth may range between 24 and 27 cm (9 to 11 in.).
- A nonskid surface on stairs and ladders is recommended.
- Use matte finishes on the steps.
- Ladder step width may be between 48 and 61 cm (19 to 24 in.).
- Ladder step separation range should be 25 to 30 cm (10 to 12 in.).

7.8 WORK AREA ARRANGEMENT

Any work area will involve the use of a number of parts, equipment, and other devices. Sanders and McCormick [29] call these items *components*. Given a user population and the specific task to be carried out, there are optimum locations for these components within the three-dimensional work area (workspace envelope). Optimum locations correspond to the best locations for positioning of these components. Theoretically, it is possible to have the optimum location for two or more components to coincide. In these instances, trade-offs have to be done to obtain the best arrangement.

Several factors determine the optimum location for a given part handled within the operator's task cycle. These include anthropometry, biomechanics, and task variables. Although biomechanics and anthropometry have been discussed extensively, we will cite these several more times in this section. In general, anthropometry defines the feasible space, and biomechanical considerations define sections in the feasible space for best location of components. On the other hand, task considerations focus on these sections. Utilizing relationships between task elements and other variables, one can find final locations for the items at hand after trade-offs.

7.8.1 Guiding Principles of Arrangement

McCormick [30] proposed a set of principles that could be used in determining the relative location of groups of components within the work area and then finding the specific location for each component within a group. He called these the *guiding principles of arrangement*. A two-tier application of these principles can lead to the development of the total work area. They are primarily subjective conventions. However, they can be used in conjunction with analytical approaches, such as link analysis and optimization techniques, in order to bring more objectivity to the process.

1. *Importance principle.* This principle claims that those components that are most important in the functioning of the system should be located in the optimal locations. This assumes that the designer in conjunction with the process engineer evaluates every component and rates each with respect to criticality in meeting system's objectives. Then the most critical items list is given the first priority in being located in the most visible and accessible locations in the work area. An example is the location of a warning label in front of a machine operator.

2. *Sequence-of-use principle.* In this case the designer would review the sequential relationship between the use of components. Then those that are always used one after the other would be located next to each other. Naturally, to evaluate sequential relationships correctly, the designer must evaluate all possible uses of the system under different missions.

3. *Functional principle.* This principle says that those components that are functionally related to each other should be located within the same area. It requires the designer to make functional groups of components. These groups may then be located in relation to each other. Components within each group can be arranged based on other principles, such as sequence-of-use and frequency-of-use.

4. *Frequency-of-use principle.* The frequency-of-use principle focuses on the absolute number of times that a component is used within a time period. Those that are used most frequently may then be located in the most accessible and otherwise best areas. This principle aims at a minimum cycle time to complete task responsibilities.

In a control panel layout investigation, Fowler et al. [31] evaluated standard military display-control panels developed under the four arrangement principles as discussed above. For each principle, three subarrangements were developed, depending on the degree of principle application (low, medium, high). Two hundred male college students used each subarrangement under each principle to carry out a simulated task. The performance measures used were number of errors made and time to complete the task. On both accounts, as well as at all levels of principle application (low, medium, high), panels arranged under the sequence-of-use principle allowed the best total system performance.

The guiding principles of arrangement call for trade-offs in many cases. For example, a component that is most frequently used may not be the component that is most important in the proper functioning of the system. Hence the designer must make decisions on the relative location of this component based on other considerations. If consistent use sequences exist among components, use of the sequence-of-use principle certainly makes sense.

7.8.2 Link Analysis

Link analysis is an objective method of developing data for component arrangement. Operational relationships between components can be symbolized by link values (in quantitative terms). For example, a value of 4 for A-B may indicate that component B has been used four times right after component A. This kind of relationship designation can be called a *sequential* [30] relationship. Hence an upper triangular matrix can be developed for expected sequential links between all components to be arranged. Another designation is that of *functional* links [30]. Functional links focus on the absolute number of times a component has been or is expected to be used per unit time (or during a representative system mission) or during a task cycle. Hence only one functional link value is associated with each component. Functional links and sequential links may then be used by a designer, ergonomist, or engineer to generate component arrangements based on objective- and task-related criteria. These data can also provide very valuable input to the guiding principles of arrangement.

7.8.3 Some General Guidelines

A higher-level arrangement scheme can be generated by a set of rules followed by other considerations, including operational links and the principles discussed in Section 7.8.1. One such set of rules is:

1. Consider visual displays first. Locate displays in upper regions of the available space. Locate critical displays first.
2. Then locate secondary displays.
3. Consider controls next. Locate controls in the lower regions of the available space. Locate primary controls first.
4. Locate secondary controls.
5. Locate other components that the operator has to use during the task cycle (parts bins, tools, etc.) based on anthropometric and biomechanical principles.
6. Locate auxiliary displays and controls.
7. Make adjustments in the layout based on additional considerations such as compatibility (see Chapter 10) and similarity with other stations.

As discussed previously, alternative designs thus obtained may be tested with and without human users for making final layout retrofits and decisions.

7.8.4 Biomechanical Considerations

As discussed in Chapter 3, biomechanics deals with determining physical stresses on the body with the objective of minimizing such stresses. Its methodologies and results may also be used to design work areas where operators are expected to apply forces and carry out other physical activities. The resulting work areas and work methods are hopefully less tiring and more compatible with human capabilities. In Chapter 3 we learned that in a seated posture human capability to apply push and pull forces far exceeds capability for force application in the up and down directions. Furthermore, Dupuis et al. [32] have shown that human beings can apply maximum pull forces on the handle of a control when the handle is located 60 to 65 cm (23 to 25 in.) in front of the seat reference point. Figures 7.11 and 7.12 provide additional biomechanical data that affect work area arrangement decisions [33]. The interpretation of these figures is that as the weight handled and the distance to the body increases, muscle fatigue sets in faster. Furthermore, as the head tilt increases, neck extensor fatigue increases. Both of these figures have the following work area design implications:

· Place heavier objects as close to the body as possible.
· Present information to the worker such that the natural rather erect posture of the head is maintained.

7.8.5 Specific Component Locations

In this section we briefly review the suggested general locations for specific components.

1. *Visual displays.* Locate these in upper areas of the work area, still maintaining a comfortable neck and head angle (not to exceed 20 degrees). Primary visual displays need to be located within ± 15 degrees of arc around the normal line of sight. Secondary displays may be located around the primary units but within ± 30 degrees of arc around the NLS. Auxiliary units would then be located around those.

2. *Hand controls.* Locate these in the lower regions of the work area. This strategy aids in two ways: keeps the elbows close to the body and minimizes the chances for blocking of view via hand and arm movements. Furthermore, it enhances accessibility of controls. If the control is too close to the body, the resulting operating posture may not allow sufficient force application. For best manipulative accuracy, design control locations for elbow angles between 80 and 120 degrees.

3. *Foot controls:* Locate these for maximum effectiveness of the foot. Figure 7.13 gives suggestions for a seated operator [34]. It is evident that a region between 60 and 90 cm (23 to 35 in.) in front of the seat reference point sloping upward is the best position for such controls. However, these areas are for foot controls that do not require substantial force application. Controls that require substantial force application need to be located well to the front with the leg in almost a straight position. Naturally, the seat must be fixed to the floor in such cases.

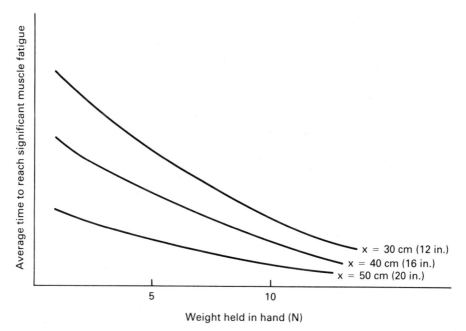

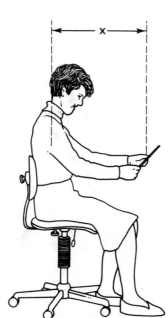

Figure 7.11 Relationship between weight handled, reach distance, and endurance time. (Adapted from D.B. Chaffin, "Localized Muscle Fatigue—Definition and Measurement," *Journal of Occupational Medicine,* 15(4), 1973, pp. 346-354.)

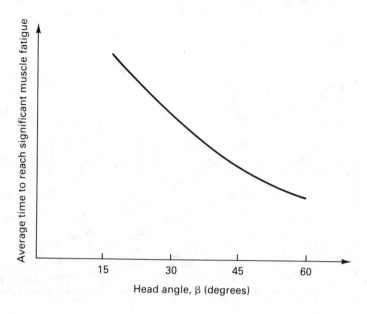

Figure 7.12 Relationship between head angle and neck fatigue. (Adapted from D. B. Chaffin, Localized Muscle Fatigue—Definition and Measurement. *Journal of Occupational Medicine*, 15(4), 1973, pp. 346–354.)

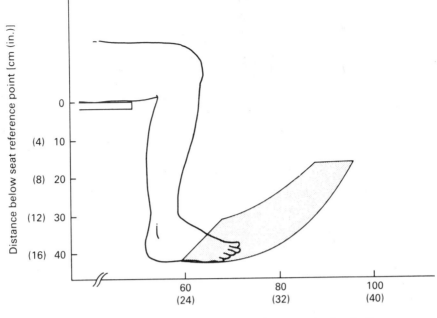

Figure 7.13 Best locations for foot controls. (Adapted from Ref. [29].)

7.8.6 Panel Design for Remote Control Operations

Remote control operations require one or more operators to work primarily with a wall-mounted instrument panel with displays and controls in addition to computer terminals and work manuals. The operators may assume a seated, standing, or sit/stand posture for the task. The primary task is to monitor system status and take action when necessary. The action may involve control activation, communication with others, and so on.

Several computer-aided heuristics have been developed to design instrument panels with displays and controls. CAPABLE, developed by Bonney and Williams [35], designs the workplace of a seated operator. Specifically, it attempts to position controls and displays automatically on several panels within reach distances. WOLAP, developed by Rabideau and Luk [36], aims at placing the more frequently used and functionally more critical components in optimal (more accessible and having better viewing angles) locations while taking anthropometric constraints into consideration. The technique used for optimization is a "minimum utility cost" procedure. This heuristic is structured to produce three quantitatively optimized solutions of the layout problem. CAPADES (Computer Aided Panel Design and Evaluation System), developed by Pulat [37] in 1980, starts out with optimal selection of components to be laid out [38] and proceeds to preparing panel layouts [39] for a standing operator. Finally, alternative layouts are evaluated via simulation. Figures 7.14 and 7.15 provide additional details.

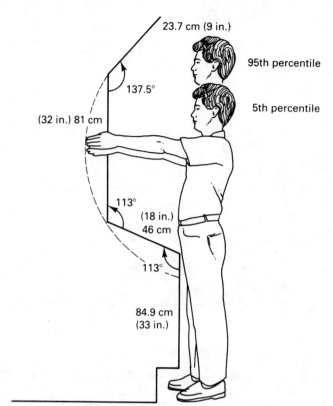

23.7 cm (9 in.)

95th percentile

137.5°

5th percentile

(32 in.) 81 cm

113°

(18 in.)
46 cm

113°

84.9 cm
(33 in.)

Figure 7.14 Panel-operator geometry in CAPADES. (After Ref. [37].)

7.9 WORK SPACE DESIGN

There are many similarities between work area design and workspace design. The latter covers multiple equipment, workstations, and people. Links between people and workstations add complexities to the picture. Other variables that are of interest are crew mobility, common displays, and voice communication between people [2]. Pulat [40–42] developed a computerized procedure to design a workspace for a crew at sit/stand operation. The process starts from work load balancing and decisions on the number of workstations. Then a general layout of the workspace is prepared, followed by the design of each workstation. The resulting model is called MAWADES (Multi-Man-Machine Work Area Design and Evaluation System). The major assumption is that the operators work with instrument panels. Figure 7.16 provides the sit/stand workplace geometry used at each station in MAWADES.

7.10 CASE ILLUSTRATION 5

The case study presented in this section is based on the redesign of the board preparation operation in the kitting function at an electronics assembly factory. Ergonomic principles received major consideration in the design and selection of equipment for combining sep-

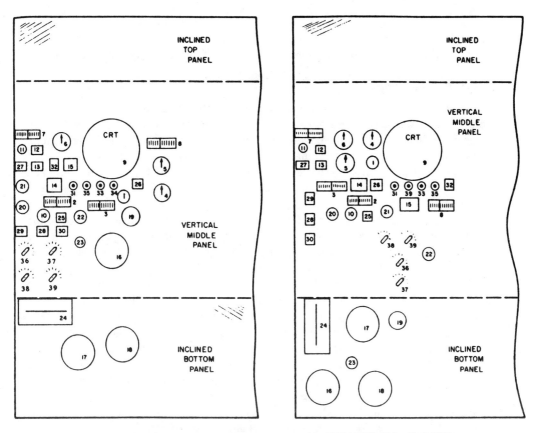

Figure 7.15 Two panel layouts generated by CAPADES. (After Ref. [39].)

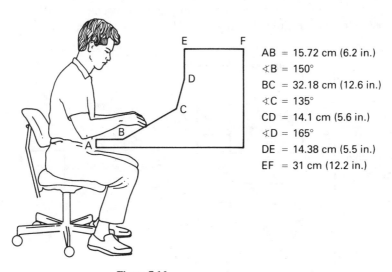

AB = 15.72 cm (6.2 in.)

∡B = 150°

BC = 32.18 cm (12.6 in.)

∡C = 135°

CD = 14.1 cm (5.6 in.)

∡D = 165°

DE = 14.38 cm (5.5 in.)

EF = 31 cm (12.2 in.)

Figure 7.16

arate work centers into one. Board prep work was previously done in four separate work centers (one center repeated four times) with extensive material handling and staging. Before presenting the improvements, some discussion of the kitting function is in order.

7.10.1 Parts Kitting

A kit is a collection of components and/or (sub)assemblies that together support one or more assembly operations for a product [43]. The kitting operations at this manufacturer are geared primarily toward the circuit board assembly (circuit pack) parts. The factory views kitting as the first shop operation, where parts are transformed from the form they are stored in and picked from the buffer zones into the form in which they will be consumed at subsequent assembly operations. A secondary function of kitting is final material verification.

Figure 7.17 gives the material flow diagram for the kitting function in each of the four kit shops prior to the modifications. Parts to be processed in kitting arrive primarily from the miniload AS/RS (automated storage and retrieval system). All arriving components are verified against the accompanying material identifier information. Parts to be rearranged only get staged until all other subkits are complete. Others take one, or two, or all three routes of prep work before getting joined with the rest of the kit at the consolidation operation. Sequencing develops an output reel of components (diodes, resistors, capacitors, etc.) sequenced in the order required by the insertion machines from single-component reels. Part preform involves bending, straightening, lead forming, and cutting. Board preparation work involves stamping, cutting circuit paths, taping, drilling, labeling, and baking. All these operations are aimed at identifying and preparing printed wiring boards for the subsequent operations. The baking operation assures good circuit connections by removing moisture from the inner layers of board material. The consolidation operation involves lot sizing, counting, stamping, and containerization. It is at this point that final material check is performed for kit completeness. A final quality control check assures kit integrity before delivery to the assembly operations. Due to the variety of operations involved, the kitting function is notably labor intensive.

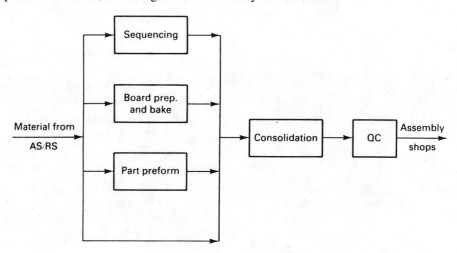

Figure 7.17 Kitting material flow.

Figure 7.18

7.10.2 Ergonomic Improvements

In this section we present a summary of the ergonomic interventions in the board prep operation with "before" and "after" statements.

Before.   Prior to centralization of the operation, all such work was performed at four different kit shops duplicating resources. This resulted in under- or overutilization of resources on a per shop basis, frequently generating employee complaints and at times, capacity bottlenecks. Each kit shop was equipped with an oven (Figure 7.18) and other prep facilities (Figure 7.19), causing excessive overall material handling with associated

Figure 7.19

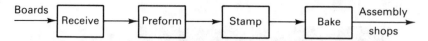

Figure 7.20 Redesigned board preparation operation.

discomfort. The distributed preparation philosophy also led to a quadrupling effect of quality problems introducing multiple sources of variability, along with job-related stress echoed in several orders of magnitude.

The baking operation required loading and unloading of ovens with batches of printed wiring boards. This job element was physically demanding in that it required bending, twisting, lifting, and carrying loads at the capability limits of the operators. Discomfort due to heat radiated from the ovens was often also noted. Information related to the suboperation requirements was developed on paper documents such as the "board prep sheet" and the "flow tag." This handwritten form of operation documents presented readability problems. Furthermore, paper documents would get lost, resulting in operation delays. Overall, the previous board prep operation presented much opportunity for improvement. Other than relieving physical discomfort, the new operation could significantly enhance material velocity by pooling resources together and organizing work in a logical task sequence.

After. Figure 7.20 gives the material flow diagram for the redesigned board preparation operation. Ninety-five percent of the boards are now being processed in this work center. The remaining 5% still need to be processed at other locations due to unique work requirements. At this central facility, boards are first prepared via bench operations (Figure 7.21), then stamped for identification, and finally, baked (Figure 7.22) with an in-line philosophy. Conveyors at knuckle height are used extensively for material transportation from one work area to another in a serpentine configuration. For several reasons, including flow rate controls and material mix concerns, operator intervention with the conveyor mechanisms is necessary at two points for an end-to-end material flow in the board prep room.

Figure 7.21

194   Chapter 7   Work Area Design

Figure 7.22

The operation is enclosed within a 52 ft by 30 ft room for additional fire controls with adequate access and egress allowance for material and personnel traffic.

Receiving and Bench Operations. At the receiving station, arriving boards are laser scanned for material tracking purposes. Then boards for the same select-ID (identifying a particular manufacturing order) are routed to a particular prep bench via a powered wheel conveyor. An instrument panel at the receiving station (Figure 7.23) identifies benches requesting work using an indicator light/pushbutton combination. Depressing the pushbutton activates the transfer conveyor and the mechanism that offloads the material to the correct bench. Bench operators request work via a pushbutton that activates an indicator light on the instrument panel. Spatial compatibility between indicator light/pushbutton positions on the instrument panel and the physical positions of benches along the conveyor

Figure 7.23

minimize work routing errors. After completing the preform operation, the bench operator pushes a button to transfer the tubs containing boards to the oven conveyor loader.

Paper form instructions for board prep operations have been automated for CRT presentation at each position upon request. This eliminated much confusion over handwritten information. Backup copies are still available for use as contingency to the video presentation.

Automated Lift and Lower Mechanisms. The baking oven operator retrieves boards from the prep conveyor using an automated elevator/lowerator mechanism (Figure 7.24). The lowerator automatically lowers as boards are stacked on it at waist height until the stack of boards rests on an orange-tined cart placed at the base. Yellow separator boards help identify boards (green in color) that belong to different select-IDs. A manual down switch is also available for bringing the lowerator down to the cart level when less than a capacity stack is loaded. When full, the operator transfers the cart to the oven conveyor loading position. For a comfortable push task, the tined truck is equipped with removable handles.

Another automated mechanism used for material handling in the board prep operation is the circuit board loading device at the front end of the oven conveyor. This machine

Figure 7.24

uses a vacuum cup pickup and push fixturing to lift the trailing edge of the boards and push them into pinch rollers to load material onto the conveyor. Initially, the elevator lift is empty and at its bottom position. If not, it can be lowered to that position via a switch. An audible alarm sounds throughout the descent of the lift device. With the elevator at its lowest position, the boards are loaded between guide rails against a vertical stop. The tined truck handles are then removed and the lift mechanism started. The elevator stops its ascent automatically when the stack of boards is in the required position for feeding. Various sizes of boards can be loaded onto the oven conveyor by this device.

Stamping and Baking. Prior to the baking operation, boards are date coded via a date code stamping machine mounted to a conveyor section attached to the loader. They are then fed into the flow-through oven via the conveyor and right-angle transfer mechanisms. After baking, boards are accumulated on an automated drop leaf stacker. A light is activated and an alarm sounds when the stacker is full and needs operator attention. At this point the operator's responsibility is to remove the boards, group them by select-ID, and dispatch them to the subsequent assembly operations. The stacker is equipped with catch arms to accumulate boards exiting the oven during stock removal. This allows an uninterrupted baking operation.

7.10.3 Conclusions

Implementing automated equipment operated by programmable controllers, all linked together for in-line continuous operation while retaining operator intervention features, resulted in notable productivity improvement. Control consoles on each bench for signaling the need for more work and for transferring finished work onto the conveyor give the operators the feeling of being a part of the automation project and in control of it. Prior to this development, it was necessary for employees to lift tubs of boards to and from portable carts, workbenches, and shelf racks and travel excessively to stage the product in kits. The redesigned work center eliminated all handling except that required to load and unload the in-line process. Now all handling is done in a standing position with no stooping, bending, or reaching. Special handling equipment complements human capability for a task safely performed.

In addition to presenting a more enjoyable, acceptable, and comfortable work environment, the new board prep operation led to better utilization of existing resources by consolidating capacity. Material velocity increased with a resulting drop in in-process inventory.

7.11 SUMMARY

Industrial work areas must be designed for the users. This means that the user characteristics should determine the specific workplace and workspace dimensions. Naturally, task parameters will also be considered in the overall design of the work area. The best design fits the worker population (adjustable design) and meets the task and production requirements.

QUESTIONS

1. Define *work area*.
2. What are the objectives of work area design?
3. Why are the functional requirements of the system an important factor in work area design?
4. Discuss visibility, hearing requirements, clearance, and access requirement considerations of work area design.
5. Discuss the various adjustability approaches in work area design.
6. What are the relative advantages and disadvantages of seated and standing workstations?
7. Discuss the critical dimensional factors in seated work areas.
8. How could design trade-offs be performed between seated and standing workstations?
9. What is the rule of thumb for work surface height?
10. List the most important three variables that affect work surface height.
11. Which posture imposes more intervertebral pressure on disks: seated or standing?
12. List several characteristics of ergonomic seating.
13. What are the guiding principles of arrangement? List and discuss.
14. Discuss the sequential and functional links.
15. Discuss the biomechanical basis of workstation design.
16. Discuss the preferred locations for displays and controls within the workspace.
17. What are the primary responsibilities of operators engaged in remote control tasks?
18. Discuss three factors that are of importance in crew station design.

EXERCISES

1. Select a workstation in your environment. Evaluate it with respect to general considerations in workstation design.
2. Select a sitting workplace in your environment. Evaluate it with respect to task requirements and specific seated work area design considerations. Also evaluate the work surface height.
3. Review your secretary's chair with respect to ergonomic requirements. What changes would you recommend?
4. Select a work area with at least eight *components*. Develop four arrangements of these components using link analysis and the *guiding principles of arrangement*. Which one would you choose? Why?
5. Select a multioperator work area in your environment. Redesign the workspace with ergonomics considerations. Discuss the trade-offs you made.

REFERENCES

1. Design of Individual Workplaces. 1972. In *Human Engineering Guide to Equipment Design*. Van Cott, H. P., and Kinkade, R. G. (Eds.). U.S. Government Printing Office, Washington, DC.
2. Thomson, R. M. 1972. Design of Multi-Man–Machine Work Areas. In *Human Engineering*

Guide to Equipment Design. Van Cott, H. P., and Kinkade, R. G. (Eds.). U.S. Government Printing Office, Washington, DC.

3. Bendix, T., and Hagberg, M. 1984. Trunk Posture and Load on the Trapezius Muscle whilst Sitting at Sloping Desks. *Ergonomics,* 27, pp. 873–882.

4. Brand, J. L., and Judd, K. W. 1993. Angle of Hard Copy and Text-Editing Performance. *Human Factors*, 35(1), pp. 57–69.

5. Niebel, B. 1987. *Motion and Time Study* Richard D. Irwin, Homewood, IL.

6. Farley, R. R. 1955. Some Principles of Methods and Motion Study As Used in Development Work. *General Motors Engineering Journal,* 2(6), pp. 20–25.

7. Squires, P. C. 1956. The Shape of the Normal Working Area. *Report 275. U.S.* Navy Department, Bureau of Medicine and Surgery, Medical Research Laboratories, New London, CT.

8. Konz, S., and Goel, S. 1969. The Shape of the Normal Work Area in the Horizontal Plane. *AIIE Transactions,* 1, March, pp. 70–74.

9. Konz, S. 1985. *Work Design.* Grid Publishing, Columbus, OH.

10. Elias, R., Cail, F., Tisserand, M., and Christman, H. 1982. Investigations in Operators Working with CRT Display Terminals: Relationships between Task Content and Psychophysiological Alterations. In *Ergonomic Aspects of Video Display Terminals.* Grandjean, E., and Vigliani, E. (Eds.). Taylor & Francis, London.

11. Crawford, B. M., and Altman, J. W. 1972. Designing for Maintainability. In *Human Engineering Guide to Equipment Design.* Van Cott, H. P., and Kinkade, R. G. (Eds.). U.S. Government Printing Office, Washington, DC.

12. Anderson, N. S., and Lee, R. 1963. Application of Human Factors in Maintenance of Electronic Systems. *4th National Symposium on Human Factors in Electronics: Men, Machines and Systems.* IEEE, Washington, DC.

13. Eastman Kodak Company. 1983. *Ergonomic Design for People at Work*, Vol. 1. Lifetime Learning Publications, Belmont CA.

14. Kvalseth, T. O. 1985. Work Station Design. In *Industrial Ergonomics: A Practitioner's Guide.* Alexander, D. C., and Pulat, B. M. (Eds.). Industrial Engineering and Management Press, Atlanta, GA.

15. Grandjean, E. 1980. *Fitting the Task to the Man.* Taylor & Francis, London.

16. Hunting, W., Grandjean, E., and Maedka, K. 1980. Constrained Postures in Accounting Machine Operators. *Applied Ergonomics,* 11, pp. 145–149.

17. Lehman, G., and Stier, P. 1961. Mensch und Gerat. *Handbuch der gesamten Arbeitsmedizin,* 1, pp. 718–788.

18. Carter, J. B., and Banister, E. W. 1994. Musculoskeletal Problems in VDT Work: A Review. *Ergonomics*, 37(10), pp. 1623–1648.

19. Redfern, M. S. and Chaffin, D. B. 1995. Influence of Flooring on Standing Fatigue. *Human Factors,* 37(3), pp. 570–581.

20. Green, A. B., and Pulat, B. M. 1984. Microcomputer Software for Workstation Design and Estimation of Energy Cost of Work. *Proceedings of the Fall IIE Conference,* Atlanta, GA, October 1984, pp. 342–349.

21. Ayoub, M. M. 1973. Work Place Design and Posture. *Human Factors,* 15(3), pp. 265–268.

22. Cox, C. F. 1984. An Investigation of the Dynamic Anthropometry of the Seated Workplace. M.Sc. Thesis. University College, London. In *Bodyspace: Anthropometry, Ergonomics, and Design.* Pheasant, S. (Ed.). 1986. Taylor & Francis, London.

23. Nachemson, A. 1974. Lumbar Intradiscal Pressure: Results from In Vitro and In Vivo Experiments with Some Clinical Implications. *7th Wissenschaftliche Konferenz der Gesellschaft Deutscher Naturforscher und Aerzte.* Springer-Verlag, Berlin.

24. Nachemson, A., and Elfstrom, G. 1970. Intravital Dynamic Pressure Measurements in Lumbar Disks. *Scandinavian Journal of Rehabilitation Medicine,* Suppl. I.

25. Andersson, G. B. J., Ortengren, R., Nachemson, A., and Elfstrom, G. 1974. Lumbar Disk Pressure and Myoelectric Back Muscle Activity during Sitting. I. Studies on an Experimental Chair. *Scandinavian Journal of Rehabilitation Medicine,* 3, pp. 104–114.

26. Kramer, J. 1973. *Biomechanische VerÄnderungen im lumbalen Bewegungssegment.* Hippocrates, Stuttgart.

27. Helander, M. G., Zhang, L., and Michel, D. 1995. Ergonomics of Ergonomic Chairs: A Study of Adjustability Features. *Ergonomics,* 38(10), pp. 2007–2029.

28. Swensen, E. E., Purswell, J. L., Schlegel, R. E., and Stonevich, R. L. 1992. Coefficient of Friction and Subjective Assessment of Slippery Work Surfaces. *Human Factors,* 34(1), pp. 67–77.

29. Sanders, M. S., and McCormick, E. J. 1982. *Human Factors in Engineering and Design.* McGraw-Hill, New York.

30. McCormick, E. J. 1976. *Human Factors in Engineering and Design.* McGraw-Hill, New York.

31. Fowler, R. L., Williams, W. E., Fowler, M. G., and Young, D. D. 1968. An Investigation of the Relationship between Operator Performance and Operator Panel Layout for Continuous Tasks. *Technical Report* 68–170. U.S. Air Force AMRL (AD-692-126).

32. Dupuis, H., Preuschen, R., and Schulte, B. 1955. *Zweckmassige gestaltung des Schlepperfuhrerstandes.* Max Planck Institutes fur Arbeitphysiologie, Dortmund, Germany.

33. Chaffin, D. B. 1973. Localized Muscle Fatigue: Definition and Measurement. *Journal of Occupational Medicine,* 15(4), pp. 346–354.

34. Human Factors Engineering. 1980. *AFSC Design Handbook, DH1-3,* 3rd ed., rev. 1. Air Force Systems Command. Andrews Air Force Base, D. C.

35. Bonney, M. C., and Williams, R. W. 1973. CAPABLE: A Computer Program to Layout Controls and Panels. *Ergonomics,* 20(3), pp. 297–316.

36. Rabideau, G. E, and Luk, R. H. 1975. A Monte-Carlo Algorithm for Workplace Optimization and Layout Planning. *Proceedings of the Human Factors Society 19th Annual Meeting,* Dallas, TX, pp. 187–192.

37. Pulat, B. M. 1980. A Computer Aided Panel Design and Evaluation System. Ph.D dissertation. North Carolina State University, Raleigh, NC.

38. Pulat, B. M., and Ayoub, M. A. 1984. A Computer Aided Display-Control Selection Procedure for Process Control Jobs?UNISER. *IIE Transactions,* 16(4), pp. 371–378.

39. Pulat, B. M., and Ayoub, M. A. 1985. A Computer Aided Panel Layout Procedure for Process Control Jobs?LAYGEN. *IIE Transactions,* 17(1), pp. 84–93.

40. Pulat, B. M. 1984. A Computer Aided Multi-Man–Machine Work Area Design and Evaluation System?MAWADES. *Technical Report.* North Carolina A&T State University, Greensboro, NC, September.

41. Pulat, B. M., and Pulat, P. S. 1985. A Computer Aided Workstation Assessor for Crew Operations?WOSTAS. *International Journal of Man-Machine Studies,* 22, pp. 103–126.

42. Pulat, B. M., and Grice, A. 1986. Computer Aided Techniques for Crew Station Design?WORG and WOLAG. *International Journal of Man-Machine Studies,* 23, pp. 443–457.

43. Bozer, Y. A., and McGinnis, L. E 1984. Kitting: A Generic Descriptive Model. *Technical Report MHRC-TR-84-04.* Material Handling Research Center, Georgia Institute of Technology, Atlanta, GA.

CHAPTER 8

Job Design

Proper job design assures job acceptance and satisfaction. These are attributes that everyone values and wishes to have in a job. This chapter dwells on the purpose and the mechanics of job design. Furthermore, various support processes are highlighted.

8.1 INTRODUCTION

Every organization has a structure. People in an organization interact with each other based on the formal and/or sometimes the informal structure. Their operational interrelationships are also defined by this system.

An industrial organization is no different. In such a system there are multiple subsystems, including manufacturing, research and development, sales, design, and customer support. Each subsystem, in turn, is composed of segments. For example, in manufacturing, one may have material planning, production planning, receiving, buffering, material preparation, assembly, testing, and shipping. Each division must be able to perform its functions effectively and on time. Furthermore, for total system effectiveness, each segment must function with the others in a synchronized manner.

Just as there are segments in a subsystem, there are functions of a segment that define how the segment performs. For example, in the test area of a factory, one may observe functions such as material receiving, test program installation, test-set preparation, testing, results management, and test-set maintenance. Segment effectiveness is largely dependent on how efficiently each function of the segment is carried out. Finally, functions break into jobs to be performed by the human operators, machines, and/or a combination of the two.

In addition to the effectiveness in design and synchronization of jobs, functions, segments, and the subsystems of an organization, the element that contributes most to the effectiveness of humans in performing the jobs is the *organizational atmosphere*. This includes variables such as motivation, morale, opportunity for growth, and rewards for achievement. From a worker's perspective, it is not hard to understand that those jobs that are motivating and offer opportunity for growth and good rewards for achievement will be more readily accepted than others.

8.2 DEFINITIONS AND QUALITY OF WORKING LIFE

A *job* is composed of a specific set of tasks that are performed by a person on an ongoing basis [1]. A *job description* details the tasks in a job. It also describes skills necessary for successful performance on the job. When an organizational structure is developed, jobs are defined along with it. It is only when people are assigned to these jobs that the organization starts functioning.

Each person working in an organization has his or her own goals, unique commitments, and other aspirations. These do not necessarily coincide with the institution's objectives. A major challenge to an organization is to offer these opportunities to its employees while meeting product or service goals. Several are:

1. Recognition for performance
2. Opportunity for further education
3. Some decision making affecting the job being performed
4. Planned career path
5. Social support at the job

Table 8.1 provides a checklist of attributes that offer a *good-quality work life*. Viewed from the workers' angle, the organization's effectiveness depends primarily on the quality of work life offered to the employees. On the other hand, the *production* view considers the worker as an element in the production system whose operational utilization must be maximized. In those instances where there is excess work force in one area of the production system, force adjustment can be achieved by moving some workers to other areas or through layoffs. Naturally, this philosophy does not lead to a work climate that is satisfying to the worker. The challenge is to find a balance between the two. Making jobs satisfying while maintaining a cost-competitive status should be the principal objective of any organization.

TABLE 8.1 QUALITY OF WORK LIFE
CHECKLIST

Job content
    Perceived value to the organization
    Feedback on results
    Stress
    Task variety
Social relations
    Interaction with others within the organization
    Interaction with others outside the organization
    Recognition of achievement
Education
    In-plant courses
    Workshops
    Conferences
    Trade shows
Environment
    Physical
        Lighting
        Noise
        Ambient temperature
    Social
Compensation
    Pay
    Benefits
Career Path
    Within the division
    Outside the division

8.3 THE JOB DESIGN PROCESS

After reviewing the purpose of job design and providing several definitions, we introduce the mechanics of job design at a macro level. Concepts presented in this section are supported by the micro-level details given in Section 8.4. The job design process may take in several levels of detail. According to one philosophy, the *laissez-faire* approach, jobs emerge based on needs without much planning. Certainly, before the Adam Smith–Charles Babbage–Frederick Taylor era (1750–1910) many jobs emerged this way [2–4]. Smith, Babbage, and Taylor wrote that managers must design each job so that when a worker reads the instruction card, he or she will be able to carry out the job with no difficulty. This approach led to *specialization,* where a worker performed a task repeatedly. Many assembly line jobs are still based on this philosophy. In the 1940s and 1960s, remedies were sought to problems caused by extreme specialization. Job enlargement [5] and job enrichment [6] concepts were developed on these grounds. *Job enlargement* proposes that several specialized tasks be grouped into a job. *Job enrichment* assumes vertical integration of jobs as opposed to job enlargement's horizontal integration. Job enrichment proposes to cover other aspects of job definition, such as planning and control, and hence provide more complete jobs.

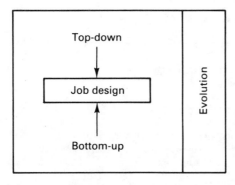

Figure 8.1   Job design process.

It is evident from the discussion above that a complete job design process must involve both a *top-down* and a *bottom-up* approach (Figure 8.1). The details of the top-down approach were discussed in Chapter 6, starting with the definition of mission (s) and operational function(s) and continuing with subfunction and task analysis. In the bottom-up approach, task groupings are prepared (with job enlargement and expansion concepts), their interactions with other jobs determined, and reward structures finalized.

It is also important to realize that the process of job design is an iterative one. Jobs evolve, changing their nature, content, and other aspects with time. Thus the designer must review job contents, structures, and interrelationships frequently to make necessary adjustments. The ergonomist's major responsibility in job design is to assure that with each definition/redefinition of jobs, human needs and organizational needs are blended in fair proportions. Furthermore, when specific work methods are designed, they should not demand work elements and motion sequences beyond human capability (see the next section).

8.4 WORK METHODS DESIGN

Up to this point, we dealt with the macro aspects of job design. At a micro level, the ergonomist is much concerned with the fit between the work methods and the worker capabilities. Several suggestions are given below to aid in this process [7–10]:

1. Work methods that require highly repetitive hand/arm movements must be evaluated for CTD (cumulative trauma disorder) risk. Use of the joints at extreme mobility ranges and force application in those postures warrant methods changes (see Chapter 3).

2. Hand-operated tools and other devices must be used with the hand in neutral position (wrist straight). These tools must be operable by either hand. They should require the functions of appropriate muscle groups, avoiding single-finger repetitive action.

3. Minimize static loading on the musculoskeletal system, including the tendons, muscles, ligaments, intervertebral disks, and bones. Static loading on such structures causes strain that may be reversed if continuous exposure has not taken place. However, the ensuing fatigue will degrade performance.

4. Do not design the work area and/or work methods to require arm use in an out-stretched or raised posture.

5. The work area or work methods should not require the head and the trunk to be inclined excessively either backward or forward.

6. Twisting arm motions should be performed with the arms bent from the elbow.

7. The preferred hand should be used during motions since it is faster.

8. Keep the arm movements within the maximum, and preferably the normal work area.

9. Avoid having both feet move simultaneously unless seated.

10. Arms should move along continuous and curved patterns.

11. Both arms should move simultaneously toward and away from the body. They should not be idle except for rest periods.

12. Avoid having simultaneous foot and arm motions when the task requires considerable attention.

13. Arm motions are more speedy and accurate than foot motions.

14. If forces are to be applied, start from a posture in which the muscle is fully extended.

15. For seated posture:
  · The hand is more powerful when in pronation (inward turn) than when in supination (outward turn).
  · Pull-down forces exceed push-up forces for the hand.
  · Push power is maximal when the hand is 50 cm (20 in.) in front of the body.
  · Pull power for the hand is greatest at 70 cm (28 in.) in front of the body.
  · The greatest push power with the legs is obtained when the knee angles range between 140 and 160 degrees.
  · Maximum bending strength is obtained at an elbow angle between 80 and 120 degrees.

16. For standing posture:
  · Pulling and pushing power are greater in the sagittal plane than in the transverse plane.
  · In general, pushing power is greater than pulling power.

17. Avoid a bent or any other unnatural posture.

18. Most frequent arm movements should be carried out with the elbows bent and near the body.

19. If support is provided to the elbows and the arms, the work surface may be elevated if the need arises.

20. Grasping force can be increased four times by changing from a fingertip grasp to a whole-hand grasp.

21. Skilled operations should not call for much force to be exerted.

22. A free rhythm is much preferred to an imposed rhythm.

23. Horizontal movements are easier to control than vertical ones.

24. Accuracy and speed of forearm movement is best within an arc of 45 to 50 degrees to each side.

25. Work requiring use of the eyes should be carried out within the field of normal vision.

26. Distribute work to various body elements in accordance with inherent capabilities.

27. Momentum should be utilized to aid motions. It should be minimized if muscular effort is required to overcome it.

28. The greater force can be applied by the middle finger and the thumb as compared with the other fingers.

29. Finger motions are the fastest. Trunk motions are the slowest.

Observing the points above during work methods design minimizes stress on the body and maximizes the effectiveness of human motion.

8.5   JOB ANALYSIS METHODS

Several job analysis and stress evaluation methodologies have been developed in the past, especially during the last decade. The purpose of these methods is to determine stress points in a job, if any, and to eliminate them or develop remedies. We review a few of these in this section [11].

Between 1950 and 1975, several physical stress checklists were developed [12,13]. These are important in identifying jobs that could be physically harmful to the musculo-skeletal system. Such surveys collect data on task activity, frequency of occurrence, and weight handled during job. They are particularly appropriate for jobs that are not repetitive, for which traditional work measurement methods are not appropriate. Koyl and Masters-Hanson's [13] GULHEMP method is a good example. This method rates job applicants and jobs on a seven-category scale: (1) general physique, (2) upper extremities, (3) lower extremities, (4) hearing, (5) eyesight, (6) mentality, and (7) personality type. The term "GULHEMP" comes from the combination of the first letters of the scale categories. Each category is also described by a finite number of levels, which vary from 1 (highest demand or full capacity) to 7 (lowest demand or low capacity). In each category the job being evaluated is assigned a representative level that best describes its demands. Similarly, an applicant's capacity is evaluated in each category. An applicant is deemed acceptable for hire if he or she is assigned levels equal to or better than job demands.

Hanman's [12] method is similar to the GULHEMP method in concept. However, it uses a different set of factors for evaluation. The three primary attributes of job demand dealt with in this method are (1) weight, (2) distance, and (3) time. In addition, environmental factors are considered (Figure 8.2). Both the GULHEMP and Hanman methods have been demonstrated to work in many industries. However, the lack of objectivity in determining job demands and applicant capacity leads to the most serious criticism of these methods.

NIOSH has developed the *physical stress job analysis sheet* to record task characteristic data for lifting tasks (Figure 8.3). As discussed in Chapter 4, using these data, AL and MPL values are calculated and matched with job requirements to evaluate the task.

A static strength prediction model developed by Chaffin et al. [14] compares the load moments produced at various body joints during manual exertion tasks with static strength

Form 1
Physical Demands Analysis Work Sheet

Job Title:

Job Location:

Physical Factors

#	Physical Factor		#	Factor	
1	1- 5		42	Far — Snellen	
2	6- 10		43	Near — Jaeger	Seeing
3	11- 25	Lifting (Pounds) — Includes pushing and	44	Color	
4	26- 50	pulling effort while	45	Depth	
5	51-100	stationary	46	Hearing	
6	100+		47	Talking	
7	1- 5		48	Other:	
8	6- 10	Carrying (Pounds) —	49	Other:	
9	11- 25	Includes pushing and			
10	26- 50	pulling effort while walking		**Environmental Factors**	
11	51-100		50	Inside	
12	100+		51	Fair Weather	Outside
13	R	Fingering	52	Wet Weather	
14	L		53	Hot °F	
15	R	Handling	54	Cold °F	
16	L		55	Sudden Temperature Changes	
17	R	Below Shoulders	56	Humid	
18	L		57	Dry	
19	R	Above Shoulders / Reaching	58	Moving Objects	
20	L		59	Hazardous Machinery	
21	R	Throwing	60	Sharp Tools or Materials	
22	L		61	Cluttered Floors	
23	Sitting		62	Slippery Floors	
24	Total Time on Feet		63	High Places	
25	Standing		64	Electrical Hazards	
26	Walking		65	Exposure to Burns	
27	Running		66	Explosives	
28	Jumping		67	Radiant Energy: (Kind)	
29	Legs Only	Climbing	68	Poor Lighting	
30	Legs and Arms		69	Poor Ventilation	
31	R	While Sitting	70	Toxic Conditions: (Kind)	
32	L		71	Wet Quarters	
33	R	While Standing / Treading	72	Close Quarters	
34	L		73	Vibration	
35	Stooping		74	Noise	
36	Crouching		75	Working With Others	
37	Kneeling		76	Working Around Others	
38	Crawling		77	Working Alone	
39	Reclining		78	Shifts	
40	Twisting		79	Other:	
41	Waiting Time		80	Other:	

Job Analyst's Name	Verified With: Foreman's Name	Date

Figure 8.2 Physical demands analysis worksheet. (From Ref. [12].)

moments obtained from tests of over 3000 U.S. workers. It predicts the population percentage capable of performing the exertion. A 75% or better reading is desired for each evaluation point.

Several job postural evaluation methods have also been developed. Corlett et al. [15] developed a procedure that requires the job analyst to make random observations on a task

Physical Stress Job Analysis Sheet

Department_____ Date _____

Job title_____ Analyst's Name _____

Task Description	Object Weight Ave | Max pounds		Hand Location				Task Freq.	AL	MPL	Remarks
			Origin		Destination					
	Ave	Max	H_{in}	V_{in}	H_{in}	V_{in}				
Load Stock	44	44	21	15	21	63	0			

Figure 8.3 NIOSH physical stress job analysis form.

during the day and record angular configurations of various body segments with the aid of a body diagram. Observing the cluster of marks, an engineer passes judgment as to the amount of postural stress imposed by the job. A drawback of this procedure is its lack of consideration of the load handled.

Another postural observation system was developed in Finland by Karhu et al. [16]. The system is called OWAS (Ovako Working Posture Analysis System). It is a method for the identification and evaluation of unsuitable working postures. OWAS is composed of two parts. The first part is an observational technique for work posture evaluation. The second part is composed of a set of criteria for workplace and work methods redesign. These criteria are based on safety and health factors, with primary emphasis placed on the discomfort caused by unacceptable working postures. A similar Swedish system is called ARBAN [17]. It is an ergonomic work analysis method that focuses on body postures and loads. It allows for:

1. Recording work methods on a videotape
2. Posture and load coding on closely spaced "frozen" situations
3. Computer evaluation of input

A body ergonomic stress measure is then calculated based on stress due to specific acts and loads handled. The peaks of resulting ergonomic stress/time curves represent heavy-load situations.

An indirect measure of job stress is a discomfort measure assessed by workers themselves. Several methods and scales have been developed for this purpose [18,19,20].A combination method is presented by Cameron [21] which makes use of a body map (front and back, left side and right side) with 22 sections, including shoulder, lower arm, wrist, hand ("pinkie" side), hand (thumb side), knee, foot, eyes, upper back, and buttocks. On this work-related-body-part discomfort (WRBPD) scale, for each body section, the worker indicates the severity of body discomfort: from "no discomfort" (—) to "intolerable" (5); frequency of discomfort: from "never" (—) to "always" (4); and duration of discomfort: from "not at all" (—) to "a very long time" (4).

Armstrong et al. [22] proposed a film analysis technique for upper extremity posture during work and subsequent evaluation for work methods improvement. Such an analysis may lead to method changes to reduce upper extremity and shoulder ailments due to undesirable benchwork. Keyserling et al. [23] developed a two-page checklist to determine the presence of ergonomic risk factors associated with the development of upper-extremity cumulative trauma disorders. A similar technique has been developed by Nordin et al. [24] for trunk flexion analysis and evaluation.

Due primarily to their short history, the biomechanical job analysis techniques discussed above have not been well validated in industry; hence they must be used with caution. However, they have opened an era for scientific evaluation of work methods that would otherwise have been unavailable to an engineer.

8.6 AUTOMATION

An important characteristic of modern business processes is continuing automation. In this section we review automation and pinpoint the ergonomist's role in it.

The term *automation* implies a degree of mechanization. A fully automated system requires no human input or effort. A system may display sections that are fully automated. Several other sections may be semiautomated. Today, very few industries are completely automated. Complete automation is feasible for continuous processing operations such as gasoline, oil, and detergent production. Partial automation is used in mass production of consumer goods, such as automobiles, cookies, cereals, light bulbs, electronic assemblies, and cigarettes.

Several factors encourage process or plant automation:

1. Increasing cost of labor
2. Difficult or hazardous operations for human beings
3. Lack of labor force
4. Desire for market penetration through reduced cost and price reduction

Several other factors discourage automation:

1. Government incentives for labor employment
2. Unavailable automation technology
3. Costs of automation
4. Lack of market for increased production
5. Social and internal forces (unions, and so on.)
6. Already low labor cost percentage in the total product cost

The best strategy is *sensible automation*. People and machines must complement each other. As discussed in Chapters 4 and 6, there are some aspects of system performance that can best be aided by human input; others must be performed by machines [25,26]. Whichever combination of person and equipment/machine makes best sense and provides the highest cost-effectiveness must be integrated into the overall process. The ergonomist's role here is to make sure that the resulting design does not tax the person extensively and that the requirements are within capability ranges. Naturally, the ergonomist starts with those tasks that are monotonous, repetitive, and outright dangerous. Manual material-handling tasks that are stressful may be the first set of tasks to be automated. These are win–win applications where the human being is already being taxed, with notable injury potential. The workers and the employers will both be happy to see these tasks being automated. It should be remembered that automation brings more responsibility to the maintenance activity. Hence maintainability of design must be an integral part of the automation effort.

8.7 MACHINE PACING

With automation, many jobs are becoming machine paced. Various estimates average around 50 million people or more working in machine-paced jobs worldwide [27]. Estimating stress on such jobs and improving working conditions, as well as identifying population characteristics that are most suitable for such jobs, are what ergonomists are interested in with respect to machine pacing. There are economic advantages to machine-paced work over self-pacing. These are automation induced and center around cost minimization and economic use of high technology. The disadvantages are related to the negative aspects of automation, such as not making best use of human capacity and economical feasibility only for high-volume production.

The challenge to the ergonomist and the engineer is to maintain the benefits of machine pacing while alleviating human disadvantages. For this purpose, researchers classified such work by the demand placed on human performance as well as by research methodology [27]. Others have identified the impact on human and performance effects. In general, it seems that the higher the perceptual and mental load associated with task performance, the higher the psychophysiological stress. Furthermore, the higher the decision freedom associated with task performance, the lower the stress level. Stress level also decreases with feedback. The more precise the feedback is, the lower the stress [28,29].

Several industrial studies have concluded that there is no statistically significant difference in psychophysiological measures and the quality of work life between machine-paced and self-paced tasks [30,31]. On the other hand, several other studies concluded that there are differences. Broadbent and Gath [32] report higher anxiety on workers paced by an assembly line in car manufacturing than those who were not. Stammerjohn and Wilkes [33] report lower job strain levels in a group associated with worker-preference lines relative to machine pacing in an inspection task. In field studies, Haider et al. [34] found higher pulse rates and longer reaction times in secondary tasks for groups of workers with paced assembly line work than in control groups with self-paced work. In a sawmill products sorting task [35], Seppala and Nieminen observed increased grading errors and heart rate, decreased blinking rate, and narrowed inspection area of the board with increased pacing. Of five baud rates (300, 600, 1200, 2400, and 4800 bits/sec) tested, Raouf et al. [36] found optimum performance at 1200. At this rate, time taken in text editing and errors were minimal. The diversity observed in field research may be due to other factors confounding the picture, as well as original stress levels prior to the additional stress imposed by the task in question.

In his studies, Salvendy[30,31] observed that those who prefer to work in a machine-paced environment are "less intelligent," more humble, more practical, more forthright, and more group dependent. Those who preferred self-paced work were "more intelligent," more assertive, more imaginative, more shrewd, and more self-sufficient.

8.8 MOTIVATION

The effectiveness of a human being in any organization is partially a function of his or her motivation. Job design process cannot ignore those job elements that are motivating to the employee. In this section we provide a review of the subject.

Motivation is a state of readiness for better performance on the job. Motivational research is the study of forces that lead to motivation. Better performance can be obtained from people who are motivated to do better, assuming equal abilities, skills, and background. A job can be motivating if it has certain qualities. These will be discussed later. Although a job may motivate people, it is also true that some people are in general more motivated than others. Furthermore, external factors such as pay may motivate people.

Much research has been carried out to discover those factors that motivate people. Perhaps the earliest psychologist to generate extensive information about drives, motives, and motivators was Abraham Maslow [37]. Maslow suggested that people have a *hierarchy of needs*. Five major needs direct people's actions. Maslow stated that:

· Man is a wanting animal whose behavior is determined by unsatisfied needs. A satisfied need no longer drives behavior.
· Man's needs are arranged in a hierarchy of importance.
· Higher-order needs differ from lower-order needs in that they are never completely satisfied.

Figure 8.4 gives Maslow's hierarchy of needs. *Physical needs* consist of the primary needs such as food, water, and sex. When physical needs are not met, other, higher-level needs

Figure 8.4 Maslow's hierarchy of needs.

become less important. Once physical needs are met, the individual's safety or security needs are activated. *Safety needs* include protection from physical harm, ill health, and economic disaster. *Social needs* include the desire to feel a part of a group; and the need to receive love. Social needs are concerned with establishing one's position relative to others in the organization. Social needs become important to a person once the physical needs and safety needs are reasonably well assured. *Achievement needs* consist of the need for self-respect, a feeling of competency. These fall into two categories: the need to have a high opinion of oneself, and the need to have others think highly of one. *Self-actualization* needs include the need to achieve one's fullest potential in terms of self-development and creativity.

Maslow's needs hierarchy is a dynamic theory. At any given moment in time, more than one need may be operating. Designers may apply these concepts in a work situation by first concentrating on basic physiological and safety needs, such as proper work–rest schedules, protection from physical harm, adequate heating and ventillation, and sufficient food facilities. These lower-order needs must be satisfied before needs higher up the scale can be considered. The designer may create a work climate where the performers may develop their potential. A variety of challenging jobs, responsibility, and autonomy are several elements that are parts of such a climate.

Fear has historically been a major motivator at the workplace. Herzberg's KITA (kick in the ass) concept [38] relates to this idea. Throughout the past four or five decades, many scientists researched the determinants of motivation. It is not the purpose of this book to review these in detail. However, one more theory will be covered, Herzberg's *motivation-hygiene theory* [39].

During the late 1950s, Herzberg and co-workers interviewed many employees in different occupations and countries for work situations that lead to satisfaction and dissatisfaction. According to the motivation-hygiene theory, the factors that produce job satisfaction (motivation) are different from factors that produce dissatisfaction. Although semantically satisfaction and dissatisfaction imply opposite extremes (a dissatisfier must also be a nonsatisfier), the opposite of job satisfaction is not dissatisfaction. Rather, it is no job satisfaction. By the same token, the opposite of job dissatisfaction is not job satisfaction. It is no job dissatisfaction.

Table 8.2 lists Herzberg's factors that lead to job satisfaction (motivation) and job dissatisfaction (hygiene). The motivator factors are related to the job itself and include such elements as achievement, recognition, responsibility, and the like. The hygiene factors are not related to the job. Rather, they are extrinsic to the job and include working conditions, salary, supervision and the like. The interesting thing about dissatisfiers, according to Herzberg, is that by removing or neutralizing things that dissatisfy people, one does not create a satisfier but eliminates a dissatisfier. In other words, the nonexistence of hygiene factors does not motivate people, but the existence of hygiene factors demotivates people.

Herzberg's theory also has design implications. Job designers may develop jobs that are *worth* doing. Satisfying work is work that is worthwhile to the performer. A worker who is recognized for his or her performance will be motivated. A garment inspector who is allowed to leave his or her mark on a finished product feels recognized.

A person who is motivated by the job and the results achieved will perform better. As better results are achieved, the person gets more motivated, especially if the task is rewarding. Hence a loop is established. If the job designer can keep the motivating aspects

TABLE 8.2 HERZBERG'S MOTIVATORS AND
HYGIENE FACTORS

Motivators	Hygiene factors
Achievement	Company policy and administration
Recognition	Supervision—technical
Work itself	Work conditions
Responsibility	Salary
Advancement	Facilities
Growth	Interpersonal relations—supervision

of the job intact and avoid dissatisfiers, the person performing the job may get great satisfaction from the job for a long time.

A closely related concept is *job enrichment*. Job enrichment enables a worker's psychological growth. All job enrichment techniques focus on the development of the worker through more responsibility and challenge. Skill utilization and development are integral parts of job enrichment. The key to job enrichment seems to be the working atmosphere that nurtures a server–customer relationship [38]. Functional relationships frequently break the desired informal links between groups. How many times have we heard the statement "That is not my job" or "My job does not cover that; you better find somebody else to help you." Hierarchical relationships promote the KITA. Usually, when a boss wants something to be done, it gets done before everything else even though it may not be important.

A supplier–customer relationship motivates the supplier to satisfy the customer to the highest degree possible. To do this, the supplier may have to learn more about his or her job, and develop better communication skills and other skills. All of these help enrich the supplier's job. Naturally, the customer is also the supplier to another customer. Hence the process grows and extends.

8.9   THE AMERICANS WITH DISABILITIES ACT OF 1990 (ADA)

President George Bush signed into law "The Americans with Disabilities Act of 1990" on July 26, 1990. Similar to the civil rights protection provided to individuals on the basis of race, sex, national origin, and religion, this law provides protection to individuals with disabilities. Many believe that ADA is the most sweeping civil rights legislation since the Civil Rights Act of 1964.

For many years, people with disabilities could not fully participate in American society. The desire for full participation led to the passage of the Rehabilitation Act of 1973 and the Education For All Handicapped Children Act of 1974. The latter was renamed The Individuals With Disabilities Education Act. The former law prohibited discrimination on the basis of disability in local programs and activities benefiting from federal financial assistance. Enforcement of this law improved program accessibility to health care, social services, transportation, housing, recreation, etc. It also began to open educational opportunities to disabled persons.

The Individuals with Disabilities Education Act required access of disabled students to regular classrooms, if appropriate, and the establishment of individualized educational programs for students with disabilities. In spite of these initiatives in the 1970s, a Louis Harris poll in 1985 indicated that the most important difficulty of people with disabilities is with employment. The Harris Poll showed that 67% of all Americans with disabilities between the ages of 16 and 24 are unemployed. Hence, although in the 1970s and early 1980s educational and vocational training opportunities for individuals with disabilities greatly improved, access to places of public accommodation (including transportation services and employment opportunities in the private sector) did not improve. This is the matter that ADA of 1990 was created to address.

There are two parts to ADA. One part deals with public accommodation, and the other part deals with employment provisions. "Public-accommodation" businesses include almost every service, retail, transportation, rental, or entertainment business. Regardless of size, public-accommodation businesses must eliminate obstacles that restrict availability or accessibility, if "it is not too expensive, disruptive or difficult, and *if* reasonable efforts have been made to comply." Reasonable changes include wider path of travel for disabled persons, grab bars in bathrooms, entrance ramps, support rails, 36-inch doorways, removal of turnstiles, etc. All public-accommodation businesses, regardless of size, were mandated to comply with the law effective January 26, 1992. If barrier removal is not readily achievable, alternative methods of providing the services must be offered, such as curbside service, home delivery, etc.

As far as employment provisions are concerned, beginning on July 26, 1992, employers with 25 or more employees were prohibited from discriminating against qualified individuals with a disability in job training, job application procedures, compensation, hiring, advancement and discharge, other terms, privileges, and conditions of employment. Beginning on July 26, 1994, ADA applied to employers of 15 or more employees. The law also applies to employment agencies, labor organizations and joint labor-management groups. Specific types of discrimination against persons with a disability include denying employment solely on the basis of the need to make "reasonable accommodation" to the specific disability observed, and not making the "reasonable accommodation" to the disability of the qualified person unless such accommodation would impose "undue hardship" on the employer. Under the law, an action imposes undue hardship if it requires significant difficulty or expense. Nature and cost of the accommodation, financial resources of the employer, and the impact on such resources determine qualification for undue hardship. "Reasonable accommodation" examples include job restructuring, modified or part-time work hours, acquisition or modification of equipment or work devices, adjustment of policies, tests, and training materials. A study showed that less than 25% of employees with disabilities need "accommodations," and almost 70% of "accommodations" cost $500 or less per disabled person.

ADA provisions apply to public services including public transportation (state and local governments, Amtrak, etc.); public accommodations operated by private entities (including transportation provided by places of public accommodation when people transportation is not the major activity, such as senior citizen centers and hotels); telecommunications services for hearing-impaired and speech-impaired individuals; and miscellaneous provisions applicable to insurers, hospitals, medical service companies, and health maintenance organizations. The ADA does not apply to the employees of the United States government. People working for the federal government are covered by a different law.

In the 1990 Revenue Reconciliation Act, Congress provided for Section 44 of the Internal Revenue Service Code, which allows an eligible small business to elect a non-refundable tax credit for barrier removal in existing buildings to specifically comply with the ADA requirements. Current maximum for this credit is $5,000 for any taxable year. Other businesses, private entities and places of public accommodation, including transportation systems under provisions of the ADA, may claim an annual tax deduction of $15,000 to remove barriers to disabled individuals at existing places of business or trade. Such barriers may be around doors or doorways, parking lots, elevators, toilet rooms, public telephones, stairs, etc. In addition, communication barriers may exist with hearing and visually-impaired individuals. Removal of such barriers may take the form of providing means of making aurally and visually delivered materials to enhance the existing sensory mechanisms.

The Americans With Disabilities Act is enforced primarily by the United States Equal Employment Opportunity Commission. There are three ways for a person with a disability to file a complaint regarding job discrimination:

1. File a complaint with the local office of the Equal Employment Opportunity Commission (There is no charge for such services.).
2. File a private lawsuit through a lawyer.
3. Do both of the above.

A valid complaint may result in the person with a disability winning back pay, "reasonable accommodation," hiring, and certain types of damages [40,41].

8.10 SUMMARY

Job design includes both macro and micro elements. Macro elements include designing of the work environment, work–rest cycles and other elements that relate to allocation of functions to machines and human beings. Micro elements of job design focus on the determination and sequencing of human job elements, including their relationships with the workplace arrangement. The overall system should also look at the motivating elements for best human performance.

QUESTIONS

1. Discuss formal and informal structure in an organization.
2. What organizational element contributes most to people's effectiveness?
3. Define a job and a job description.
4. Why do organizational goals and individual goals not necessarily match?
5. List and discuss the two primary points of view for maximum system effectiveness.
6. What was the main reason for job specialization during the early twentieth century?
7. Discuss job enlargement and job enrichment.
8. Discuss the top-down and the bottom-up approaches to job design.

9. What is the ergonomist's major responsibility in job design?
10. List and discuss four micro-level job design considerations.
11. What is the major objective of physical stress evaluation methods?
12. What is the "physical stress job analysis sheet" used for?
13. What type of operations lend themselves to complete automation?

14. List and discuss the factors that encourage automation.
15. How could an ergonomist reduce stress on a machine-paced task?
16. What is motivation?
17. Discuss Maslow's hierarchy of needs.
18. What is the significance of KITA?
19. Discuss Herzberg's motivation-hygiene theory.
20. How could a designer make a job worth doing?

EXERCISES

1. Write the description of your secretary's or receptionist's or laboratory technician's job based on your observations. Compare the write-up with the department record. Discuss any differences.
2. Select an organization in your environment. Study it. List and discuss improvement ideas.
3. Go to a local industry and select two bench assembly jobs. Evaluate each with respect to the suggestions given in Section 8.4. Propose improvements.
4. Investigate the reasons for an automation project at a local industry. Do they seem sensible? Discuss. Also list the intangible benefits of the project.
5. Conduct an interview with several students to find out what motivates them. Compare your findings with the discussion in Section 8.8. List and discuss the main commonalities and differences.

REFERENCES

1. Davis, L. E., and Wacker, G. J. 1987. Job Design. In *Handbook of Human Factors*. Salvendy, G. (Ed.). Wiley-Interscience, New York.
2. Babbage C. 1965 (orig. published in 1835). *On the Economy of Machinery and Manufactures*, 4th ed. Augustus M. Kelley, New York.
3. Taylor, F. W. 1911. *The Principles of Scientific Management*. Harper & Row, New York.
4. Smith, A. 1970 (orig. published in 1776). *The Wealth of Nations*. Penguin, London.
5. Conant, E. H., and Kilbridge, M. D. 1965. An Interdisciplinary Analysis of Job Enlargement: Technology, Costs and Behavioral Implications. *Industrial and Labor Relations Review*, 18, pp. 377–390.
6. Herzberg, F. 1966. *Work and the Nature of Man*. World, Cleveland, OH.
7. Grandjean, E. 1980. *Fitting the Task to the Man*. Taylor & Francis, London.
8. Kvalseth, T. O. 1985. Work Station Design. In *Industrial Ergonomics: A Practitioner's Guide*. Alexander, D. C., and Pulat, B. M. (Eds.). Industrial Engineering and Management Press, Atlanta, GA.

9. Niebel, B. 1987. *Motion and Time Study*. Richard D. Irwin, Homewood, IL.

10. Salvendy, G. (Ed.). 1982. *Handbook of Industrial Engineering*. Wiley, New York.

11. Chaffin, D. B., and Andersson, D. B. 1984. *Occupational Biomechanics*. Wiley, New York.

12. Hanman, B. 1959. Clues in Evaluating Physical Ability. *Journal of Occupational Medicine*, 1, pp. 595–602.

13. Koyl, F. F., and Masters-Hanson, P. 1973. Physical Ability and Work Potential (unpublished contract report). Manpower Administration, U.S. Dept. of Labor, Washington, DC.

14. Chaffin, D. B., Herrin, G. D., Keyserling, W. M., and Garg, A. 1977. A Method for Evaluating the Biomechanical Stresses Resulting from Manual Materials Handling Tasks. *American Industrial Hygiene Association,* 38, pp. 662–675.

15. Corlett, E. N., Madeley, S. J., and Manenica, I. 1979. Postural Targeting: A Technique for Recording Working Postures. *Ergonomics,* 22(3), pp. 357–366.

16. Karhu, O., Harkonen, R., Sorvali, P., and Vepsalainen, P. 1981. Observing Working Postures in Industry. *Applied Ergonomics,* 12, pp. 13–17.

17. Holzman, P. 1982. ARBAN: A New Method for Analysis of Ergonomic Effort. *Applied Ergonomics,* 13, pp. 82–86.

18. Corlett, E. N. 1990. Static Muscle Loading and the Evaluation of Posture. In *Evaluation of Human Work: A Practical Ergonomics Methodology.* Corlett, E. N., and Wilson, J. R. (Eds.). Taylor & Francis, London, pp. 542–570.

19. Borg, G. A. V. 1982. Psychophysical Basis of Perceived Exertion. *Medicine and Science in Sports and Exercise*, Vol. 14, pp. 377–381.

20. Krawczyk, S., and Armstrong, J. S. 1993. Psychophysical Assessment of Simulated Assembly Line Work: Combinations of Transferring and Screw Driving Task. in *Proceedings of the Human Factors and Ergonomics Society 37th Annual Meeting*, Seattle, WA. Human Factors and Ergonomics Society, Santa Monica, CA, pp. 803–807.

21. Cameron, J. 1995. The Assessment of Work-Related-Body-Part Discomfort: A Review of Recent Literature and a Proposed Tool for Use in Assessing Work-Related-Body-Part Discomfort in Applied Environments. In *Advances in Industrial Ergonomics and Safety VII.* Bittner, A. C., and Champney, P. C. (Eds.). Taylor & Francis, London, pp. 173–180.

22. Armstrong, T. J., Foulke, J. I., Joseph, B. S., and Goldstein, S. A. 1982. Investigation of Cumulative Trauma Disorders in a Poultry Processing Plant. *Journal of American Industrial Hygiene Association,* 43(2), pp. 103–115.

23. Keyserling, W. M., Stetson, D. S., Silverstein, B. A., and Brouwer, M. L. 1993. A Checklist for Evaluating Ergonomic Risk Factors Associated with Upper Extremity Cumulative Trauma Disorders. *Ergonomics*, 36(7), pp. 807–831.

24. Nordin, M., Ortengren, M., and Andersson, G. B. J. 1983. Measurement of Trunk Movement during Work. *Spine,* 8, pp. 66–74.

25. Woodson, W. E., and Conover, D. W. 1966. *Human Engineering Guide for Equipment Designers,* 2nd ed. University of California Press, Berkeley, CA.

26. Alexander, D. C. 1986. *The Practice and Management of Industrial Ergonomics.* Prentice Hall, Englewood Cliffs, NJ.

27. Salvendy, G. 1981. Classification and Characteristics of Machine Paced Work. In *Machine Pacing and Occupational Stress.* Salvendy, G. and Smith, M. J. (Eds.). Taylor & Francis, London, pp. 5–12.

28. Karasek, R. A. 1979. Job Demands, Job Decision Latitude, and Mental Strain: Implications for Job Redesign. *Administrative Science Quarterly,* 24, pp. 285–308.

29. Knight, J. L., and Salvendy, G. 1981. Feedback Effects in Internally and Externally Paced Tasks. Cited in *Machine Pacing and Occupational Stress.* Taylor & Francis, London, pp. 5–12.

30. Knight, J. L., Geddes, L. A., and Salvendy, G. 1980. Continuous Unobtrusive Performance and Physiological Monitoring of Industrial Workers. *Ergonomics,* 23, pp. 500–506.
31. Knight, J. L., Salvendy, G., Geddes, L. A., Jans, K., and Smitt, E. 1980. Monitoring the Respiratory and Heart Rate of Assembly Line Workers. *Medical and Biological Engineering and Computing,* 18, pp. 797–798.
32. Broadbent, D. E., and Gath, D. 1981. Symptom Levels in Assembly Line Workers. In *Machine Pacing and Occupational Stress.* Salvendy, G. and Smith, M. J. (Eds.). Taylor & Francis, London, pp. 243–252.
33. Stammerjohn, L. W., Jr., and Wilkes, B. 1981. Stress/Strain and Linespeed in Paced Work. In *Machine Pacing and Occupational Stress.* Salvendy, G. and Smith, M. J. (Eds.). Taylor & Francis, London, pp. 287–293.
34. Haider, M., Koller, M., Groll-Knapp, E., Cervinka, R., and Kundi, M. 1981. Psychophysiological Studies on Stress and Machine Paced Work. In *Machine Pacing and Occupational Stress.* Salvendy, G. and Smith, M. J. (Eds.). Taylor & Francis, London, pp. 303–309.
35. Seppala, R. and Nieminen, K. 1981. Workers' Behavioural and Physiological Responses and the Degree of Machine Pacing in the Sorting of Sawmill Products. In *Machine Pacing and Occupational Stress.* Salvendy, G. and Smith, M. J. (Eds.). Taylor & Francis, London, pp. 319–327.
36. Raouf, A., Hatami, S., and Chaudhary, K. 1981. CRT Display Terminal Task Pacing and Its Effect on Performance. In *Machine Pacing and Occupational Stress.* Salvendy, G. and Smith, M. J. (Eds.). Taylor & Francis, London, pp. 347–354.
37. Maslow, A. H. 1954. *Motivation and Personality.* Harper, New York.
38. Herzberg, F. 1987. One More Time: How Do You Motivate Employees? *Harvard Business Review,* 65(5), September-October, pp. 109–120.
39. Herzberg, F., Mausner, B., and Snyderman, B. 1959. *The Motivation to Work,* 2nd ed. Wiley, New York.
40. Allen, J. G. 1993. *Complying with the ADA: A Small Business Guide to Hiring and Employing the Disabled.* John Wiley, New York.
41. Parry, J. (Ed.). 1992. *The Americans with Disabilities Act Manual: State and Local Public Accommodations.* American Bar Association, Washington, DC.

_____CHAPTER 9

Design of the Work Environment

The two preceding chapters focused on work area design and job design. In this chapter we complement the work systems design process by discussing the design of the physical work environment. Optimal structuring of the work environment helps to improve comfort and performance and reduce potential risks due to environmental factors.

9.1 INTRODUCTION

The human–machine system functions in an environment composed of other people, machines, equipment, and other energy sources. Furthermore, the person brings personal experiences and other background to the system, as do other system elements. All these establish a *context* for the system to function within. Hence the behavior of a human–machine system must not be evaluated against present time events only. The history of the system may significantly affect the way the system operates.

The two major components of the work environment that affect the behavior of a human–machine system are the physical environment and the social environment [1]. Elements of the social environment include isolation, task pressures, group dynamics, and the like. Although human performance has been studied and reported under a variety of environmental factors, the ones that most concern an industrial ergonomist are the physical factors that exist in industrial environments, such as illumination, noise, vibration, and ambient temperature (heat and cold). Hence it is not our purpose to discuss social environmental factors in this chapter.

Our discussion starts with the visual environment and the elements associated with it. Then the auditory environment is presented. Vibration and ambient temperature follow. In each, special emphasis is placed on how human performance is affected by the elements discussed, with the objective of verifying the limits within which people perform best. The designer may then take the necessary steps to control the physical environment for most effective functioning of the human–machine system.

9.2 THE VISUAL ENVIRONMENT

Adequate light must be present for the human visual system to function effectively. A person's ability to perform depends, to an extent, on the amount and the quality of light. Our perception of the environment may also be affected by lighting conditions. In general, an outdoor environment does not present much of a problem in this respect. However, the ergonomist must target well-designed lighting systems for indoors. Naturally, glare may be a problem outdoors. Indoors, there are additional parameters of a lighting system that one should consider beyond the number, type, and location of luminaries which affect human performance. However, before we discuss these, some background in photometry is necessary.

9.2.1 Photometry

Photometry deals with the measurement of light, usually with an electronic device. Three aspects of light are of most relevance to an ergonomist:

1. Illumination
2. Luminance
3. Reflectance

Illumination is defined as the amount of light falling on a surface. In international measurement units (the SI system), illumination is measured by *lux*. One *lux* is equal to 1 lm/m². The "lm" stands for luminous flux, or *lumens*, which is equal to 1/683 W (watts) for light with a wavelength of 555 nm. Illumination is measured by:

$$\text{illumination} = \cos\theta \frac{I}{d^2}$$

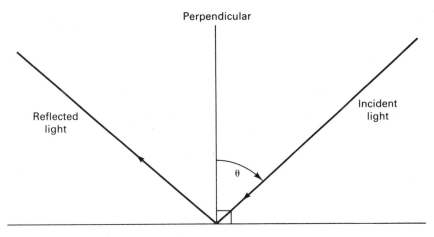

Figure 9.1 Reflectance relationships.

where,

θ is the angle between a line perpendicular to the surface on which illumination is being calculated and the direction of light (Figure 9.1)

I is the intensity of illuminating source

d is the distance between the illuminated surface and the source that illuminates it

When the light is impinging on the surface at a right angle, the formula reduces to

$$\text{illumination} = \frac{I}{d^2}$$

Luminance is the amount of light emitted in a given direction by a luminous source. It is measured in candelas per square meter (cd/m^2) in the SI system. For a perfectly diffuse reflecting system, in SI units, luminance is calculated as

$$\text{luminance} = \frac{\text{illuminance} \times \text{reflectance}}{\pi}$$

Reflectance is defined as the ratio of light reflected from a diffuse surface to the incident light. Konz [2] provides the reflectance properties of different materials given as percents (Table 9.1). Luminance is measured using a photometer. An illumination meter that is attached to a photometer may be used to measure the illumination on a surface.

9.2.2 Lighting System Design

Proper selection and placement of light sources are the two important elements of a good visual environment design process. The discussions in this section and in Section 9.2.3 provide readers with directions for the best decisions in this respect. Two major types of

TABLE 9.1 REFLECTANCE VALUES

Object	Reflectance (%)
Mirrored glass	80–90
White matte paint	75–90
Aluminum paint	60–70
Newsprint, concrete	55
Dull brass	35
Cast, galvanized iron	25
Black paint	3–5

Source: Adapted from Ref. [2].

lighting are general lighting and supplemental lighting. The purpose of *general lighting* is to provide general illumination in public spaces and in workspaces. *Supplemental lighting* is used to support general lighting at specific locations for enhanced visibility on tasks that demand it. General lighting can be broken down further into direct and indirect lighting. *Direct lighting* provides the most luminous flux to work areas; however, such luminaries may be sources of glare. *Indirect lighting* eliminates this problem; however, the amount of light that falls on an object is less. Hence for the same amount of lighting with a direct option, more powerful luminaries may have to be used. Lenses may solve this problem. In fact, in a study on parabolic direct lighting, the lensed indirect lighting system resulted in better productivity and preference by two-thirds of the workers [3]. The following should be considered in lighting systems design [4]:

1. Avoid locating direct light sources within the visual field of workers.
2. Avoid using glossy paint on machines or tables and in workplaces.
3. Align fluorescent tubes perpendicular to the line of sight.
4. Use diffuse light to provide the best working atmosphere.
5. Use more lamps, each of lower power, rather than using a few high-powered luminaries.
6. Avoid locating luminaries within 30 degrees of the NLS (normal line of sight).
7. Avoid flickering light sources.

Special-purpose lighting is also available for tasks such as inspection. Stroboscopic lighting, spectrum-balanced lights, polarized light, transillumination, black light, cross-polarization, and other techniques are used to enhance surface projections and indentations, thickness changes in material, surface scratches, opacity changes, color changes, and the like [5]. The purpose of each special lighting technique is to improve the detectability of certain visual targets (e.g., defects).

9.2.3 Types of Luminaries

There are two types of artificial light sources. The *incandescent* lamp produces light by electrical heating of a filament or by combustion of gases within a medium. The *gas dis-*

charge lamp produces light by passing an electric current through a gas. The most popular is the incandescent lamp. Its low cost, availability in numerous sizes, configurations, and bases, good color rendering qualities, and use convenience make it the most widely sought light source. However, since light is produced by heating a filament, most of the energy generated is in the infrared (nonvisible) region of the spectrum. Hence incandescent lamps are the least efficient sources of light. In actuality, lighting engineers like to use the term *efficacy* rather than *efficiency*. "Efficacy" refers to the amount of lumens per watt.

Related to incandescent lamps are tungsten-halogen or "quartz" lamps. These lamps operate at higher temperatures and contain iodine or bromine in addition to a fill gas. The primary advantages of quartz lamps are higher efficacy, smaller decline in output over the lamp life, and longer lamp life.

The second major type of light source is the gas discharge type, which includes fluorescent lamps, metal halide lamps, mercury lamps, and low- and high-pressure sodium lamps. In terms of efficacy, low-pressure sodium lamps are best at 100 to 180 lm/W and mercury lamps are worst at 50 to 55 lm/W. On the other hand, mercury lamps have the longest expected life, about 9 to 12 years, and low-pressure sodium lamps have the least, about 4 or 5 years. The designer must also look at cost and color rendering variables for all lamps to make the best trade-off.

9.2.4 Performance Effects

The two major variables that affect the visibility of a target, and hence visual performance, are *target size* and *contrast* between the target and the background. If these can be adjusted, notable visibility gains are possible for minor improvements. Many other variables also affect visual performance. The most important are discussed in this section.

1. *Amount of light.* In general, as the amount of light increases in an environment, detectability of objects also increases. Since the measure of the amount of light is illumination, recommendations can be given in terms of how much light is needed for different types of visual activity. Table 9.2 gives a set of recommendations. However, there is a limit here. Light that is too intense obscures details and presents glare problems. The suggestions in Table 9.2 are for medium-aged users with reflectance of task background between 30 and 70% and important speed or accuracy demands. A 33% reduction in these values can be made for younger users, reflectance of task background greater than 70%, and unimportant speed or accuracy requirements. On the other hand, a 33% addition should be made for critical demands in task speed or accuracy, less than 30% reflectance of task background, and users over 55 years of age [6]. Age is an important factor here. Weale [7] estimates that the retinal illuminance of a 60-year-old is about one-third that of a 20-year-old. This is why the elderly require more light in their environments.

2. *Glare.* Glare is caused by brightness that exceeds the adaptation level of the eyes. It may cause annoyance and loss of visual abilities. Glare can be sourced on a reflected light (indirect) or caused by light sources in the field of vision (direct). A light source located within 40 degrees of line of sight will cause direct glare if it is intense enough. A glossy paint may cause indirect glare. Indirect glare from a video display terminal screen may also affect the performance of a data-entry operator. There are no established relationships between amount of glare and the degree of performance loss that is applicable to all

TABLE 9.2 RECOMMENDED ILLUMINATION
LEVELS FOR INTERIOR LIGHTING

Activity type	Illumination level (lx)
Rough orientation	75
Occasional rough visual tasks	150
Rough assembly	320
Rough toolmaking	550
Office work—simple	750
Bookkeeping—small character size	1,500
Difficult inspection	1,500
Technical drawing	2,200
Precise assembly work	5,000
Prolonged difficult visual task	7,500
Precise and delicate visual work	11,000
Very special visual tasks—extremely low contrast and small object size	15,000

Source: Adapted from Refs. [5], [6], and [8].

situations. In general, we would like to eliminate glare, whether it is direct or indirect. Sanders and McCormick [8] propose several methods of reduction of glare:

· Reduce the luminance of light sources.
· Position them away from the line of sight.
· Use light shields.
· Set windows at some distance from the human activity.
· Construct canopies above windows.
· Use diffuse light.
· Use surfaces that diffuse light.
· Reposition work areas and light sources to minimize reflections.
· Provide adequate level of general illumination.

3. *Target/background luminance ratio.* Visual performance as well as comfort may be affected by the ratio of target luminance to the luminance of the area immediately surrounding the target. Guth [9] showed that a relative target/background ratio of 5:1 significantly impairs detectability of fine detail and visual comfort. Although many studies recommend a ratio of 3:1, several studies have concluded that these ratios are not valid. Until doubts are cleared, however, our recommendation will focus on the ratios noted above.

4. *Brightness contrast.* In general, visual effectiveness increases as brightness contrast between the target and the background increases. However, there are limitations, as discussed in item 3. Contrast can be defined as

$$C = \frac{L_t - L_b}{L_t}$$

where L_t is the luminance of the target and L_b is the luminance of the background. Brightness contrast is one of those variables that helps visual performance the most.

For tinted window glass, federal standards call for a minimum of 70% luminous transmittance. Freedman et al. [10] showed that under luminous transmittance 70% or below, target visibility through automobile rear windows depends largely on the type of target. Furthermore, detection probability decreases with age.

5. *Duration in the visual field.* The more time that one has to view an object, the more details can be reviewed and recognized. It is natural for performance to be affected positively by longer viewing time.

6. *Movement of target.* Visual performance declines with the movement of target. Fewer details can be observed and less time is available for observations. Visual acuity suffers loss with movement of target. For angular velocity of 50 degrees per second, acuity is 56% of normal. For angular velocity of 150 degrees per second, acuity is 19% of normal [11]. Burg [12] also showed that visual acuity declines rapidly once angular velocity of target exceeds 60 degrees per second. Hence, if possible, designers should minimize the movement of target on tasks that require fine detail identification.

7. *Color of target.* The color sensitivity curves presented in Chapter 4 indicate that the eye is more sensitive to hues in the middle range of the visible spectrum in relation to hues in the opposite extremes. Hence, if properly selected, color of the target may help enhance visual performance, although the effect will be small. Color may also have psychological effects. Table 9.3 gives a summary [13].

8. *Size of target.* Another important variable that affects visibility is size. Much can be gained by increasing the size of the target. If this is not possible, increase the contrast between the target and the background.

9. *Position of target in the visual field.* Closely linked to the distribution of the rods and the cones, position of the target in the visual field will make a difference in terms of discrimination of detail. Maximum effectiveness is achieved when the image falls on those regions of the retina where the appropriate photoreceptors are densely populated, the fovea for day vision and ± 20 degrees off the fovea for night vision.

TABLE 9.3 PSYCHOLOGICAL EFFECTS OF COLORS

Color	Response
Yellow	Warm, cheerful, pleasing
Green	Cool, comfortable, calming
Blue	Cool, protective, calming, slightly depressing
Violet	Slightly warm, calming
Purple	Rich, protecting, may be depressing
Fluorescent red	Warm, stimulating, exciting
Brown	Warm, comfortable, rich, substantial
Gray	Neutral, calming, slightly hard
White	Neutral, sterile, clean, fresh, stark

Source: Adapted from Ref. [13].

10. *Direction of lighting.* Direction of lighting is important because if the light is not coming in the correct angle, glare and shadows may be created. In both cases, performance will be negatively affected.

9.3 THE AUDITORY ENVIRONMENT

Another physical environment within which the human–machine system operates is the auditory environment. Warning signals, the high-pitched sound of a sawmill, the banging sound of a press, and the "clicking" of the keyboards are all examples of the environment that is in question. The main element of the auditory environment that we will be concerned with in this section is *noise*. In simple terms, noise has been defined as *unwanted sound*. Hence noise is sound that creates a negative psychological effect on a person. This effect frequently manifests itself in the form of distraction, annoyance, and frustration. Furthermore, ergonomists are interested in the effect of sound on living tissue, especially on the hearing mechanisms. Both research and industrial experience tell us that exposure to sound may not be desirable or healthy. Duration of exposure and sound intensity emerge as the two important variables here. These factors are investigated in more detail later.

Many researchers also view noise as an environmental stressor. In this respect, noise has a differential effect on performance, as will be discussed later. The important point is that some sound is always present in our environment. Whether it is viewed as beneficial or detrimental depends on many factors. In this section we first review the physiological effects of noise. Psychophysical attributes of sound are covered next. Then our emphasis will shift to performance effects. Noise control is an important element of managing possible ill effects. This topic is summarized next. Finally, we provide some concluding remarks.

9.3.1 Physiological Effects of Noise

As discussed in Chapter 4, the two physical attributes of sound that are of interest to an ergonomist, *intensity* and *frequency*, are also important in detailing the effects on physiology and well-being. Intensity is the element that is of most concern. Sound intensity is measured in decibels (dB) on the A-scale, similar to the human ear's weighting of certain frequencies. Table 9.4 gives the sound intensity in various environments.

Instantaneous sound pressure level (intensity) is measured by a portable sound level meter. A noise dosimeter, on the other hand, calculates a time-weighted average for noise exposure on an 8-hour basis. Operators may carry a noise dosimeter in their pockets for the duration of their work. At the end of the shift, an LED readout or printed copy of total exposure with instantaneous exposure data can be obtained. Noise dosimeters were developed to aid in meeting regulations of the Occupational Safety and Health Administration. The OSHA permissible noise exposure limits were published in 1971 [14]. According to these data, a worker can be exposed to a sound intensity of 90 dB(A) for an 8-hour period. If exposure intensity is increased beyond this level, exposure duration must be reduced. Table 9.5 gives the OSHA permissible noise exposure levels.

The roots of controlled exposure to noise lie in the evidence that it can cause hearing loss. After exposure to sufficiently intense noise, temporary hearing loss occurs, which lasts

TABLE 9.4 SOUND PRESSURE LEVEL

Source	Intensity [dB(A)]
Quiet residence	42
Dictation	67
Conference speaking	70
Quiet factory	76
Loud shouting	82
18-in. automatic lathe	87
Wire drawer	89
Sawmill	90
Chain saw	105
Pneumatic bore hammer	120
Rifle shot	130

several hours or days. Chambers et al. [15] showed that even a 25-minute exposure to welding noise between 87 and 100 dB(A) may cause a temporary threshold shift. With extended exposure to noise, the amount recovered becomes less and less. At some point in time, no recovery occurs. This is called *permanent hearing loss*.

Noise-related deafness is caused by slow and progressive degeneration of the sound-sensitive cells in the inner ear. The louder and more extensive the noise, the more the damage. It is also known that high-frequency noise is more deafening than low-frequency noise. Intermittent noise is also known to be more harmful if the intensity is sufficient. A single explosion may also damage the hearing mechanism permanently.

Hearing measurement is conducted by *audiometers*. The most common is the one that tests hearing capability at various frequencies. The threshold for hearing, which is the lowest intensity that can just be heard, can be determined at each frequency in the audible spectrum.

Noise-induced hearing loss must be distinguished from age-related hearing loss. Noise deafness starts at around 4000 Hz [16]. A pure-tone audiogram of a person who has experienced permanent noise-related hearing loss shows a dip around 4000 Hz. Figure 9.2 demonstrates this. In this figure, curve C_2 shows the experience of a person who has been exposed to noise for a longer period. Raja and Ganguly [17] confirmed noise-induced hearing losses at around 4000 Hz in a railcoach factory in India. At 4000 Hz, statistically sig-

TABLE 9.5 OSHA PERMISSIBLE NOISE EXPOSURE

Duration per day (hr)	Sound level [dB(A)]
8	90
6	92
4	95
3	97
2	100
1	105
0.5	110
0.25 or less	115

Source: Ref. [14].

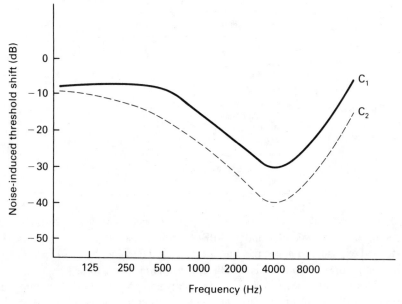

Figure 9.2 Noise induced threshhold shift versus exposure to noise and test frequency. (After Ref. [16].)

nificant differences were noticed with respect to noise level and duration of exposure.

The threshold of hearing rises with age. Loss of hearing is more evident in the higher frequencies of the audible range. Age-related hearing loss is more pronounced in men than in women. The audiogram of an elderly person does not show the characteristic dip at around 4000 Hz.

Other points related to noise-induced hearing loss are [4]:

1. Alternating between quiet and noisy environments produces less damage.
2. The time of recovery is about 10% longer than the duration of exposure.
3. The amount of threshold shift is greater with increasing intensity of noise.
4. The amount of threshold shift is proportional to the duration of exposure.
5. The time taken for recovery is more if the intensity and duration are higher.
6. Some people are more sensitive than others to noise.

9.3.2 Psychophysical Evaluations

In addition to the physical elements of noise, psychophysical attributes such as loudness (discussed in Chapter 4) have been used in the past to evaluate the psychological experience of human beings. Such indices define the subjective experience due to physical elements of sound, such as the intensity and the frequency. Loudness is the most widely used subjective attribute of sound, and the most widely used loudness indices are the *phon* and the *sone*. Other indices, such as the *equivalent sound level* (L_{eq}) developed by the Environmental Protection Agency (EPA) [18], the *perceived noise level* (PNL) proposed by Kryter [19], the *day-night level* (L_{dn}) used by the EPA, and DENL (day–evening–night level) proposed by Kryter [20], have all been linked to the annoyance effects of noise (see the next section). We will not discuss these indices in detail since most have been developed in the context of community response/exposure to aircraft noise. However, some of these indices may be applied in an industrial setting, with caution, to evaluate the annoyance effects of noise objectively.

9.3.3 Performance Effects of Noise

We will investigate the performance effects of noise along several pathways.

1. *Annoyance effects.* The annoying characteristics of noise are quite obvious. Since noise is sound that is not wanted, it is expected that it will be annoying. In general, unlocalized, unpredictably intermittent, and loud noise is more annoying. McCann [21] found that at 50 dB(A), intermittent noise is more detrimental than continuous noise to vigilance performance. Annoyance may bring about irritation, discomfort, displeasure, and complaints. Indoors, noise may be more annoying than when it is heard outdoors. High-frequency noise is more annoying than low-frequency noise. All of the expected effects on people of annoying noise lead to the conclusion that performance may be affected. Reducing the level of noise helps reduce annoyance; however, due to the social and individual factors involved, it is difficult to predict when problems will be least.

2. *Distraction effects.* Noise may distract a person. A sudden loud noise may be startling. It is not hard to predict that in these situations performance will worsen.

3. *Effects on alertness.* Since noise is an environmental stress agent, it can be predicted that some tasks will be positively affected by noise and some negatively affected. Hockey [22] claims that tasks that are already lacking the motivating elements to the performer, simple and routine tasks, are expected to show improvement as a result of higher levels of continuous noise. Eschenbrenner [23] has shown detrimental effects on difficult tasks requiring high levels of concentration and perceptual capacity.

4. *Interference with communication.* Noise may also affect speech intelligibility. Speech comprehension depends on the loudness of voice and the level of background noise. The following summarizes what is known in this respect:

- For familiar information, speech comprehension is unimpaired when the background noise level is at least 10 dB(A) below the level of the speaking voice.
- For unfamiliar conversation, words, and signals, the difference should be a minimum of 20 dB(A).

In an office environment, the background noise level should not exceed 55 to 60 dB(A). This may be a problem if offices are located next to busy streets.

5. *Attention effects.* Closely related to the distraction and annoyance effects, noise may negatively affect attention [24,25]. The effect of noise on time sharing is that it moves attention resources away from low-priority items toward high-priority items. Broadbent [26] called this the "funneling of attention."

6. *Productivity effects.* Relatively few studies suggest that noise may create more accidents and lower accuracy. Speed of performance does not seem to be affected [27,28].

Introduction of ear protection devices seems to cancel the negative effects. Younger and inexperienced workers seem to be more susceptible to noise.

7. *Effects of music on performance.* Music has been presented in many factories as a means of enhancing productivity. In some instances, music has produced beneficial effects. Still in some trials, it has not been well received. Most workers like music, but some do not. There is some evidence that boring or repetitive tasks are positively affected by music, especially if the task rhythm corresponds well to the rhythm of the music. On jobs that require heavy concentration and attention, music may be an undesirable distraction. In planning for music in the occupational environment, consider that [5]:

· Music will interfere with oral communication.

· Music is not recommended in background noise levels of 70 dB(A) or above.

· Good-quality presentation systems need to be used.

· The employees should have input to the presentation schedule.

· The type of music should be selected by the employees.

9.3.4 Noise Controls

Both engineering controls and administrative controls are needed for proper management of noise.

Engineering controls.   Engineering controls of noise focus on three points:

1. *Control at the source.* Reduction of noise at the source is the choice of control; however, this may be prohibitively expensive. One may use softer materials in place of harder ones to reduce the effects of impact forces. Worn components create unnecessary noise. These should be replaced periodically. Belt drives are quieter than cogwheels. Noise created by vibrating plates may be reduced by stiffening them, making them curved, and using nonresonant materials [4]. Using sound-absorbing material on the inside and outside of machine surfaces also helps.

2. *Control along its path.* Enclosures with porous linings around noise sources will help reduce the transmission of noise to the workers. A housing of suitable material may reduce noise by 20 to 30 dB. Special sound-reflecting material may also help.

3. *Control at the receiver level.* This is the least preferred method since people generally do not like to wear extra equipment or clothing. Hearing protection equipment comes in all sizes and costs. Earplugs block the outer ear passages. Earmuffs cover the entire ear. Protective helmets enclose the head, including the ears. Earplugs are cheap and reduce noise by as much as 30 dB. Earmuffs provide 40 to 50 dB protection; however, they are more expensive. Helmets provide more protection. They are the most expensive of the three. As a general rule, use earplugs if noise level is between 90 and 120 dB(A). Above these values, one may use earmuffs unless a helmet is necessitated by the extremes of noise.

Administrative controls.   Administrative controls focus on exposure management. Rotating employees between noisy and quiet environments and education are within the realm of administrative controls. Administrative controls do not compare favorably with engineering controls. Our preference is to control noise by engineering means.

9.3.5 Concluding Remarks

In this section we provide several conclusions and recommendations with respect to noise management [29]:

1. Target for a maximum 8-hour weighted noise exposure of 85 dB. This will assure maintenance of well-being.
2. For best signal intelligibility, increase the signal-to-noise ratio.
3. Engineering controls of noise are greatly preferred to administrative controls.
4. Control noise at its source to the extent possible.
5. The speech reception range is between 50 and 80 dB(A).
6. Noise affects rehearsal, hence retention of information.
7. Extremes of noise may lead to extremes of judgment.
8. Noise increases chances for aggressive behavior.
9. Conduct audiometric screening of employees once every 2 to 3 years.

9.4 VIBRATION

Vibration has been defined in many ways: a series of reversals in velocity [8], mechanical oscillations [4], and particle motion around an equilibrium [5] are a few definitions proposed. In this book we define it as rhythmic or random particle oscillation. Vibration is different from sound. Sound is carried in the air; vibration is carried in solid structures that are in contact with each other. Vibration leads to sound. Examples of vibrating structures are an earth-moving vehicle, a power screwdriver, a jackhammer, and a utility cart being pushed.

Vibration is characterized by its frequency and intensity. Frequency of vibration refers to the frequency of oscillations per second and is measured in terms of hertz (Hz). Intensity is measured in a variety of ways, such as peak amplitude, peak displacement, peak velocity, and acceleration. Physiological reaction to vibration is frequency and intensity dependent. Performance effects are frequency, axis, duration, body support, and age dependent. Table 9.6 gives vibrating (vertical) characteristics of various vehicles [30].

There are several types of vibration. *Sinusoidal vibration* [8] is one or many sine wave forms at different frequencies. Its main characteristic is regularity. Sinusoidal vibration is encountered primarily under laboratory conditions. The second type of vibration, *random vibration*, occurs irregularly and is unpredictable. Random vibration is characteristic of vibrating equipment in the real world.

One may also talk about *whole-body vibration* versus *localized vibration*. Whole-body vibration is transmitted to the body mostly through the buttocks, as in the case for an operator sitting in a vibrating seat. Whole-body vibration may also be experienced when vibration enters the body through stiff arms and legs. Localized vibration may occur to the hands and legs when they are supporting vibrating tools, such as powered hand tools and pedals.

TABLE 9.6 VERTICAL VIBRATING CHARACTERISTICS OF VARIOUS VEHICLES

Vehicle	Maximum frequency (Hz)	Acceleration (m/s^2)
Car	1–2	0.5–1.0
Agricultural tractor	3–5	0.8–2.5
Tractor with trailer	3–4	0.8–4.2
Caterpillar tractor	9–12	0.9–1.6
Truck	3–4	0.8–2.0
Earthmoving vehicle	3–4	1.0–1.5

Source: Ref. [30].

9.4.1 Physiological Effects

The physiological effects of vibration will be investigated along the lines of whole-body and localized vibration. Before going into these areas, we would like to discuss several more terms that are of importance.

Any object has its own natural frequencies of vibration. These depend on its mass and structure. At certain frequencies, the amplitude of vibration may be reduced. This phenomenon is called *attenuation*. At still other frequencies, at resonating frequencies of the object itself, the amplitude of vibration is increased. This is called *resonance*. Hence when an object is vibrated at its own vibrating frequencies, maximum amplitute vibration will occur. Radke [31] claims that when the ratio of input frequency to the resonant frequency of the body part is greater than 1.414, attenuation will occur. Otherwise, amplification occurs.

Whole-body vibration. The body does not possess the characteristics of a rigid structure. Different parts of the body have different resonance characteristics.

1. *Vertical vibration.* Rasmussen [32] and Coermann [33] have shown that the resonance characteristics of different body segments under vertical vibration are different. Table 9.7 gives a summary. Human complaints and symptoms due to whole-body vertical

TABLE 9.7 RESONANCE CHARACTERISTICS OF THE BODY SEGMENTS UNDER WHOLE-BODY VERTICAL VIBRATION

Body segment	Resonance frequency (Hz)
Abdominal mass	4–8
Spinal column (axial)	10–12
Head (axial)	20–25
Eyeball (intraocular)	30–80
Lower arm	16–30
Hand grip	50–190
Lumbar vertebrae	4
Shoulder girdle	4–5

Source: Refs. [32] and [33].

vibration are in line with the data shown in the table. There is severe pain in the chest between 4 and 10 Hz. Complaints of backache are dominant between 8 and 12 Hz. Headache and eye strain occur in the range of 10 to 20 Hz. There is abdominal pain in the range of 4 to 10 Hz. Speech is most affected within 15 to 20 Hz. There is breathing difficulty between 4 and 8 Hz.

In general, there is a feeling of fatigue and discomfort between the frequencies of 4 and 9 Hz. The ISO (International Standardization Organization) [34] has developed vibration exposure limits for various types of symptoms. Figure 9.3 gives the fatigue-decreased proficiency boundaries. Each line in this figure represents the upper limits of vibration exposure before fatigue onset.

The most severe reactions in whole-body vertical vibration occur between 4 and 8 Hz. If the acceleration is sufficient (1.5g or above), there may be internal hemorrhage and damage to the internal organs. Whole-body vertical vibration also affects the spine. Research shows that the possibility of disk problems increases by a factor of 4 in truck drivers and by a factor of 2 in car driving [35]. However, Bobick et al. [36] showed that the back muscle function may not be compromised by whole-body vibration. In two studies supported by the U.S. Bureau of Mines, the authors could not show an effect of random, low-intensity whole-body vibration on maximum back extensor strength and back endurance. On the other hand, Bobick et al. [37] concluded, after research with whole-body vibration exposure to a typical underground mine haulage vehicle, that the heart rate, systolic blood pressure, mean blood pressure, and overall subjective discomfort increase significantly. A person tolerates vibration better when standing than while sitting. While standing, the legs absorb some of the effects. Women, in general, experience more discomfort than men do at the same exposure and intensity level.

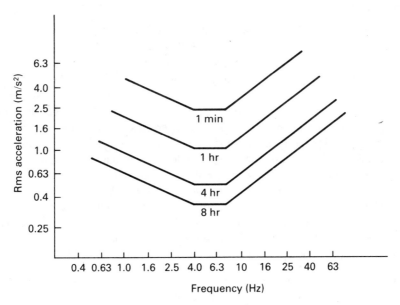

Figure 9.3 ISO proficiency boundaries for vertical vibration. (After Ref. [34]).

2. *Horizontal vibration.* Horizontal vibration, in the fore-and-aft direction, also may affect human physiology. The critical frequencies here are between 2 and 3 Hz. At these frequencies, if the intensity is sufficient, damage to the soft tissue may occur. Again, a person standing with the legs slightly bent and relaxed will be at much less risk than will a seated person. However, a suspension seat may absorb most of the effects of vibration.

3. *Lateral vibration.* Compared with vertical and horizontal vibration, effects of lateral vibration are the least known. This type of vibration is the least common. Naturally, fatigue effects may be expected in addition to performance effects.

Localized vibration. Another term for this type of vibration is *segmental vibration*. We are more specifically concerned with the vibration effects on the hands and arms. *Raynaud's phenomenon* (vibration-induced white finger disease—VWFD) is a vibration-induced circulatory disorder of the hands. Frequencies in the range of 25 to 150 Hz and accelerations from 1.5 to 80g are most commonly associated with this disorder. People with this condition experience severe vascular spasms and associated pain when they expose their hands to cool or cold environments. They also lose grip power and may not have endurance in sustained-holding tasks. For short exposures (30 min), upper limit of acceleration for safe exposure to vibration to the hands extends into the 250g range (at 250 Hz) and is a linear function of the frequency range. However, for longer exposures (4 to 8 hours), maximum acceleration limit drops to 20g at similar frequencies. There is also some evidence that vibration to the hands (40 Hz and below) may cause degeneration of the bones, joints, and tendons, possibly leading to arthritis in the wrist, elbow, and sometimes in the shoulder [4]. Research [38] with chain saws identified the acceleration in the firing frequency as the primary risk factor for the development of VWFD.

9.4.2 Performance Effects

In addition to possible tissue damage, vibration may lead to performance effects. Vibration degrades fine manipulative ability and skilled muscular control between 5 and 25 Hz. Visual functions may also be degraded. Under vibration, visual acuity is poorer and the visual field becomes blurred and unsteady. The greatest visual effect is in the range 10 to 30 Hz. Vibration also affects speech, especially between 13 and 20 Hz. Speech intelligibility drops as a result of effects on the chest wall, vocal cords, and tongue and mouth.

Vibration is also seen as a nuisance. It creates an annoyance. As a result of this, mental processing performance may be affected. According to Poulton [39], low-acceleration, 5-Hz vibration may keep people alert on long-term monitoring tasks. Low-frequency vibration (1 to 2 Hz) with sufficient amplitude may lead to motion sickness. Some people are more prone than others to motion sickness. Motion sickness leads to severe physiological reactions and performance effects. One may reduce the effects of motion sickness by learning to hold the head still. Vibration also affects eye–hand coordination and the ability to write. The head tends to rotate during vibration. This may have a marked effect on vision [40].

9.4.3 Vibration Controls

Both engineering and administrative controls are necessary to reduce the effects of vibration. Employees could be rotated to control exposure. Vibration could be reduced at the source by proper equipment maintenance, mounting equipment on pads or springs, using materials that generate less vibration, and equipment modifications to reduce speed and acceleration [5]. The main effort has been devoted to preventing vibration from reaching the human being. Spring suspensions between vibrating sources and seats or platforms, as well as rubber or vinyl floor mats, help in this respect. Tools can be designed to reduce the vibration transmitted to the hands.

9.4.4 Combined Effects of Noise and Vibration

It is important for the reader to realize that the combined effect of multiple physical stressors may be different from the algebraic sum of the individual effects, whether performance or physiological variables are in question. In a recent study, Notbohm and Gros [41] studied performance in a five-choice reaction task under six conditions of environmental stress: white noise of 75 and 100 dB(A) alone and in combination with sinusoidal whole-body vibration at 4 and 8 Hz (acceleration of 3 m/s^2). Reaction time was shortest with combined stress of 8-Hz vibration and 100-dB(A) noise, whereas reaction response accuracy was best with noise alone. The observed subtractive effect at 8 Hz and 100 dB(A) could not be confirmed with any combination of 4-Hz vibration and noise, or with 4-Hz vibration and noise alone.

9.5 AMBIENT TEMPERATURE

The temperature of a workplace may significantly affect performance in addition to well-being. Added heat and humidity beyond the heat generated by the body during physical work can lead to a notable performance decrement and may place the health of a worker at risk. Cold conditions, especially when coupled with high wind, may lead to physical harm to uncovered flesh and will definitely lower output and increase errors. Both conditions also give rise to the potential for unsafe acts.

Hot conditions in industry may arise from smelting, molding, steaming, boiler operations, extruding, drying operations, and work with heat-producing chemical reactions. Exposure to cold may be due to work outside during winter, to work under conditions of refrigeration or cold storage, and to work in poorly heated buildings. Normal body temperature is 37 degrees Celsius (98.6 degrees Fahrenheit). The body is most efficient when heat balance is maintained. With physical work, the body generates heat; this extra heat must be dissipated. Those conditions that allow the body to maintain its thermal balance define a *comfort zone* for the worker. The American Society of Heating, Refrigerating, and Air-Conditioning Engineers (ASHRAE) [42] sponsored the development of an *effective temperature* (ET) scale, a variant of which is published in Figure 9.4. The ET scale considers the combined effects on the body of air velocity, humidity, and dry-bulb temperature. According to Figure 9.4, the comfort zone lies between 19 and 26 degrees Celsius, assuming 50% humidity and 0.2 m/s (0.64 ft/sec) air velocity.

There are several variables that affect the comfort zone definition:

· Humidity
· Air velocity

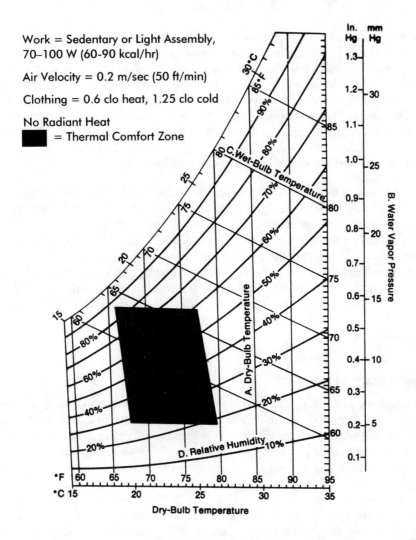

The dry bulb temperature and humidity combinations that are comfortable for most people doing sedentary or light work are shown as the shaded area on the psychometric chart. The dry bulb temperature range is from 19° to 26°C (66° to 79°F), and relative humidities (shown as parallel curves) range from 20 to 85 percent, with 35 to 65 percent being the most common values in the comfort zone. On this chart the ambient dry bulb temperature (A) is plotted on the horizontal axis and indicated as parallel vertical lines; water vapor pressure (B) is on the vertical axis. Wet bulb temperatures (C) are shown as parallel lines with a negative slope; they intersect the dry bulb temperature lines and relative humidity curves (D) on the chart. In the definition of the thermal comfort zone, assumptions were made about the work load, air velocity, radiant heat, and clothing insulation levels. These assumptions are given in the top left corner of the chart.

Figure 9.5 Thermal comfort zone (From Ref. [5]. Reprinted courtesy of Eastman Kodak Company.)

- Work load
- Clothing
- Radiant heat

As humidity increases, the limits of the comfort zone shift downward (i.e., at 80% humidity, the limits are 18.5 and 24 degrees Celsius). As air velocity increases, comfort limits shift upward. As work load increases, comfort limits shift downward. As clothing insulation value increases, limits move downward. Finally, with an increasing radiant heat load, comfort zone limits shift toward lower values.

The only variable in the list above that is not self-explanatory is the clothing insulation value, the clo. Table 9.8 gives clo values for various items of clothing [43,44].

Closely related to thermal comfort is thermal balance, which a person seeks with regard to his or her environment. In thermal equilibrium, net heat exchange (loss or accumulation), ΔS, is zero. Heat loss is mainly through the evaporation, E, of perspiration from the skin. People will also either add or lose heat due to convective, C, or radiative, R, heat transfer. *Convection* refers to heat exchange between skin and air due to a temperature difference between the two. *Radiation* is due to the difference between a person's average skin temperature and the temperature of surfaces in the environment. Finally, in the heat balance equation, we have M, which stands for metabolic heat generation. Hence the final equation is

$$\Delta S = M \; \Delta C \; \Delta R - E$$

where Δ indicates net gain or loss (+ or −).

TABLE 9.8 INSULATION VALUE (CLO) OF VARIOUS ITEMS OF CLOTHING

Clothing item	Clo
T-shirt	0.09
Briefs	0.06
Shirt	
Long sleeved (heavy)	0.29
Short sleeved	0.25
Sleeveless blouse, light skirt, and sandals	0.3
Long light trousers and open-neck short-sleeved shirt	0.5
Light business suit (full jacket)	1.0
Light vest	0.15
Heavy sweater	0.37
Heavy jacket	0.49
Heavy trousers	0.32
Boots	0.08
Heavy business three-piece suit, long underwear, wool socks, shoes, long-sleeved shirt	1.8

Source: Adapted from Refs. [43] and [44].

There also exist ISO standards concerned with the ergonomics of the thermal environment. ISO 7243 and ISO 7933 are for the assessment of hot environments. ISO 7730 and ISO DIS 10551 provide methods for assessing moderate environments. ISO TR 11079 provides an analytical method to evaluate cold environments [45].

9.5.1 Physiological Effects

Heat stress. Heat stress requires additional effort on the part of the body to maintain thermal equilibrium. The most important physiological reactions due to increased ambient temperature beyond the comfort zone are:

· Vasodilation
· Increase in heart rate
· Decrease in blood pressure
· Increased skin temperature
· Initial decrease and then increase in core temperature

Acclimatization to heat is possible to a certain extent; however, it takes time. It takes several weeks to get heat adapted. A heat-acclimatized body needs less liquid in volume but more frequent drinks. Salt tablets may also help in the process. A staged heat-acclimatization process is recommended. A worker should spend only 50% time in heat at the initial stages. Exposure may then be raised by 10% each day.

Normal core temperature, 37°C (98.6° F), may easily be exceeded by heat accumulation due to convection, radiation, and metabolic heat generation. A rise in core temperature to 37 3°C (99.2°F) may impair performance. A core temperature of 38.5°C (101.3°F) makes a person uncomfortably hot. Collapse or heat stroke is possible at 40°C (104°F). Few people will survive a core temperature of 43°C (109.4°F).

In addition to the ET scale, the WBGT (wet-bulb globe temperature) index can be used to evaluate jobs carried out in heat, or design jobs based on a hot environment. The following formulas summarize the steps involved [46]: If the difference between radiant temperature and air temperature is negligible (mostly indoors),

$$WBGT = 0.7NWB + 0.3GT$$

Otherwise (mostly outdoors),

$$WBGT = 0.7NWB + 0.2GT + 0.1DBT$$

where,

WBGT is the wet-bulb globe temperature, °C.

NWB is the natural wet-bulb temperature, °C; this is the temperature of a wet wick exposed to natural air currents

NWB = WB for air velocity greater than 2.5 m/sec (8 ft/sec), where WB is the psychometric wet-bulb temperature; NWB = 0.1DBT + 0.9WB for air velocity between 0.3 and 2.5 m/s.

GT is the globe temperature, °C.

DBT is the dry-bulb temperature, °C.

In addition to this methodology, there are direct-reading devices for WBGT. The WBGT index has been accepted as a composite measure of thermal environment. WBGT threshold values exist for evaluating tasks carried out in hot environments for continuous work over an 8-hour period. For example, with an air velocity of less than 1.5 m/s, the threshold that should not be exceeded is 30°C for light work (metabolic rate less than 198 kcal/h), 27.8°C for moderate work (metabolic rate between 198 and 301 kcal/h) and 26°C for heavy work (metabolic rate above 301 kcal/h) [29]. These values should be interpreted with caution. They are not clearcut values, merely expert advice.

The Occupational Safety and Health Administration recommends the limits given in Table 9.9[47]. As can be observed, these are very close to the limits discussed. Although there is excellent agreement between researchers with respect to industry limits due to heat stress for continuous work over a period of 8 hours, the effects of intermittent work in hot environments have not been clearly established. Hence the data above should be applied to these situations with extreme caution.

NIOSH [48], as early as 1972, specified a 1-hour time-weighted WBGT that could not be exceeded without compensatory measures. This was to assure that the body's core temperature would not be elevated by more than 1°C. The NIOSH [49] revised criteria for the recommended standard occupational exposure to hot environments proposed heat stress limits as given in Figure 9.5. The REL (recommended exposure limit) applies to the heat-acclimatized worker, and the recommended alert limit (RAL) applies to the unacclimatized worker.

Others have also made exposure limit recommendations based on physiological criteria. Grandjean [4] offers the following limits for working in heat for an entire workday:

1. Heart rate (daily average) of 100 to 110 per minute
2. Rectal temperature of 38°C
3. Evaporated sweat of 0.5 liter per hour

An index that is closely related to the WBGT index is the Botsball (BB) index [50]. Beshir et al. [51], after field studies, observed that BB readings were highly correlated (0.96) with the WBGT index. Since the Botsball index uses a special type of thermometer that combines the effects of air temperature, wind speed, humidity, and variables related to

TABLE 9.9 SUGGESTED MAXIMUM WET-BULB GLOBE TEMPERATURES FOR VARIOUS WORKING CONDITIONS [°F (°C)]

Air velocity	Work intensity (kcal/h)		
	Light (≤ 200)	Moderate ($201 < \leq 300$)	Heavy (> 300)
Low: < 1.5 m/s (< 300 ft/min)	86 (30)	82 (27.7)	79 (26.1)
High: $\geq$ 1.5 m/s ($\geq$ 300 ft/min)	90 (32.2)	87 (30.5)	84 (28.9)

Source: Adapted from Ref. [47].

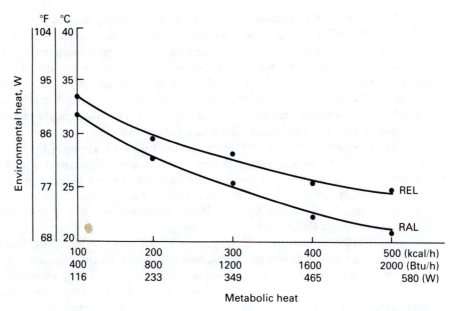

Figure 9.5 Recommended heat-stress exposure limits (REL) and alter limits (RAL). (After ref. [49].)

radiation, the authors concluded that BB readings from this simple device can be converted into WBGT values using the following equation:

$$WBGT = 1.01BB + 2.6$$

Then heat exposure limits using the WBGT index (such as NIOSH limits) can be applied. ·

Several other heat evaluation indices have been developed in the past with limited application success. Some of these are the heat stress index [52], the operative temperature [53], and the Oxford index. Since these are not widely used in industry, additional details will not be presented·here.

Working in hazardous environments may require special protective equipment to be worn. Such equipment may present significantly increased thermal burden and early onset of physical fatigue. Research with microclimate cooling during rest at a constant WBGT of 31°C shows that periodic circulation of chilled liquid over torso may double work capacity and reduce heat stress [54].

Cold stress.   Physiological reaction of the body to cold stress takes the form of vasoconstriction, piloerection (erection of hair), shivering (to generate heat internally), and tension in the muscles. Body core temperature initially rises and then shows a continuous decline. A body core temperature of 32°C (89.6°F) produces muscle rigidity and unconsciousness. There is severe cardiovascular stress at 29.5°C (85.1°F), and death occurs around a core temperature of 27°C (80.6°F).

The major concern regarding body exposure to cold is hypothermia and subsequent death. Furthermore, *frostbite* of the face or extremities may occur due to exposure to cold. The most useful thermal index for cold is the *wind-chill index*. This index shows the combined effect of air velocity and temperature on cold sensations as well as possibility for frostbite. Table 9.10 provides a sample wind-chill index range. Exposure to cold at equivalent temperatures of $-32°$ C ($-25.6°$F) may result in cold injury to exposed flesh. Cold may also lead to vibration injury syndromes and aggrevate preexisting arthritic conditions. Furthermore, tissue damage to skin is possible when it comes in contact with very cold surfaces.

9.5.2 Performance Effects

Heat stress. Several studies concluded that after around 27 to 30°C ET, performance on physical tasks deteriorates. On tasks that require mental performance, efficiency increases with increasing ET up to about 31.5°C. After this point, there is a continuous decline in performance [55,56]. Tasks requiring perceptual motor skills show a performance decrement in the 30°–33°C WBGT range [57]. Improvement in performance up to around 29.4°C ET is attributable to the alerting effects of heat and increased neural conductivity. Performance decrement after this point is due to sweating, drowsiness, lassitude, and subjective feelings of discomfort. Sweating may also affect visual performance and reduce grip capability. Even moderate heat stress (27°C) has been shown to significantly affect performance in a vigilance task during driving [58].

Cold stress. The most researched index with respect to the effects of cold on performance is the hand skin temperature (HST). One can put more clothes on and hence protect the body under cold conditions, but if manipulative skills are required, wearing gloves does not help. Therefore, several scientists have been concerned with measuring hand skin temperature and predicting the point below which manipulative performance decrement can be predicted. Fox [59] summarized the literature, which indicates that the critical HST for tactile sensitivity is near 8°C (46.5°F). The corresponding figure for manual dexterity is between 12 and 16°C (52.6 to 60.8°F).

With increasing degrees of cold, fine manipulative skill is lost first. Then degradation in finger–wrist action follows. Finally, gross movement ability is affected. Impairment in performance due to cold is commonly said to be due to loss of cutaneous sensitivity, changes in characteristics of synovial fluid at the joints, shivering, loss of muscle strength, and distracting effects of cold sensations. It is possible to protect hands in cold and maintain some dexterity by using gloves with the tips of fingers removed. Sundheim and Konz [60] also showed that the bare hand can be kept somewhat warm with extra clothing on the body.

9.5.3 Combined Stressors

Performance effects of combined stressors seem to be highly variable and a direct function of the level of the stress agent and the specificity of the conditions used in the study. Bell [61] and Dean and McGlothlen [62] showed that combined effects of heat and noise are sometimes smaller and sometimes additive. Van Order et al. [63] showed that cold in combination with a command-and-control task may produce significant changes in behavioral response patterns.

TABLE 9.10 WIND-CHILL INDEX

Air velocity m/s (ft/s)	Measured dry-bulb temperature [°C (°F)]							
	10(50)	4(39)	-1(30)	-7(19)	-12(10)	-18(-1)	-29(-20)	-40(-40)
	Equivalent temperature [°C (°F)]							
Calm	10(50)	4(39)	-1(30)	-7(19)	-12(10)	-18(-2)	-29(-20)	-40(-40)
2.2(7)	9(48)	3(37)	-3(26)	-9(16)	-14(7)	-21(-6)	-32(-25)	-44(-47)
6.7(21)	2(35)	-6(21)	-13(8)	-21(-6)	-28(-18)	-36(-33)	-50(-58)	-65(-85)
8.9(28)	0(32)	-8(17)	-16(3)	-23(-9)	-32(-25)	-39(-38)	-55(-67)	-71(-96)
13.4(43)	-2(28)	-11(12)	-19(-2)	-28(-18)	-36(-33)	-44(-47)	-62(-80)	-79(-110)
15.6(50)	-3(26)	-12(10)	-20(-4)	-29(-20)	-37(-35)	-46(-51)	-63(-81)	-81(-114)
17.9(56)	-3(26)	-12(10)	-21(-6)	-29(-20)	-38(-37)	-56(-69)	-73(-99)	-82(-116)
	Little danger			Dangerous			Very dangerous	

Source: Adapted from Ref. [5].

9.5.4 Controls

Both engineering and administrative controls are recommended. Engineering controls focus on environmental control (i.e., removing the heat or cold source or minimizing their effects). After an extensive review of the literature, Ramsey [64] proposed that the thermal environment should be kept within -30 and $+30°C$ WBGT (-22 and $+86°F$ WBGT). Administrative controls concentrate on job rotation, frequent breaks to heat hands (in cold), clothing management, and acclimatization of workers.

9.6 SUMMARY

The visual environment, noise, vibration, and ambient temperature are the physical environmental stressors that are most important in industry. For best human performance and minimal effects on the body, their levels must be kept within comfortable ranges. Ergonomic focus on the design of the physical environment aims at accomplishing this objective.

QUESTIONS

1. How does the context affect human–machine system performance?
2. Briefly discuss the two major components of the work environment that affect the behavior of human–machine systems.
3. What is the objective of specifying physical environment limits for human–machine systems?
4. List and discuss the aspects of light that are most relevant to an ergonomist.
5. What are the different types of lighting available? Discuss each.
6. Discuss the advantages and the disadvantages of incandescent and gas discharge lamps.
7. What are the two most important factors of the visual environment that affect visual performance?
8. How does amount of light affect visual performance?
9. How does the color of target affect visual performance?
10. What is noise? How may a loud theme from Bach not be perceived as noise?
11. How does noise intensity affect human physiology?
12. Discuss the mechanics of age-related and noise-induced hearing loss.
13. How does noise affect alertness?
14. Discuss the "funneling of attention" due to noise.
15. How can noise be controlled?
16. What is vibration?
17. What are resonance and attenuation?
18. What are the ISO-accepted critical frequencies due to whole-body vertical vibration?
19. Discuss Raynaud's phenomenon.
20. Discuss the performance effects of vibration.
21. How can vibration be controlled?

22. Define the human comfort zone due to ambient temperature.

23. How do humidity, clothing, and air velocity affect the comfort zone?

24. What are the physiological limits of heat exposure?

25. Discuss the WBGT index.

26. Discuss the physiological limits due to cold stress.

27. How does performance get affected by heat stress?

28. What is hand skin temperature? What is its importance?

EXERCISES

1. Using a photometer, take several illumination readings in your environment. Compare the results with recommended levels. Discuss differences.

2. Using a sound-level meter, take five readings of sound pressure level in your environment. Compare results with recommended maximums. Discuss differences.

3. Select a noise source in your environment. Investigate ways of reducing the noise level. List your alternatives in the rank order of cost of implementation. Then list them again in the rank of user preference. Discuss differences.

4. Assume that a job requires 285 kcal/h of energy expenditure and is done under 1 m/s air velocity. The average WBGT reading in the immediate environment is 30°C. Would you consider this job to pose unacceptable stress? Show your work.

5. Assume that light rays from a mercury lamp of 50 W are impinging on a surface at an angle of 50 degrees with the surface. The perpendicular distance between the surface and the light source is 2 m. Would you recommend that a reading task be carried out on this surface? Support your answer by work.

REFERENCES

1. Bailey, R. W. 1982. *Human Performance Engineering.* Prentice Hall, Englewood Cliffs, NJ.

2. Konz, S. 1979. *Work Design.* Grid Publishing, Columbus, OH.

3. Hedge, A., Sims Jr., William R., and Becker, F. D. 1995. Effects of Lensed-Indirect and Parabolic Lighting on the Satisfaction, Visual Health, and Productivity of Office Workers. *Ergonomics.* 38(2), pp. 260–280.

4. Grandjean, E. 1980. *Fitting the Task to the Man.* Taylor & Francis, London.

5. Eastman Kodak Co. 1983. *Ergonomic Design for People at Work.* Lifetime Learning, Belmont, CA.

6. RQQ. 1980. Selection of Illuminance Values for Interior Lighting Design. *Journal of Illuminating Engineering Society*, 9(3), pp. 188–190.

7. Weale, R. A. 1963. *The Aging Eye.* H.K. Lewis, London.

8. Sanders, M. S., and McCormick, E. J. 1993. *Human Factors in Engineering and Design,* 7th ed. McGraw-Hill, New York.

9. Guth, S. K. 1958. Light and Comfort. *Industrial Medicine and Surgery,* 27, pp. 570–574.

10. Freedman, M. Zador, P. and Staplin, R. 1993. Effects of Reduced Transmittance Film on Automobile Rear Window Visibility. *Human Factors,* 35(3), pp. 535–550.

11. *Design Handbook.* 1977. DH 1-3. Air Force Systems Command. Andrews Air Force Base, DC.
12. Burg, A. 1966. Visual Acuity as Measured by Static and Dynamic Tests: A Comparative Evaluation. *Journal of Applied Psychology,* 50(6), pp. 460–466.
13. Woodson, W. E. 1981. *Human Factors Design Handbook.* McGraw-Hill, New York.
14. Federal Register. 1971. 36(105), May 29.
15. Chambers, R. M., Fernandez, J. E., and Marley, R. J. 1989. Noise Exposure of Plumbers in New Home Construction: A Case Study. In *Advances in Industrial Ergonomics and Safety I.* Mital, A. (Ed.). Taylor & Francis, London, pp. 559–566.
16. Melnick, W. 1979. Hearing Loss from Noise Exposure. In *Handbook of Noise Control.* Harris. C. (Ed.). McGraw-Hill, New York.
17. Raja, S., and Ganguly, T. 1986. Noise as a Problem of Work Environment in a Railcoach Factory. In *Trends in Ergonomics/Human Factors III.* Karwowski, W. (Ed.). Elsevier, Amsterdam, pp. 407–414.
18. Environmental Protection Agency. 1974. *Information on Levels of Environmental Noise Requisite to Protect Public Health and Welfare with an Adequate Margin of Safety.* EPA 550/9-74-004. Washington, DC.
19. Kryter, K. 1970. *The Effects of Noise on Man.* Academic Press, New York.
20. Kryter, K. 1985. *The Effects Noise on Man,* 2nd ed. Academic Press, Orlando, FL.
21. McCann, P. H. 1969. The Effects of Ambient Noise on Vigilance Performance. *Human Factors,* 11, pp. 251–256.
22. Hockey, G. 1978. Effects of Noise on Human Work Efficiency. In *Handbook of Noise Assessment.* Van Nostrand Reinhold, New York.
23. Eschenbrenner, A. J., Jr. 1971. Effects of Intermittent Noise on the Performance of a Complex Psychomotor Task. *Human Factors,* 13(1), pp. 59–63.
24. Jones, D. M., Smith, A. P., and Broadbent, D. E. 1979. Effects of Moderate Intensity Noise on the Bakan Vigilance Task. *Journal of Applied Psychology,* 64, pp. 627–634.
25. Davies, D. R., and Jones, D. M. 1984. Effects of Noise on Performance. In *The Noise Handbook.* Tempest, W. (Ed.). Academic Press, London.
26. Broadbent, D. 1976. Noise and the Details of Experiments: A Reply to Poulton. *Applied Ergonomics,* 7, pp. 231–235.
27. Kerr, W. A. 1950. Accident Proneness and Factory Departments. *Journal of Applied Psychology* 34, pp. 167–170.
28. Cohen, A. 1976. The Influence of a Company Hearing Conservation Program on Extra-auditory Problems in Workers. *Journal of Safety Research,* 8, pp. 146–162.
29. Jones, D. M., and Broadbent, D. E. 1987. Noise. In *Handbook of Human Factors Engineering.* Salvendy, G. (Ed.). Wiley, New York.
30. Dupuis, H. 1974. Mechanische Schwingungen, sowie: Messung und Bewertung von Schwingungen und Stossen. In *Ergonomie,* Vol. 2. Schmidtke, H. (Ed.). Carl Hanser Verlag, Munich.
31. Radke, A. O. 1957. Vehicle Vibration: Man's New Environment. *Paper 57-A-54.* American Society of Mechanical Engineers, December 3.
32. Rasmussen, G. 1982. Human Body Vibration Exposure and Its Measurement, *Technical Review.* Bruel and Kjaer, Vol. 1, pp. 3–31.
33. Coermann, R. 1963. The Mechanical Impedance of the Human Body in Sitting and Standing Position at Low Frequencies. In *Human Vibration Research.* Lippert, S. (Ed.). Pergamon Press, Elmsford, NY.
34. International Organization for Standardization. 1974. Guide for the Evaluation of Human Exposure to Whole Body Vibration, *ISO 2631.* ISO.
35. Chaffin, D. B., and Andersson, G. 1984. *Occupational Biomechanics.* Wiley, New York.

36. Bobick, T. G., Gallagher, S., and Unger, R. L. 1989. Effects of Random Whole-Body Vibration on Back Strength and Back Endurance. In *Advances in Industrial Ergonomics and Safety I*. Mital, A. (Ed.). Taylor & Francis, London, pp. 537–545.

37. Bobick, T. G., Gallagher, S., and Unger, R. L. 1989. Pilot Subjects Evaluation of Whole-Body Vibration from an Underground Mine Haulage Vehicle. In *Advances in Industrial Ergonomics and Safety I*. Mital, A. (Ed.). Taylor & Francis, London, pp. 521–528.

38. Hutton, S. G., Paris, N. and Brubaker, R. 1993. The Vibration Characteristics of Chain Saws and Their Influence on Vibration White Finger Disease. *Ergonomics*, 36(8), pp. 911–926.

39. Poulton, E. C. 1978. Increased Vigilance with Vertical Vibration at 5 Hz: An Alerting Mechanism. *Applied Ergonomics,* 9, pp. 73–76.

40. Poulton, E. C. 1972. *Environment and Human Efficiency*. Charles C Thomas, Springfield, IL.

41. Notbohm, G., and Gros, E. 1988. Combined Effects of Whole-Body Vibration and Noise on Performance in a Multiple-Reaction-Time Task. In *Trends in Ergonomics/Human Factors V*. Aghazadeh, F. (Ed.). Elsevier, Amsterdam, pp. 505–512.

42. American Society of Heating, Refrigerating, and Air-Conditioning Engineers. 1977. *ASHRAE Handbook and Product Directory*. ASHRAE, Atlanta, GA.

43. Fanger, P. O. 1970. *Thermal Comfort, Analyses and Applications in Environmental Engineering*. Danish Technical Press, Copenhagen.

44. Sprague, C., and Munson, D. 1974. A Composite Ensemble Method for Estimating Thermal Insulating Values of Clothing. *ASHRAE Transactions,* 80(1), pp. 120–129.

45. Parsons, K. C. 1995. International Heat Stress Standards: A Review. *Ergonomics*, 38(1), pp. 6–22.

46. Rohles, F. H., and Konz, S. A. 1987. Climate. In *Handbook of Human Factors*. Salvendy, G. (Ed.). Wiley, New York.

47. Occupational Safety and Health Administration. 1974. *Recommendation for a Standard for Work in Hot Environments*. OSHA, Washington, DC.

48. NIOSH. 1972. Criteria for a Recommended Standard Occupational Exposure to Hot Environments. *HSM 72-10269*. NIOSH, Washington, DC.

49. NIOSH. 1986. Criteria for a Recommended Standard Occupational Exposure to Hot Environments. Revised Criteria 1986. *HSM, DDHS (NIOSH) 86-113*. NIOSH, Washington, DC.

50. Botsford, J. H. 1971. A Wet Globe Thermometer for Environmental Heat Measurement. *American Industrial Hygiene Journal*, 32, pp. 1–10.

51. Beshir, M. Y., Ramsey, J. D., and Burford, C. L. 1982. Threshold Values for the Botsball: A Field Study of Occupational Heat. *Ergonomics*, 25(3), pp. 247–254.

52. Belding, H. S., and Hatch, T. F. 1955. Index for Evaluating Heat Stress in Terms of Resulting Physiological Strains. *Heating, Piping and Air Conditioning*, August, pp. 129–136.

53. American Society of Heating, Refrigerating, and Air-Conditioning Engineers. 1985. ASHRAE *Handbook, 1985 Fundamentals*. ASHRAE, Atlanta, GA.

54. Constable, S. H., Bishop, P. A., Nunnely, S. A., and Chen, T. 1994. Intermittent Microclimate Cooling During Rest Increases Work Capacity and Reduces Heat Stress. *Ergonomics*, 37(2), pp. 277–285.

55. Mackworth, N. H. 1961. Research on the Measurement of Human Performance. In *Selected Papers on Human Factors in the Design and Use of Control Systems*. Sinaiko, H. W. (Ed.). Dover, New York, pp. 174–331.

56. Wilkinson, R. T., Fox, R. H., Goldsmith, R., Hampton, I. F. G., and Lewis, H. E. 1964. Psychological and Physiological Responses to Raised Body Temperature. *Journal of Applied Physiology* 19, pp. 127–291.

57. Ramsey, J. D. 1995. Task Performance in Heat: A Review. *Ergonomics*, 38(1), pp. 154–165.
58. Wyon, D. P. 1996. Effects of Moderate Heat Stress on Driver Vigilance in a Moving Vehicle. *Ergonomics*, 39(1), pp. 61–75.
59. Fox, W. 1967. Human Performance in the Cold. *Human Factors,* 9(3), pp. 203–220.
60. Sundheim, N., and Konz, S. 1990. Keeping Bare Hands Warm with Extra Clothing on the Body. In *Advances in Industrial Ergonomics and Safety II.* Das, B. (Ed.). Taylor & Francis, London, pp. 927–932.
61. Bell, P. A. 1978. Effects of Noise and Heat Stress on Primary and Subsidiary Task Performance. *Human Factors,* 20, pp. 749–752.
62. Dean, R. D., and McGlothlen, C. L. 1965. Effects of Combined Heat and Noise on Human Performance, Physiology, and Subjective Estimates of Comfort and Performance. *Proceedings of the Annual Technical Meeting.* Institute of Environmental Science, New York, pp. 55–64.
63. Van Orden, K. F., Benoit, S. L., And Osga, G. A. 1996. Effects of Cold Air Stress on the Performance of a Command and Control Task. *Human Factors*, 38(1), pp. 130–141.
64. Ramsey, J. D. 1990. Working Safely in Hot Environments. In *Advances in Industrial Ergonomics and Safety II.* Das, B. (Ed.). Taylor & Francis, London, pp. 889–896.

CHAPTER 10

Selection and Design of Displays and Controls

This chapter opens a two-unit section that deals with design of equipment and information aids. Chapter 10 covers a classic ergonomics focus area, design and selection of displays and controls, and Chapter 11 dwells on the design of consumer products and information aids, such as forms, labels, and video displays. These two chapters provide information that supports the earlier chapters on work systems design.

10.1 INTRODUCTION

The purpose of this chapter is to present design and selection considerations for commonly used displays and controls. The discussion that follows is based largely on the foundation established in earlier chapters. Displays and controls are the main interface points of the worker with machines and other equipment. Their ergonomic design assures effective use by human beings. In addition, errors in use are minimized.

A display must present timely data to a user. The data presented must be in an immediately usable form. Assuming that all other use conditions (visual functions, physical environment, task variables) are favorable, the user perceives and processes the data presented.

An action may or may not follow. If an action is required, it may take the form of activating/deactivating or adjusting a control to a desired setting. Hence a control device, such as a pushbutton or knob, accepts human output as its input and generates energy as input to a machine or equipment.

We begin our discussion with display design. The objective is to have the worker acquire proper information at the right time with no confounding or confusion. Ergonomic design of displays helps this process to a great extent.

10.2 DISPLAYS

A display is a human-made means of presenting information. A computer screen is a display, a tachometer on a machine is a display, a page in a book is a display. It is the interaction of people with displays that is of most interest to the ergonomist.

People acquire information from their environment either directly (looking at a road, hearing the humming of a machine) or indirectly [1]. Indirect sensing is the type relevant to display need, and hence design. In this case the person is presented with information that cannot be detected by the sensory mechanisms. Some transformation must take place. An example is the fluid temperature in a plating tank. The eyes or ears cannot sense temperature, and due to the chemical and burning hazard, the skin senses are not used. Therefore, sensing probes placed in the tank detect the temperature and after transducing the level, the exact temperature may be displayed continuously on a display mounted on an instrument panel. Naturally, there must be a reason for collecting and displaying these data. In this case, plating is effective only within a certain range of solvent temperature, which must be monitored by the worker.

10.3 TYPES OF DISPLAYS

Displays present data to a user to be converted into information. The data may be static or dynamic. An example of a static display is a page in a machine manual. The content of the page is not expected to change within the foreseeable future. Hence such displays present data that are not expected to change over time. Although the manual may be updated from time to time, the content does not change during a particular issue of the manual. On the other hand, a dynamic display presents data that are expected to change over time. The indicator of solvent temperature is a good example. Figure 10.1 presents the breakdown for visual displays with additional details. Both display types need close scrutiny of the ergonomist during the design stage. However, due to their complexity (involving more design parameters), dynamic displays present more challenges.

10.4 DETECTABILITY OF SIGNALS

The prime function of a display is presentation of data in a clear (unambiguous) format that is detectable by human beings. Detectability is the first consideration since the idea behind

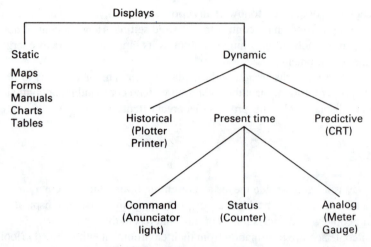

Figure 10.1 Types of visual displays.

using a display is to make otherwise undetectable data available to the person. Conditions that make it easier to detect a signal are:

1. *Increased duration of signal.* Human sensory systems do not respond immediately to signals. Patterson [2] showed that the minimum signal duration for short-burst flight deck warning applications should be 100 ms with 25-ms rise and 25-ms fall times. Munson [3] found that pure auditory signals "build up" in terms of detectability in 200 to 300 ms. Naturally, in normal-use conditions, signal duration should be increased many times to allow the user to carry out multiple activities.

2. *Enhanced signal-to-noise ratio.* As we discussed earlier in the section on signal detection theory, the higher the signal-to-noise ratio, the more detectable the signal is. Deatherage [4] claims that a good rule of thumb for auditory displays is that the signal intensity at ear entrance should be halfway between the masked threshold of the signal in noise and 110 dB(A).

3. *Multichannel presentation of data.* In addition to improving detectability, presentation of information to multiple channels (visual, auditory, tactual, etc.) enhances processing of information in background noise.

4. *Multichannel monitoring of data.* Perception is more efficient when data are monitored by multiple input channels. A visual image of a newscaster is more effective when it is accompanied by sound than otherwise. Multiple-channel communication is also more natural for people.

5. *Optimal signal presentation rate (neither too slow nor too fast).* The inverted U relationship between performance and stress backs this hypothesis. Naturally, the best rate of signal presentation depends largely on the task demands and other existing stress.

Table 10.1 lists some other methods of increasing signal detectability, with emphasis on human detection performance [5].

TABLE 10.1 FACTORS THAT ENHANCE
SIGNAL DETECTABILITY

1. Artificial signals to be responded to
2. Two operators monitoring
3. Task duration only 20 minutes
4. Rest periods between monitoring intervals
5. Only one monitoring task
6. Feedback to operator
7. Fresh operator

10.5 VISUAL DISPLAYS

Once a decision is made as to which sense modality to present the signals and the coding
dimensions have been determined (see Chapter 4), the next step is either to design a display
or to use one that already exists, to achieve the information transfer function. Whether a
new display is being designed or one that exists is being purchased, before actual production use, the engineer must exercise several ergonomic judgments. This section is devoted
to such considerations along several display categories. The purpose is to make the display
readable, understandable, signals detectable, and various design parameters compatible
with learned expectations and research results. We first present design considerations along
the lines of qualitative and quantitative displays. Then a section on electronic data display
follows. Then suggestions with respect to display installation and location are presented.
Finally, cognitive factors are analyzed.

10.5.1 Quantitative Displays

Quantitative displays present quantitative information on some rather frequently changing
variable or on a static variable for application purposes later. There is a requirement that
such devices display data with a certain precision [1]. Therefore, each quantitative display
can be read up to a certain level of accuracy with interpolation following that.

   Three generic types of quantitative displays are:

1. A counter or digital display
2. A moving scale with fixed indicator
3. A moving indicator with fixed scale

   Figure 10.2 gives symbolic descriptions of these displays. Various forms of each can
be observed, depending on such factors as the manufacturer, space availability, and economics. Table 10.2 presents information as to various characteristics of each. For accurate
reading, counters outperform other types. Grether and Baker [6] showed that scales are less
efficient in displaying quantitative information. On the other hand, for relative deviation
from a certain value, a pointer–scale combination is better. Sinclair [7], after a review of
literature, concluded that in general, analog displays with moving pointers and fixed scales
are superior to those with fixed pointers and moving scales. Such displays are also better

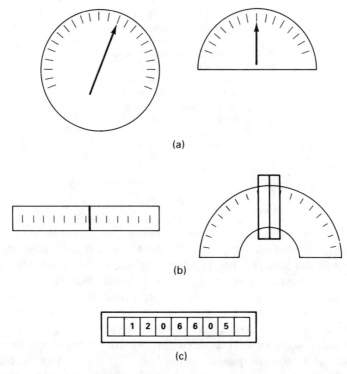

Figure 10.2 Various quantitative displays: (a) fixed scale, moving pointer (b) fixed pointer, moving scale; (c) counter.

with respect to compatibility between their movement and the associated control element's movement (see Section 10.9.3). Specifically, if a pointer moves rather than a scale, there is less confusion of the movement relationship between the control element and the display indicator. However, the disadvantage is in space and illumination requirement. Displays with fixed pointers (markers) and moving scales require less panel space. The illumination requirement on such devices is only on the portion displayed. Furthermore, a larger variable range can be used with such displays.

Scale design. The specific features of quantitative displays to be discriminated affect the effectiveness of visual discriminations. In this respect, scale design emerges to

TABLE 10.2 CHARACTERISTICS OF QUANTITATIVE DISPLAYS

	Counter	Moving pointer, fixed scale	Moving scale, fixed pointer
Quantitative reading	Good	Medium	Medium
Qualitative reading	Poor	Good	Medium
Illumination requirement	Good	Poor	Good
Check reading	Medium	Good	Medium
Deviation from a reference	Poor	Good	Poor

be the most important step to be considered. Scale design itself includes the design of many parameters, including the pointer (marker), numerical progression, feature size, and the like. These are discussed next.

1. *Numerical progression.* Quantitative scales list numerical values for each major scale marker. The specific progression of these defines the ease of use for a worker. Certain progression schemes are easier to use than others. Progressions by ones, tens, and fives are easiest to use, in that order. Decimals on scales are not recommended. Graduation intervals in fours, threes, and sixes are the worst.

2. *Design of pointers.* The most important variable here is the distance between the pointer and the scale surface. Having the pointer close to the scale surface minimizes visual parallax. Pointers should have a tip angle of 20 degrees. Whitehurst [8] recommends that pointers should meet but not overlap the smallest graduation mark on the scale surface. A uniform color throughout the pointer is also recommended.

3. *Scale markers.* The desired reading precision as well as the range of the parameter values to be displayed defines the number of scale markers to be included in the design. Naturally, if this number is very large, a fixed pointer, moving scale may be considered. Scale markers should not be crowded on the display surface. Figure 10.3 gives some suggestions for viewing conditions under normal illumination. For low illumination, a multiplier of 2.5 must be applied to the width of markers.

4. *Scale length.* As discussed previously, this parameter is determined largely by the variable range (e.g., temperature range for the solvent) to be displayed and the design characteristics of each scale marker.

5. *Size of scale markers.* The marker size recommendations given by Figure 10.3 are for a viewing distance of 71 cm (28 in.). These must be adjusted for other viewing distances by the factor "viewing distance [cm (in.)]/71 (28)." Hence a linear proportioning of the suggested size is appropriate.

Alphanumeric display. On many occasions, alphanumeric data must be displayed to the operator (instructions, etc.). These may be integrated into a static display (labels, maps,

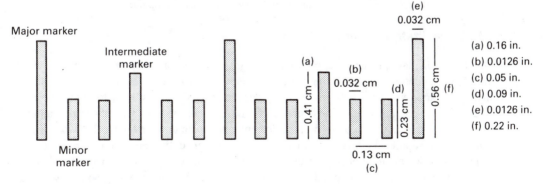

Figure 10.3 Quantitative display scale marker design guidelines. (Adapted from Ref. [1] and [6].)

etc.). Several characteristics of such data lend themselves to more effective use by humans:

1. *Character height.* Peters and Adams [9] recommend the following relationship between the character height and variables that affect it:

$$H \text{ (character height, cm)} = 0.000866D + K_1 + K_2$$

$$H \text{ (character height, in.)} = 0.0022D + K_1 + K_2$$

where,

D is the viewing distance in cm (in).

K_1 is a correction factor for illumination and viewing conditions

K_2 is a correction factor for importance of the message conveyed by the display (see Table 10.3 for details).

2. *Character orientation.* Orient numbers and characters in the upright position.

3. *Character width/height ratio.* For numerals, the recommended ratio is 3:5. For capital letters, the recommendation ranges between 1:1 and 3:5.

4. *Stroke width/height ratio.* This is the ratio of the thickness of stroke to the height of characters. For black characters on a white background, the recommended range is 1:6 to 1:8 [1]. Berger [10,11] suggests that for white characters on a black background, the range should be 1:8 to 1:10.

Legibility.   *Legibility* refers to discriminability of the details of a display. Legibility may be enhanced by optimal contrast between characters and the background, using type fonts that are easier to see, minimizing glare from the display surface, minimizing visual parallax, optimizing character dimensions with respect to the viewing angle, and optimal design of the physical elements of a display.

Readability.   Readability of a display goes beyond legibility. Legibility of data is the minimum requirement for an effective display. Helander [12] lists other factors that enhance the readability (comprehension) of a display, such as relevancy, location, prominence, emphasis on important words or phrases, simplicity, and clarity. While legibility drives for distinguishability of details, readability aims at understandability of contents.

10.5.2 Qualitative Displays

Qualitative displays present an approximate reading on a variable of interest. Normally, the variable is a continuously changing one; however, the user is not interested in exact readings. Only approximate values or ranges are desired. An example is the working temperature of a production machine. The operator may have to take action only if the temperature exceeds a certain value. Furthermore, the designer may wish to warn the worker when this range is being approached. Hence a display may be designed that shows three temperature ranges: cold, warm, and hot (see Figure 10.4). If a mechanical indicator is to be used as a qualitative display, a moving pointer, fixed scale is best. This will provide the user with a "feel" for deviation from normal benchmarks.

TABLE 10.3 PARAMETER VALUES FOR CHARACTER SIZE

	Above 1.0 fc/ favorable reading	Above 1.0 fc/unfavorable reading Below 1.0 fc/favorable reading	Below 1.0 fc/ unfavorable reading
Nonimportant markings ($K_2 = 0$)	$K_1 = 0.15$ cm (0.06 in.)	$K_1 = 0.4$ cm (0.16 in.)	$K_1 = 0.66$ cm (0.26 in.)
Important markings [$K_2 = 0.19$ cm (0.075 in.)]			

Source: Adapted from Refs. [1] and [9].

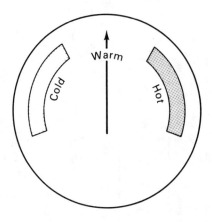

Figure 10.4   Qualitative display.

Sometimes, displays that are primarily quantitative in nature are used for check reading. In this case a logical arrangement of such displays may help the perceptual needs of the worker. Figure 10.5 gives one such arrangement, called the *extended line design*. The case of check reading borrows from the *gestalt* principles [13,14]. *Gestalt* implies wholeness. Several gestalt principles have been stated by the gestalt psychologists. These include *continuity, good closure, common fate,* and *proximity*. As stated in Chapter 4, humans possess a tendency to link isolated sensory inputs to lead to wholeness. The gestalt principles define the elements of wholeness. Hence things that are close to each other are perceived to belong to the same group (see Figure 10.6). In the case of check reading, the principle of good continuity guides the perception.

A group of qualitative displays is known as *status indicators*. These are units that present two-state information concerning the status of a system, such as go/no-go, on/off, working/idle, and normal/abnormal. All applicable display design principles must be carefully exercised during their design. A special class of status indicators is *warning lights*. Their ergonomic design assures that the data presented will be correctly recognized and interpreted in a minimum amount of time. Several design parameters for warning lights are:

1. *Color.* Utilize color coding. Reynolds et al. [15] recommend the following colors in the order of eliciting fast human response: red, green, yellow, and white.

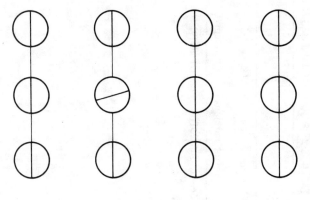

Figure 10.5   Display arrangement for check reading.

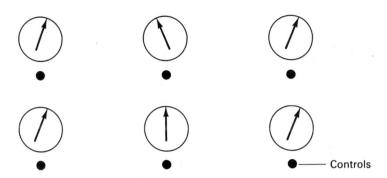

Figure 10.6 Law of proximity in action.

2. *Flash rate.* A warning light should get the attention of an operator immediately. In this respect, a flashing light is more effective than one that is steady. Several researchers investigated the best flash rates. The general consensus seems to focus on the range between 3 and 10. Heglin [16] suggests a flash rate of 4 per second as optimum.

3. *Brightness.* A warning light should get attention without glare. Brightness may be adjustable if the external lighting is expected to be dimmed.

4. *Multichannel presentation.* For high urgency of viewing under visual noise, an auditory signal may be added to the visual signal for extra attention-getting capability.

5. *Contrast with background.* Good brightness contrast and good color contrast help the attention-getting characteristics of a warning light.

Nowadays, most American and European passenger cars are equipped with a center high-mounted stop lamp (CHMSL) to improve the rear lighting and signaling systems. This is a direct result of research that showed such an arrangement may reduce rear-end collisions by promoting faster perception of relevant information by the driver of the trailing vehicle [17]. This was shown to be true by accident statistics on U.S. highways [18].

10.5.3 Symbolic Displays

Symbols may be effective in transmitting information to a user. Since there is no written message in symbols, such displays require minimum time for comprehension. Kline and Fuchs [19] showed that the average distance at which standard symbolic signs could be identified was about two times that of text signs for young, middle-aged, and elderly observers. Another advantage of symbols is that they do not have to be translated for users from different countries. Symbols are very popular in displaying warning messages and are widely used in traffic and public displays.

In addition to simplicity, a symbol should be appropriate for its referent (what it stands for). The method of measuring the appropriateness of a symbol is by measuring errors in comprehension. Zwaga and Easterby [20] developed a standard for evaluating symbols submitted for ISO's (International Standards Organization) acceptance. According to this standard, every symbol to be approved by ISO as an international symbol must go

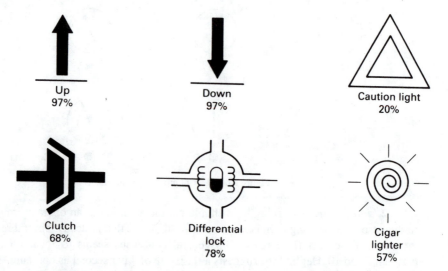

Up
97%

Down
97%

Caution light
20%

Clutch
68%

Differential
lock
78%

Cigar
lighter
57%

Figure 10.7 Sample construction machine symbols with percent correct comprehension. (Adapted from Ref. [12].)

through testing in six countries, with the average comprehension (correct identification of referent) being at least 66%. Helander [12] lists several machine symbols in Figure 10.7 considered by the Society of Automotive Engineers, with the corresponding percent correct responses.

10.5.4 Electronic Display Design

Recently, many electronic methods of information display have been introduced. LEDs (light-emitting diode), dot matrix character generators, and liquid crystals are examples. These save notable panel space; however, the ergonomist must be cognizant of their potential for reduction in legibility. Examples of such displays are segmented numbers such as those found in calculator displays and gas pump displays.

Several design suggestions for electronic displays are [5,21,22]:

1. For most accuracy, use a dot matrix character style of 7×7 or 7×9. Gould [25] found that legibility is better with a larger matrix size.

2. Lines describing characters should be sharp, not diffused, and with equal brightness throughout. Korge and Krueger [26] observed a shift of accommodation toward the resting position due to blurred characters. Gomer and Bish [27] detected stronger and more clearly defined evoked potentials in the brain due to images of higher resolution than with images of lower resolution.

3. For changing data, the character image should not persist too long. Otherwise, images will be blurred and confused.

4. Choose vertical characters. Plath's [28] research showed that slanted numerals lead to more identification errors than do vertical numerals.

5. Dot spacing should be between 0.4 and 0.6 mm (0.0157 to 0.0236 in.). This offers best character segment continuity.

6. Distance between characters should be between 1.1 to 1.4 times the stroke width. This helps continuity between characters with minimum intercharacter confusion.

7. Minimize reflections on the screen. Reflections cause distraction. Details may also be lost due to reflections.

8. Character contrast with the background should be adjustable. This helps visual effectiveness under varying external illumination levels.

10.5.5 Video Display Terminals

Nowadays, many factory processes are being controlled by computers and automatic identification equipment, such as bar-code scanners, radio-frequency devices, and transponders. Successful marriage of the two may provide real-time shop floor control via data displayed through video display terminals (VDTs), printers, or closed-loop automatic control. Office environments are also being equipped with word processors and other computer peripherals to enhance white-collar productivity. All such equipment requires efficiently functioning human–machine systems.

Proper design of the physical characteristics of video display terminals makes a significant contribution to the effectiveness of the human–machine interface. Table 10.4 gives specific characteristics of acceptable designs. Field surveys of VDT tasks show that many attributes of the hardware may be important in defining comfort in use. After studying such tasks at five organizations, Stammerjohn et al. [29] concluded that a number of factors related to the design of the video display terminals are bothersome, including screen readability, reflected glare, flicker, and screen brightness. The authors observed a significant positive correlation between employees' rating of screen flicker, reflected glare, and visual function complaints. Smith et al. [30] experienced similar observations during a survey conducted in a large newspaper company. Work-associated symptoms were significantly related to poor visual clarity and readability of the VDT screens. Dainoff et al. [31], working with two VDT models, observed more complaints with the old model than with the new model. The new model displayed a number of improvements in physical design. In a survey of data entry, CAD, payment, and drawing tasks using VDTs, Laubli et al. [32] found that high luminance contrasts between screen and the environment were associated with increased visual discomfort. Furthermore, increased oscillating luminance of the screens was linked to higher incidences of eye irritation requiring use of eyedrops.

Laboratory studies along these lines do not show as clear effects as the field studies do. Hedman and Briem [33] did not observe any notable changes in visual functions (accommodation, near-point convergence, etc.) due to VDT work compared with a control group. Gyr et al. [34] confirmed that a 3-hour reading task did not lead to any differences in visual functions whether a VDT or a normal printed text was used. Gould and Grischkowski [35] observed 20 to 30% faster proofreading performance with hard-copy material as opposed to VDTs with good photometric qualities. A similar effect was also observed by Kruk and Mutter [36]. In this experiment, subjects read continuous text from either printed material or from a television screen (videotext). The videotext subjects read 28.5%

TABLE 10.4 RECOMMENDED VDT CHARACTERISTICS

Hardware Characteristics

1. Resolution is best with 9 to 10 scan lines per character. This assures good image quality.
2. Choose a character color from the middle of the chromaticity diagram. Orange and green characters are recommended.
3. Luminance of the screen should be greater than 25 mL (millilambert).
4. Character regeneration rate of 60 cycles per second or more is desired.
5. The visual angle subtended at the eye of each character must be between 15 and 20 minutes of arc. At a viewing distance of 71 cm (28 in.), the minimum character height is 4.6 mm (0.18 in.). The visual angle in terms of minutes of arc can be calculated as follows:

$$\text{visual angle (min)} = \frac{(57.3)(60)L}{D}$$

Where L is the size of object measured perpendicular to the NLS (normal line of sight) and D is the distance from the eye to the object.
6. Surfaces adjacent to the scope should have a matte finish.
7. Scopes may be hooded for minimum indirect glare.
8. A minimum contrast of 88% with the background is recommended.
9. Use uppercase letters for headings and titles. Lowercase characters may be used for other text.
10. Scan lines should not be visible to the operator.
11. Optimum dot size seems to be between 0.8 and 1.2 mm (0.0315 to 0.047 in.). Dots should be large enough and spaced so that they fuse together.
12. A dot matrix size of 7×9 or larger is recommended.
13. The width/height ratio of characters should be between 0.7 and 0.8.
14. Characters should be equally focused and well defined at every point on the screen.
15. The cursor should be easily visible and should not affect the legibility of characters.
16. The cursor should blink at a rate of about 2 to 3 Hz.
17. Cursor control should be easily accomplished by either the right or left hand.
18. If an alarm is to be used to alert the operator, both auditory and visual alarms should be used. The auditory signal must be more than 200 Hz in frequency and less than 1000 Hz. A volume control is recommended. However, it should not shut off the alarm entirely.

Other

1. Orient VDTs to minimize indirect glare.
2. The CRT surface and the NLS should not make an angle greater than 30 degrees.

Source: Adapted from Refs. [23][and [24].

more slowly than the book subjects. A follow-up experiment with single and double spacing between text revealed a marked increase in reading speed with large interline spaces. In hopes of identifying the cause of reading speed difference between CRTs and printed material, Gould et al. [37] studied 10 variables, including experience with CRT, display orientation, character size, font, and polarity. They could not identify a single variable explaining the difference. The authors' tentative conclusion was that the differences are due to a combination of variables, centering on the image quality of characters, which was later shown to be true.

10.5.6 Other Considerations in Visual Display Design and Use

Many other factors affect the effectiveness of visual display use. One of these is the location of the display in the visual field. As discussed in Chapter 4, the worker's visual field must be taken into consideration in locating displays. Furthermore, normal line of sight and optimum locations for primary, secondary, and auxiliary displays come into play (see Chapter 7), which also determine desired locations for displays.

Another factor to be considered is display orientation. The need to present data as close to 90 degrees to the NLS as possible requires that displays be angled with respect to expected head and eye orientation. Naturally, many trade-offs are considered in obtaining best designs. The seated operator's workspace presents more challenges in this respect.

As discussed earlier, the message presented by a display must be understandable and require minimum reading time. Symbols may be effective in transmitting data to the user. Traffic signs have been designed on this basis for many years. Many office and production machines are also being equipped with symbols. These symbols must be relevant to their referents and present minimum confusion probability [20]. People judge change more accurately and quickly with line and bar graphs. Proportion is best judged by pie charts [38].

Recently, Wickens and Carswell [39] showed that the concept of proximity compatibility principle (PCP) can be applied to display design, including electronic displays. According to this concept, displays relevant to a common task or mental operation should be located close together in perceptual space.

Other visual display design considerations include [5]:

1. Avoid shadows on display surfaces. Provide adequate illumination on the display for reading.
2. Label displays clearly. Labels are especially useful during training and personnel changes.
3. Remove unused displays. These present clutter and space problems.
4. A circular or semicircular scale is better than vertical or horizontal scaled displays unless space is a problem.
5. Scale numbering should increase in a clockwise direction on a circular or semicircular scale, upward on a vertical, and to the right on a horizontal scale.
6. Multiscale, multipointer, nonlinear scale displays should be avoided [40].

10.6 AUDITORY DISPLAYS

Auditory displays are used to present data to the hearing sense. Several examples of auditory displays are horns, buzzers, alarms, and sirens. As in visual display design, many considerations with respect to the human auditory mechanisms drive the design of auditory displays. Concepts such as detectability, discriminability, and identification apply to such displays, also. Sorkin [41] provides several recommendations relative to the level, interference, and spectral considerations in designing auditory displays, as listed in Table 10.5.

TABLE 10.5 AUDITORY DISPLAY DESIGN RECOMMENDATIONS

1. Increase signal-to-noise ratio for detectability. Signal levels of 8 to 12 dB above the masked threshold will provide full detectability.
2. For maximum detectability, the signal duration should be at least 300 ms. If the signal has to be shorter than this, the intensity should be increased.
3. Rapid response to a signal can be achieved by levels 15 to 18 dB above the masked threshold.
4. To minimize operator annoyance, the signal level must be less than 30 dB above the masked threshold.
5. Interference is at its maximum when the signal tone is near the frequency of the interfering tone.
6. Interference spreads to additional frequencies as the intensity of interfering tone increases.
7. When it is below the signal frequency, the effect of the interfering tone is maximum, as opposed to above.
8. For absolute discriminations, keep the signal frequency at 1000 to 4000 Hz.
9. For auditory coding, use five intensity levels or less.
10. Four to seven levels should not be exceeded for frequency coding.
11. Four prominent frequency components are recommended for a signal to minimize masking effects and maximize the number of different distinct signal codes that can be generated.
12. At least four of the first 10 harmonics should be prominent. The prominent frequency components for signals should be in the range 1000 to 4000 Hz.
13. To acknowledge warning, provide the signal with a manual shutoff.

Source: Adapted from Ref. [41].

The recommendations in the table are all based on research results. For example, in a simulated driving task, Fidell [42] observed that the effectiveness of signals was a function of their detectability. Signals about 15 to 16 dB(A) above the masked threshold were detected by subjects almost consistently. The *masked threshold* of a signal is the level required for 75% correct detection when presented to an observer in a two-interval task (one interval contains noise only, and one has noise plus signal, with random occurrence of signal in one of the two intervals). Patterson [2] studied cockpit signals in terms of their annoyance. Even with notable aircraft noise, Patterson recommended a level of 30 dB(A) above masked threshold to be the upper limit for auditory signals.

Complexity of signals has also been studied. In a NASA report, Cooper [43] suggested a maximum of four or five signal levels for effective discrimination. Deaterage [4] also recommends a range between 4 and 7 for intensity and frequency coding of auditory signals.

Auditory signals are especially useful as warning signals or alarms. Several design recommendations for warning signals are as follows [41]:

1. Warning sounds should have distinct temporal and spectral patterns.
2. Warning signals can be reinforced by speech output.
3. Keep signal frequency to within the range 150 to 1000 Hz.
4. Warning signals should use harmonically regular frequency components rather than inharmonic spectra. Lower-priority signals should have most of their energy in the first five harmonics. Higher-priority signals demanding immediate action should have more energy in harmonics 1 through 10 than in others.
5. A small number of inharmonic components may make a high-priority signal very distinctive.
6. Use a modulated signal (1 to 3 times/sec) or beep sounds (1 to 8 beeps/sec).

Perceived urgency of the warning signal can be manipulated by various acoustic and temporal parameters, such as speed and harmonic content, to match the urgency of the relevant situation [44,45].

Other considerations in auditory display design include [4,5]:

1. Establish compatibility between the signal and its referent (wailing signal for emergency, increasing pitch for increasing referent values). The same signal should designate the same information at all times.
2. Avoid extremes of auditory dimensions. Where feasible, avoid steady-state signals.
3. If the distance to the listener is great (over 330 m or 1000 ft), use high intensity and frequencies below 1000 Hz.
4. If sound must bend around obstacles and pass through partitions, use frequencies below 500 Hz.

10.7 SPEECH INTELLIGIBILITY

An area that is related to auditory displays is speech intelligibility. Nowadays, many auditory displays are using speech output to get more attention from the operator. Hence ergonomic aspects of speech output gain importance. *Speech intelligibility* is the ability to recognize spoken sounds correctly. Speech is defined by a nonscientist as a combination of vowels and consonants. Speech scientists use a smaller unit called a *phoneme*. This is the smallest sound unit that may change the meaning of a word. A collection of phonemes make syllables, words, and sentences. Spoken speech generates about 12 phonemes per second. Liberman et al. [44] showed that human beings can understand speech at phoneme rates of 30 to 35 per second. Two characteristics that help the ergonomist evaluate sound effects also aid in speech intelligibility evaluations. Frequencies help evaluate phonemes and, consequently, words and sentences, and intensities determine signal-to-noise ratios.

The everyday speech spectrum spans an approximate range of 100 to 8000 Hz. Speech spectrograms are especially useful in determining the frequencies generated by any utterance. Vowels emphasize frequencies around 800, 2200, 3000, and 4200 Hz. These correspond to the resonant frequencies of the vocal cavities (pharynx, oral cavity, and nasal cavity). The problem is in the consonants. There is no simple relationship between consonant phonemes and associated spectrographic patterns.

A related concept in understanding speech is the signal-to-noise (S/N) ratio. This ratio is computed simply by subtracting the intensity (dB) of noise from the intensity of the speech signal. For example, if the average intensity of speech is 65 dB and average intensity of noise is 55 dB, the S/N ratio is + 10. If the numbers were reversed for the signal and the noise, the S/N ratio would be − 10.

Designers of speech communication systems evaluate a speech channel (telephone, radio) on the basis of intelligibility. French and Steinberg [45] developed an intelligibility calculation method called the *articulation index* (AI). This index has also been standardized by the American National Standards Institute. The AI focuses on the most important frequency region for speech communication (200 to 5000 Hz). It is a measure of the signal-to-noise ratio within this region. The more that the speech spectrum differs from that of noise, the larger the resulting AI. Maximum AI can be 1.0 and minimum 0. For a given communication channel, AI may assume any value between these two extremes.

Simplified methods of calculating AI segment the region 200 to 5000 Hz into several (between 10 and 20) bands and assign weights to these bands based on relative contribution of each to intelligibility under noisy conditions. Then within each band, the dB difference between peak speech level and the noise level is calculated. A zero value is assigned to the difference if the noise intensity is greater than the signal's. The resulting decibel differences are then multiplied by the weighting factors and summed across all bands. The resulting value is the AI. An AI value of 0.3 is quite unsatisfactory. A value of 0.5 may be acceptable for limited vocabulary communication. Sentence comprehension is excellent at an AI of 0.7 or above.

Several variables help speech intelligibility. Familiarity with the message, predictability of the contents, enhanced S/N ratio, binaural (two ears) listening, and grammatically correct messages are more understandable than otherwise. Designers must take these into consideration when developing communication systems.

10.8 TACTUAL DISPLAYS

Tactual displays are used as warning devices and used extensively for blind persons' perceptual needs. However, touch senses seem to be suitable for transmitting only a limited number of discrete stimuli. Mechanical vibrations, thermal energy, and electrical impulses may be transmitted as warning signals to attract attention. If mechanical vibrations are transmitted, an amplitude of 0.0004 cm (0.00016 in.) is sufficient. It is not recommended to apply continuous pressure to the skin senses since they are highly adaptable. The "let-go" threshold for electrical impulse is 16 mA. Hence a voltage of 10 to 12 mA will definitely attract attention without startling the worker. If thermal energy is to be applied, the pain threshold must be known. As skin temperature decreases, the amount of thermal energy to be applied increases. For example, the pain threshold corresponding to a skin temperature of 20° Celsius (68°F) is about 420 mcal/s-cm^2 (0.155 in.^{2}). This value drops to about one-fourth at a skin temperature of 40° Celsius (104°F).

Tactual displays have also been used as a substitute for seeing. The most widely known tactual display for reading printed material is braille printing. Braille print is composed of raised "dots" formed by the use of all possible combinations of six dots. Distance between dots, and height and diameter of dots are some of the parameters to be considered for making them discriminable to the touch.

10.9 CONTROLS

A *control* is a device (mechanical, electromechanical) that converts human output to machine input. Hence it serves as an interface between a machine/device and a user. Ergonomic design of controls assures effective use with minimum errors.

Control performance varies from a simple depression of a pushbutton in order to start a machine to manipulating multiple levers, toggle switches, and other devices (Figure 10.8) to control a subsystem (e.g., distillery column A) in the control room of a chemical processing plant. The former task requires little skill, time, and other resources for completion.

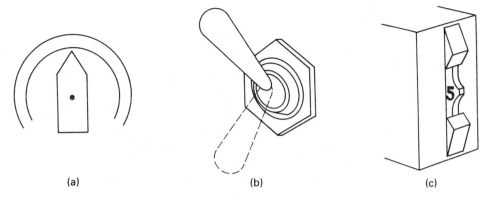

Figure 10.8 Various controls: (a) rotary selector switch; (b) toggle switch; (c) detent thumbwheel.

The latter is a much more complex task requiring many hours of training and equipment interfaces. Most control tasks fall in between these two extremes.

An important characteristic of a control task is that the type of control used (or selected) depends on the type of system response desired. Hence for a two-state response (on–off), one would use a control device that may be set at two states. On the other hand, if the system is expected to take on many values on a continuum (e.g., frequency setting on a radio), a continuous-action control such as a knob would be the best to use.

Many parameters define the use effectiveness of controls. Naturally, these are a function of human capabilities and limitations, task characteristics and other variables (environmental, etc.). The primary design variables affected are type, size, shape, material, spacing, and resistance of controls (Figure 10.9).

10.9.1 Control Design, Selection, and Arrangement

Over the years, controls have been subject to research with respect to identifying characteristics that help users achieve the control task in minimum time with minimal effort and

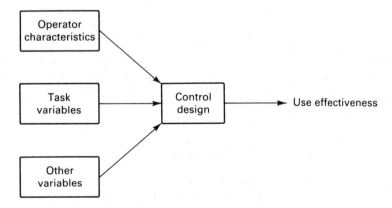

Figure 10.9 Control use-effectiveness affectors.

Chapter 10   Selection and Design of Displays and Controls   **265** errors. As a result, many general rules have been developed [46,24], some of which are listed in Table 10.6. In this section some background theory is presented.

The three variables that are most important in the design and operation of controls are resistance, speed, and accuracy. Control resistance is a function of maximum isometric forces that can be applied by the hand, arm, foot, and leg. Such forces have been experimentally determined by Caldwell [47], Rohmert [48], and Katchmer [49]. Naturally, the amount of resistance should be such that a majority of users can overcome it. Control speed depends on several factors, such as the type of control (hand, foot; discrete or continuous adjustment), amount of resistance, and shape/size variables, as shown by Conrad and Hull [50] and Bradley [51]. Control accuracy also depends on these factors in addition to others, such as motion range, type of coupling, and environmental factors. The effectiveness of a control task also depends on factors such as the specific arrangement of controls [52], consistency with learned expectations [53], and feedback on operation [54]. Other relevant factors are discussed in the subsequent sections.

10.9.2 Control Characteristics

Several concepts that apply to controls and control use are described here. The objective is to sensitize the designer and the user to design and to use the most effective controls.

Control coding. The purpose of control coding is effective identification—controls that are easy to identify result in accurate and fast operation. Proper control coding also reduces training time. The six most common control coding methods are labeling, color, location, shape, size, and texture coding. Several of these methods may be combined for maximum effectiveness. Many variables affect decisions concerning the coding

---

TABLE 10.6 GENERAL RULES FOR DESIGNING, SELECTING, AND ARRANGING CONTROLS

---

1. Locate and orient controls so that their motion and location are compatible with the movement and location of associated display element or system response.
2. Functionally related controls are candidates for combining to reduce intermediate moves and panel space.
3. Control responses should be distributed to the limbs as equally as possible.
4. The hands should be assigned those controls that require precise setting.
5. Controls requiring large and continuous application of force should be assigned to the feet.
6. A positive indication of control activation should be provided.
7. The associated elements must be designed such that when a control is activated, the user will obtain feedback that a system response has been achieved.
8. Controls should be designed to withstand abuse. This is particularly important in emergency conditions.
9. Select multirotation controls when precise settings are required over a wide range of adjustments.
10. Select discrete adjustment controls when the system is to be adjusted for discrete positions or values over a continuum.
11. Select controls that can easily be identified based on touch or visual clues.
12. When precise adjustments over a continuum are required, continuous adjustment controls are the choice.
13. Control surfaces must prevent the activating hand from slipping accidentally.
14. Frequently used and critical controls must be located within easy reach of the user.
15. Control movements should conform to user expectations.
16. In designing and locating controls, consider the probability of accidental activation. Minimize this probability.
17. The least capable user must be able to apply the forces necessary to activate controls.

Source: Adapted from Refs. [46] and [24].

method, including the total number of controls to be coded, available space, illumination, and the existing methods of coding.

The simplest coding method is labeling. Well-designed labels will enhance control operation performance. As a minimum, the designer should indicate the following on labels:

1. The function being controlled
2. The positions to which the control can be set

Labeling may require a large amount of panel space. It should be used sparingly. Labels may be placed on the control, or preferably very near to it, and they should be placed in a horizontal direction, to enhance reading. Groups of controls may be enclosed within a border and the whole group may be identified with a label. The alphanumeric character details discussed previously as a function of the reading distance also apply to label design.

Color coding is effective when certain colors can be associated with specific implications, such as red controls as emergency controls. Once color coding is adopted, it should be standardized across all controls. If many functions are to be discriminated based on color, the number of color combinations may exceed the levels suggested, as indicated in Chapter 4. For most effective use, always combine color coding with other coding methods.

Location coding refers to locating controls that refer to the same function in the same area. Size, shape (see Figure 10.10) and texture coding help identify controls without the use of vision. This is especially important in skilled performance. Futhermore, tactile identification of controls in addition to visual identification facilitates error-free operation. Size coding calls for a particular size control referring to a particular function or group of functions. Shape coding indicates shaping of controls relative to the functions they serve. Texture coding is a function of the surface finish of the control material. A caution: Size, shape, and texture coding are much less effective when the user is wearing gloves.

Control resistance. Controls always require some force to be applied for use. Hand controls require less activation force relative to foot controls. Although the designer would like to minimize the resistance of controls to maximize the population capable of use, resistance is a desirable feature of controls since to some degree it helps to overcome

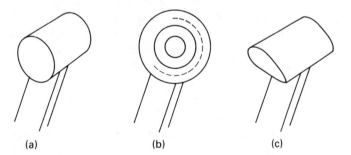

(a)   (b)   (c)

Figure 10.10   Shape-coded knobs for U.S. aircraft; (a) reverse power; (b) landing gear; (c) landing flap. (From *AFSC* Design Handbook, DH1 1-3, Human Factors Engineering, Air Force System Command, 1990.)

the possibility of accidental activation. Therefore, there is definitely a trade-off for the designer to consider here.

Several types of control resistance are [46]:

1. *Elastic resistance.* Another term for this type of resistance is *spring loading.* It varies with control displacement. Elastic resistance offers more counter force as control displacement increases. It applies force toward the null position of the control when the control is released. Hence it helps the user to identify the null position via touch senses and make changes around it. Elastic resistance also helps the user to make quick changes in the direction of control movement.

2. *Frictional resistance.* The two elements that make up frictional resistance are static and sliding friction. At the initiation of movement, static friction is at its maximum. When the control starts to move, it drops sharply, allowing sliding friction to take over. Static friction tends to hold the control in its null position, hence reducing the likelihood of accidental activation. Sufficient static friction also allows hands to rest on the control without activating it. Sliding friction is independent of control displacement or velocity.

3. *Viscous-damping resistance.* Viscous damping resists quick movements. It varies directly with control velocity and is independent of displacement or acceleration. It assists the operator in making smooth control movements.

4. *Inertial resistance.* Inertial resistance varies directly with control acceleration. It too assists the operator to make smooth control movements. It resists sudden changes in control velocity. Inertial resistance makes it difficult to make small and precise adjustments quickly.

For maximum speed of movement, control resistance should be minimum. Resistance for hand controls should not be less than 0.9 kg (2 lb). For controls that make use of the arm and the hand, the recommended minimum resistance is about 5 kg (11 lb) [55,56].

Control/response ratio. This concept applies to continuous controls, not to controls that can be set at discrete settings. It is the ratio of the distance of movement of the control to that of the movement of the system element being controlled (pointer, etc.). For rotary knobs, the C/R ratio is the reciprocal of the system element movement (in inches or centimeters) for one complete revolution of the knob. For linear controls, such as levers, that control linear displays, the C/R ratio is defined as the ratio of the control's linear displacement to the display element displacement. For controls that involve rotational displacement, affecting linear displays, the C/R ratio is defined as

$$C/R = \frac{\dfrac{a}{360} \times 2\pi L}{\text{display movement}}$$

where a is the angular control movement in degrees and L is the length of the control lever arm.

The case for the C/R ratio is rooted in the following: In making a control move with continuous controls, the user makes two distinct adjustments:

1. A rapid move (slewing action) by which the controlled system element is moved as close to the final desired position as possible.
2. A fine-adjustment move, where the final adjustments are made for final positioning of the system element at the desired position.

Figure 10.11 gives the behavior of these two elements with respect to the C/R ratio. Furthermore, the behavior of total control movement time is plotted. It is obvious that at the optimal C/R ratio, the total control operation time is minimum. The optimum C/R ratio for knobs ranges between 0.2 and 0.8. For levers, it is within the range 2.5 to 4.0. The ratios are affected by display size, tolerance, and time delay.

Control spacing. Control spacing helps minimize accidental activation. Furthermore, it is a good way to aid the functional grouping process and use the correct control. Many variables affect control separation [24], including the body member being used, the type of use, control type, blind reaching requirements, and personal equipment, such as gloves. Table 10.7 gives minimum and preferred separation recommendations for various types of controls on a panel or machine.

Accidental activation. In any use situation, there is some potential for controls to be activated accidentally. In one case, a worker with a rather large abdomen was reaching for a part when his belly accidentally activated a toggle switch, which in turn started the downward movement of a cutting blade, severing his hand from the wrist. Several methods exist to minimize the possibility of accidental activation [46]:

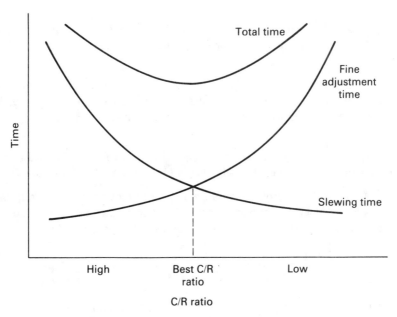

Figure 10.11 C/R ratio (Adapted from W. Jerkins and M.B. Connor, ''Some Design Factors in Making Settings on a Linear Scale,'' *Journal of Applied Psychology*, 33, 1949.

Chapter 10   Selection and Design of Displays and Controls   269

TABLE 10.7 RECOMMENDED SEPARATION FOR VARIOUS CONTROLS

Control	Minimum separation [mm (in.)]	Preferred separation [mm (in.)]	Separation measurement
Pushbutton			Between closest edges
One finger—random (a)	12 (0.47)	51 (2)	
One finger—sequential (b)	6 (0.24)	25 (1)	
Different fingers—(a) and (b)	12 (0.47)	12 (0.47)	
Toggle switch			Between centers
(a)	20 (0.78)	51 (2)	lever tips
(b)	12 (0.47)	25 (1)	
Different fingers—(a) and (b)	16 (0.63)	20 (0.78)	
Rotary selector switch—pointer type			Edge to edge
One hand—random	25 (1)	51 (2)	
Two hands—simultaneous	78 (3)	130 (5)	
Discrete thumbwheel	6 (0.24)	9 (0.35)	Nearest edges
Knobs			Edge to edge
One hand—random	25 (1)	51 (2)	
Two hands—simultaneous	78 (3)	130 (5)	
Cranks and Levers			Edge to edge
One hand—random	51 (1)	102 (4)	
Two hands—simultaneous	78 (3)	130 (5)	
Handwheels	78(3)	130(5)	Edge to edge
Pedals			Between centers
One foot—random	203 (8)	254 (10)	
One foot—sequential	152 (6)	203 (8)	

Source: Adapted from Refs. [5] and [46].

1. *Covering.* Protective covers or guards over controls may aid in the process. However, this method is ineffective, or the guards are disabled if the control is used very frequently.

2. *Locking.* Locking into position is another way of reducing the possibility of accidental activation. A control that is locked into position will have to be operated in two different directions. Since this takes time, the locking method is ineffective for frequently used controls.

3. *Resistance.* A very effective method even for frequently used controls is resistance, as discussed before. Using the proper kind and amount of resistance is essential.

4. *Location.* Controls can be located in such a manner that the sequence of operations is not conducive to accidental activation. Separating important controls from others is an example. Naturally, this process increases control operation time. Hence a trade-off must be sought.

5. *Recessing.* Controls can be positioned on recessed surfaces to minimize protrusion from other surfaces.

6. *Orientation.* The direction of movement for a control may be such that accidental movements are least likely to occur in that direction. If the hand and arm movements at a bench take place in a fore-and-aft direction, the operation of a toggle switch may be such

that it is activated in a right-left direction. Here care should be exercised not to violate recommended direction-of-motion relationships.

The designer should keep in mind that most methods of design to minimize accidental activation run counter to designing controls for physically impaired users [57]. For example, subjects suffering from arthritis or a muscular disease need controls on consumer products that require the least amount of force and one-stage activation (preferably push) instead of two-stage (e.g., pull and lift).

Feedback on operation. As discussed before, feedback is a very important element of skilled performance. Similarly, the designer should exercise sound judgment in building such dimensions into controls. A small light next to a control that lights up when the control is activated will provide the needed feedback. However, the light takes extra space. Hence tactual or auditory feedback from the control itself may be other choices. An audible "click" or resistance building and falling sharply when activation occurs are good methods of providing feedback on control operation.

10.9.3 Compatibility

The term *compatibility* establishes a relationship between two things or concepts. In the ergonomics arena it refers to relationships between stimuli and human responses that are consistent with expectations. Many applications of this concept exist in the area of displays and controls. The bases for such relationships are rooted in previous experiences, culture, and training. As discussed before, population stereotypes are affected notably by such tendencies. For example, we have been raised with the concept that *red* means danger or that things identified with this color may require immediate attention; hence an assembly accompanied by a red tag signals special attention to a production worker. A recent study [58] proved the importance of previous exposure experience on the strength of SR (stimulus-response) compatibility evaluations.

Sanders and McCormick [1] propose that there are three types of compatibility:

1. *Spatial compatibility.* The two major elements of spatial compatibility that apply to displays and controls are the physical similarity of these devices and their physical arrangement in a work area. In design, the second element is much more prevalent than the first. Perhaps full location compatibility is established when a control is associated with a display right above it. In the case of a burner-control arrangement (on two planes), Chapanis and Lindenbaum [56] experimented with four arrangements and found that the arrangement given in Figure 10.12 gives the minimum errors in operation. Displays and controls laid out with spatial compatibility considerations minimize control operation time and the number of times an incorrect control is used.

2. *Compatibility of movement relationships.* The second type of compatibility includes the movement relationships between controls and moving display elements, movement of display elements without any related response, and movement of controls and the related system response (e.g., turn steering wheel clockwise for right turn). Both errors and operation time will be minimized by adherence to the compatibility of movement relationships.

In the case of rotary controls and rotary displays on the same plane, it seems that a clockwise turn of the control is associated with a clockwise movement of the pointer (Figure 10.13). For moving scales with fixed pointers, Bradley [59] postulated that the expectation is that the scale should rotate in the same direction as the control knob. In addition, the scale numbers should increase from left to right and the control turn clockwise to increase settings.

With rotary controls and linear displays in the same plane, different relationships emerge depending on the position of the control relative to the display. Figure 10.14 indicates that for a similar movement of the display element (upward), the control should turn counterclockwise if it is to the left of the display, and in a clockwise direction if it is to the right of the display. Warrick [60] solved this dilemma by postulating that the indicator of a linear display should move in the same direction as the nearest point on the control. The reader is referred to Ref. [1] for additional information on other movement relationships, including relationships between those devices that are located on different planes.

3. *Conceptual compatibility.* This type of compatibility relates to the intrinsic relationship between items or concepts. Examples are a skull and crossbones denoting danger, increasing pitch of a flash designating a charging unit, a wailing signal denoting emergency, and an airplane shape on a map denoting an airport.

Andre and Wickens [61] showed that display-control consistency (similarity of display-control layout between designs) may be a critical factor in addition to compatibility within designs in defining the strength of human interface. There are also ethnic differences in compatibility associations. For example, Courtney [62] observed differences between the "Western" and Hong Kong residents' compatibility expectations.

10.9.4 Specific Controls and Keyboards

Many controls that are in use today may benefit from ergonomic design. Specific suggestions with respect to commonly used controls, such as a hand pushbutton, foot pushbutton, rotary selector switch, detent thumbwheel, toggle switch, knob, continuous adjustment thumbwheel, handwheel, crank, foot pedal, lever, ball controller, rocker switch, rotary disk,

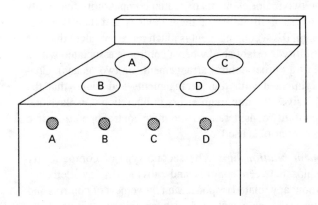

Figure 10.12 Control-burner arrangement giving fewest use errors. (After Ref. [56].)

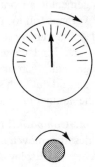

Figure 10.13 Compatibility of movement relationships for rotary displays and controls.

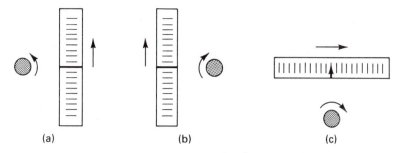

Figure 10.14 Warrick's rule.

joystick, handle, and slide are given in Refs. [21] and [63]. Appendix A provides a summary. The designer of specific controls should exercise these recommendations during design, and the engineer who has a need must screen available controls with respect to these recommendations. However, the use of keyboards is increasing nowadays, especially in the service industry. Hence this section is devoted specifically to the ergonomics of keyboard design.

Keyboard layout. An important element of keyboard design is the arrangement of keys on the board. The best arrangement optimizes data-handling performance with such devices. Along these lines, several arrangement schemes have been developed. The most widely used scheme in the Western world is the QWERTY arrangement, named after the leftmost six keys in the top row of letters. There are no detailed accounts as to how this layout was developed; however, it conforms well with the general guidelines for keyboard layout development (use frequency of character digraph—sequences of two letters—to alternate most keystrokes from hand to hand; assign least frequently used letters to the bottom row and to the ring and little fingers; and minimize successive keystrokes by the same finger). Figure 10.15 gives the various keyboard arrangements that are discussed in this section.

On the other hand, the QWERTY layout has several disadvantages [64]:

1. It slightly overloads the left hand (57%).
2. Skipping of rows is common in frequently used sequences.
3. Too much typing is done using the top row of keys. The home row is used very little.
4. Certain fingers are more loaded than others.

Many efforts to improve the QWERTY layout led to another layout that has endured. The DVORAK simplified layout, also accepted by the American National Standards Institute, was developed on the basis of loading the right hand more than the left hand, most use of the home row, minimum finger motions from row to row, and alternate hands striking letters occurring together. Yamada [65] claims that the DVORAK layout is easier to learn, faster to use, and more error free than the QWERTY layout. Kinkead [66] estimated the speed enhancement at 2.6%; however, this may be a conservative number. The upper limit of the speed advantage is estimated at 10%.

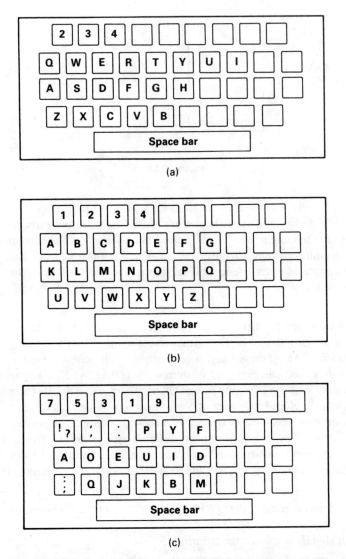

Figure 10.15 Various keyboard layouts: (a) QWERTY; (b) alphabet; (c) DVORAK.

As an alternative to the above, an alphabetical layout has also been developed. The objective was to make this layout available to the inexperienced user. However, it seems that the memory and visual search requirements introduced by the alphabetical layout overcome the expected improvements. Research with low-skilled subjects lead to no performance differences between the alphabetical and the QWERTY layouts.

Although the DVORAK layout seems to perform better than the QWERTY layout, the population experience with the QWERTY layout makes it extremely difficult for the DVORAK layout to gain acceptance in English-speaking countries. Minor variations of the QWERTY layout are in use in other countries, such as the AWERTY layout in France and the QWERTZ layout in German-speaking countries.

One of the major shortcomings of the standard (linear) keyboard is the static load it places on the hands and the forearm. This load is characterized by continuous pronation and use of hands in ulnar deviation. In order to solve these problems, changes in geometry of the keyboard with adjustable and split sections have been proposed [67,68,69]. Some of these have been field tested with successful results [70] in terms of productivity, reduced amount of errors, and user acceptance. Gerard et al. [71] evaluated the Kinesis Ergonomic Computer Keyboard and documented fast learning rate as well as reduced muscular activity required for use. Figure 10.16 provides two examples of ergonomic keyboards.

Figure 10.16 The Maltron® keyboard (top); the Apple® adjustable keyboard for the Macintosh® computer (bottom).

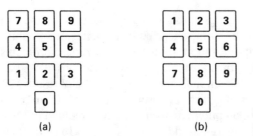

Figure 10.17 Numeric keypad layouts: (a) calculator; (b) telephone.

A related topic is the arrangement of numeric keypads. Figure 10.17 gives two commonly used layouts: the calculator layout and the telephone layout. Research with these two arrangements [45,72] led to a significant advantage in acceptance and accuracy of the telephone layout over the calculator layout.

Key characteristics. Greenstein and Arnaut [73] suggest the following as the general recommendations on key characteristics:

· *Diameter:* 13 mm (0.51 in.)
· *Spacing:* 19 mm (0.748 in.) center to center (Hoffman et al. [74] showed that best inter-key——i.e., edge-to-edge—spacing is equal to finger pad size, about 10mm.)
· *Shape:* square with slight surface concavity
· *Activation force:* 0.25 to 1.47 N
· *Total displacement:* 1 to 6 mm (0.039 to 0.236 in.)
· *Feedback:* kinesthetic, tactual, and auditory

Key functions. Whether each key or some keys will have variable or fixed functions is another consideration. A fixed-function key performs the same function in all conditions. On the other hand, a variable-function key performs various functions, depending on how it is programmed. Fixed-function keys restrict the keyboard to a specific use. In addition, the space requirements increase with such keys. The advantage is less cost. Having two or more functions assigned to a key saves space and allows flexibility; however, errors in use may increase.

10.9.5 Speech Recognition

A unique way of accepting data into computers is *speech recognition*. Speech input of process data has been an alternative to other modes of data entry for many years. However, it was not used in industrial environments on a notable scale until after the mid-1980s, due to such factors as technical problems and user resistance. Although today's technology is still not perfect, speech recognition systems, equipment, and boards are currently used in factories, with minimal problems. Use of this technology can be justified on the basis of labor time savings and enhanced data accuracy. It is the preferred mode of communication in hands-busy/eyes-busy tasks, especially where human judgment is involved. It can also be incorporated as an element in a multimodal communication design.

In the past, the major roadblocks to effective and error-free use of speech input technology have been ambient noise and operator frustration. Ambient noise reduces the S/N ratio. Operator frustration is due largely to a change in his or her voice over time and related recognition problems, as well as limited capability of the hardware to distinguish utterances such as "m" and "n." These problems have not been totally solved yet, but, significant technological advances have been made. Robust recognition algorithms and hardware as well as use of a phonetic alphabet (Table 10.8) minimize these problems. Today, there are systems in the market that span the continuum from isolated word (discrete), speaker-dependent technology to continuous, speaker-independent technology. Normally, a speech input device functions between the user and a host computer that processes the incoming data.

Twenty commandments of speech technology development

Designers of voice input devices for industrial use must be cognizant of 20 issues during design. Many of these are ergonomics related and aim at the minimization of operator frustration.

1. *Function in factory noise.* There are many processes in a factory that generate noise. The overall result is a general background noise. In addition, local noise coming from the processes in the immediate environment confound the issue. The system must be able to function in these environments.

2. *Robust against drift.* The system should not be affected by minor variations in speech tone or quality. Such drift complicates template matching in speech pattern processing(Figure 10.18).

3. *Multiple encoding.* This is a capability to recognize an utterance and send a corresponding but different code to the host computer. For example, the device may recognize a "yes" utterance but only send a "Y" code to the host computer.

4. *Buffers.* There should be a buffer between the user input data and processing of the data by the host computer. It is possible that the user will change his or her mind about

TABLE 10.8 PHONETIC ALPHABET

A	Alpha	N	Nine
B	Beta	O	Opera
C	Charlie	P	Pappa
D	Dog	Q	Quick
E	Echo	R	Reno
F	Face	S	Sign
G	Goggles	T	Target
H	Hotel	U	Unit
I	Index	V	Victor
J	July	W	Whiskey
K	Kilometer	X	X-ray
L	Loss	Y	Young
M	Mike	Z	Zebra

a specific data element in the middle of an input stream. The user should be able to make the correction and send the corrected data set to the host.

5. *Editing.* Closely linked to the point made above, there should be multiple edit modes in the system. An entire input data element or part of it or an entire data set must be editable.

6. *Multisensory prompting.* A CRT prompts the user visually for the next data. The same can be accomplished via voice output. In a factory or any other industrial environment, the operator may be expected to walk around. In this case, auditory prompting would help significantly.

7. *Remote data entry.* Closely linked with point 6, the system must be able to interface with remote data-entry technologies, such as radio frequency or infrared, for data input while the user is away from the recognition device. In this case the user may carry a two-way radio and communicate with the system continuously.

8. *Interruption allowance.* The system must allow for a data input sequence to be stopped at any point for an unexpected interruption. After the interruption, it should start from the point left off.

9. *Feedback.* This is necessary for the user to evaluate data before they are sent to the host. An individual datum may also be evaluated before buffering. A good way to accomplish this is through the auditory channel (i.e., repeat the spoken word back to the user).

10. *Multi-tasking.* The system must be capable of processing several different tasks simultaneously. While a new data set is being created, the previous data set may be sent to the host.

11. *Customization.* The user must be able to develop application programs easily, depending on the case involved. The utilities for this purpose must be flexible so they can be applied to any data collection situation.

12. *Redundant data entry.* Until all problems are resolved, the user must be able to use several data input modes. If there is a recognition problem with any utterance, the system must allow for that utterance to be entered via a keyboard or another device.

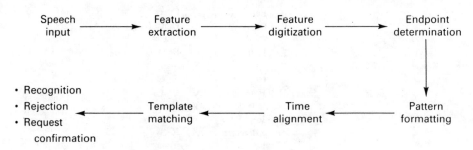

Figure 10.18 Speech pattern processing algorithm.

13. *Help messages.* Help messages must be available through visual and auditory modes. The system must allow the developer custom design help messages for any application and any data input point within the application.

14. *Continuous recognition.* The user must not be limited to pauses between utterances. The natural speaking pattern and speed must be accepted by the system.

15. *Gender-independent recogniton.* The hardware should not distinguish between males and females in terms of recognition accuracy. The designer must not limit use to a particular gender.

16. *Multiple host communication modes.* The user must be able to communicate with the host in several different ways. On-line communication and batch communication are two possible ways.

17. *User experience level.* Both novices and experts must be able to use the system without frustration. Novices may expect extensive help and edit utilities, whereas experienced users may not.

18. *Control of valid vocabulary.* The user must be able to define the valid vocabulary at each data input point. The system must be able to limit input data to those defined. This helps improve recognition accuracy and reduce security worries.

19. *Branching capability.* Custom development of an application must consider branching capability, including conditional branching. Depending on the conditions observed, the user may make one of several possible decisions, and the ensuing data type and number of data elements to be entered may change. All this will increase the flexibility of application development.

20. *Queueing of input data.* Data must be queued at the voice input device if the host is down. When the host is up, it may query for any queued data. Alternatively, the speech input device may sample the host for availability to accept data.

The above are only the most important elements of an acceptable device that is capable of functioning between a human being and a host computer. A whole host of other factors may be considered to improve acceptability.

10.10 CASE ILLUSTRATION 6

A voice input application was developed at a manufacturer's facility to enhance the process check function in the kitting operation. In this section we provide a brief review of this application and presents its results. The process check function that is in question occurs primarily in the forming and consolidation areas.

Process check before voice I/O. Process quality data were collected via a paper–pencil medium prior to the new method. An operator walked between various workstations, and based on a sampling scheme, collected the necessary data using a paper form and a pencil. The data included defect types and sources. Handwritten data on the paper form were then analyzed to generate process control charts as well as other information.

Process check after voice I/O. The improved process made use of a speech recognition system composed of various hardware and software elements to allow quality data to be spoken into a microphone while allowing the operator to walk freely between operations (Figure 10.19). The equipment that supports the application is comprised of a Westinghouse Voice System Series 100, consisting of the voice system controller (VSC) hardware and software, together with a workstation module, a Swintek two-way radio structured to communicate with the workstation in radio-frequency mode, a computer, and communications interfaces (this entity to be called hereafter the *host*), and the data collection system (DCS), comprised of a processor and various peripherals.

The software elements of the application consist of the VSC-resident operating system, which is a Westinghouse proprietary product, the UNIX® operating system, and the custom code residing on the host and in the DCS software. Both the operators and application developers interface with these software. The latter develop applications using a network-based structure defining prompts, acceptable utterances, data strings, and various other utilities for each node on the network. The nodes are then linked according to the precedence relationships between them. The software is ready for use after defining users and completing the application generation (error checks) (Figure 10.20).

The operator (user) is stepped through a voice template creation scheme (training—discrete and continuous). The templates are then linked to the user ID. Data collection may then start after a noise sampling process. Data transmission between the VSC and DCS is

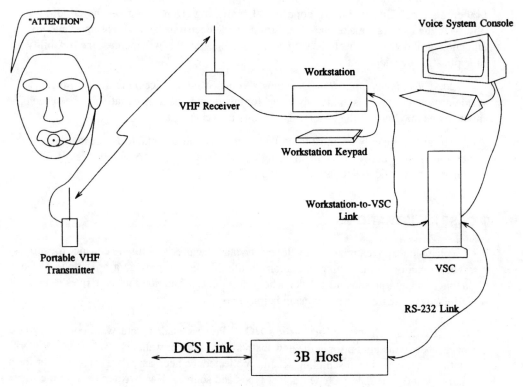

Figure 10.19 Voice I/O process check architecture. (From Pulat and Stafford.)

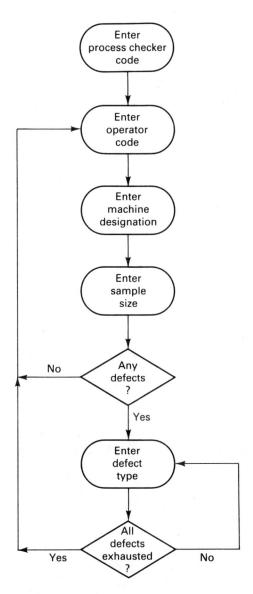

Figure 10.20 Section of the kitting application.

accomplished through the communication software, which resides on the host. This module acts primarily as a traffic controller for data transactions between the VSC and DCS.

Discussion.   In general, the factory's experiences with voice I/O have been positive. One of the observations is that some people accept the system much faster than others. There are still some who cannot work with it at all. Some became expert users in a particular application after half an hour of use experience, some took 2 to 3 hours to reach the same level of expertise, and some dropped from training. The key factor seemed to be the extent of realization that the system is totally under the control of the user. Those who develop use confidence faster become further motivated for better performance.

With experienced users, first-pass recognition accuracy is 97%. The audio repeat function, as well as error correction utilities, allow one to reach perfection in the second pass. The new method of process checking in the kitting operation reduced task cycle time, improving productivity. The users also feel that they are an integral part of automation in the area.

10.11 DISPLAY/CONTROL SELECTION

There are many types of displays and controls in use today. Each has its own use characteristics. Hence it is important for the designer to select the best displays/controls for a given job based on task requirements. In addition to the subjective matching methods (designer matches job requirements and display/control characteristics), Pulat and Ayoub [75] proposed an analytic method, called UNISER (unit selection routine) for selecting displays and controls (units) for process industry. Required inputs are a description of the various information items to be displayed by the machine and the actions to be performed by the operator. In developing the algorithm, a database was created using the available literature as well as surveying a number of experts in the field. A set of selection rules that were developed drive the process.

Figure 10.21 gives the flowchart of UNISER. The heuristic calculates a utility value for each unit based on task requirements. The display selection subroutine selects and recommends the display with the highest utility value for the task together with other displays which tie for the same value. Similarly, the control selection subroutine ranks controls according to their utility values and recommends the one with the highest value, along with three alternatives that follow the recommended control in the rank order.

Two clusters of dimensions have been considered for control selection. The first cluster is task specific. Six main dimensions were identified for this cluster:

· Control settings
· Adjustment precision
· Force required
· Range of adjustments
· Control dimension
· Setting time

The second cluster consists of non-task-specific dimensions. Some of these are well documented by Chapanis and Kinkade [41]. The ones considered most important and thus incorporated into UNISER are:

· Likelihood of accidental activation
· Space requirements for location and operation of control
· Effectiveness of check reading to determine control position when member of a group of similar controls

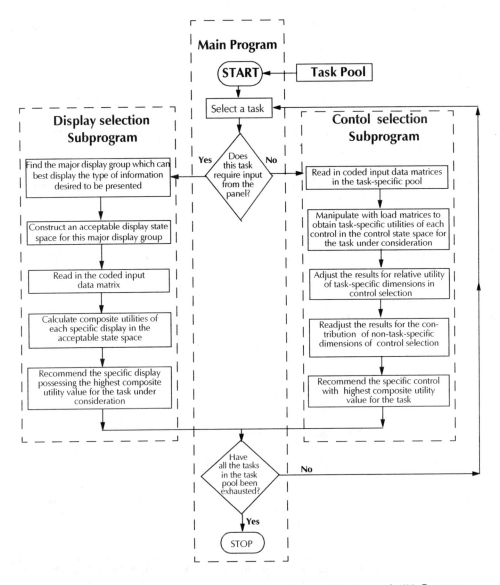

Figure 10.21 UNISER algorithm. (Reprinted from B. M. Pulat and M. A. Ayoub. ''A Computer-Aided Display-Control Selection Procedure for Process Control Jobs.'' IIE Transactions. 16(4). Reprinted with permission of the Institute of Industrial Engineers, 25 Technology Park/Atlanta, Norcross, GA 30092. Copyright © 1984.

- Effectiveness of operating control simultaneously with similar controls in an array
- Effectiveness of identifying control position other than visually
- Effectiveness of coding
- Effectiveness of visually identifying control position

In performing the trade-offs between task-specific dimensions, a database was developed by surveying a panel of judges in order to show:

- The relative merits of each control at the levels of each dimension. To achieve this, assuming a ratio scale between 0 and 100%, each judge was asked to enter percentage values for each control across the levels of the six task-specific dimensions.
- The relative contributions of task-specific and non-task-specific dimensions to the selection procedure (in percentage).
- The importance of each task-specific dimension in selecting a particular control. The sum of the overall ratings (weights) assigned to the task-specific dimensions is 6.0.

The judges were selected from among those who have carried out research in controls or manual control tasks. Participants in the survey were from the United States, Great Britain, and West Germany. The results of the survey are given in Tables 10.9 and 10.10. These data are used in UNISER to calculate composite utilities for controls.

The utility values for the candidate controls are calculated through the following relationship:

$$CU_j = TSU_j + NTSU_j$$

where

CU_j is the composite utility

TSU_j is the task-specific utility

$NTSU_j$ is the non-task-specific utility

of the jth control. In an open form, the relationship is as follows:

$$CU_j = \frac{\alpha}{6} \sum_{i=1}^{6} W_{ji} \sum I_{ik} L_{jik} + \frac{1 - \alpha}{7} \sum_{l=1}^{7} M_{jl}$$

where

$1 - \alpha$ is the percentage contribution of non-task-specific dimensions to control selection.

M_{ji} is the weight accumulated by the jth control on lth non-task-specific dimension.

α is the percentage contribution of task-specific dimensions to control selection.

W_{ji} is the importance rating on the ith task-specific dimension for jth control.

I_{ik} assumes either a 0 or a 1. It takes the value of 1 if the kth level of dimension i has been selected by the user to portray the task requirement most accurately. Otherwise, it assumes the value 0.

L_{jik} is the relative importance of the kth breakdown of the ith task-specific dimension for the jth control.

TABLE 10.9 MEAN RATINGS ON TASK SPECIFIC DIMENSIONS OF CONTROL SELECTION

	Control settings		Adjustment precision			Force required			Range of adjustment/discrete action			
	Disc	Cont.	Gross	Int.	Fine	Small	Med.	Large	$x = 2$	$10 \geqslant x > 2$	$24 \geqslant x > 10$	$x > 24$
Foot pushbutton	95.0	5.0	87.0	7.5	5.5	17.0	41.5	37.5	91.8	0.2	0.0	0.0
Foot pedal	28.7	71.3	51.1	35.5	13.4	14.7	48.7	36.6	62.8	35.0	2.2	0.0
Keyboard	98.5	1.5	34.2	9.9	59.9	90.5	6.5	3.0	64.3	7.4	9.6	18.7
Detent thumbwheel	90.0	10.0	21.7	46.3	32.0	73.0	24.5	2.5	17.2	62.3	16.0	4.5
Rotary selector switch	75.0	25.0	19.3	42.0	38.7	67.3	31.2	1.5	17.2	45.5	23.6	13.7
Hand pushbutton	98.0	2.0	78.0	10.5	11.5	67.0	27.3	5.7	90.3	7.3	2.1	0.3
Legend switch	99.5	0.5	76.0	7.5	16.5	72.8	23.2	4.0	88.7	10.3	1.0	0.0
Toggle switch	96.0	4.0	75.2	11.8	13.0	65.5	28.5	6.0	73.9	21.3	3.2	1.6
Ball controller	3.7	96.3	12.4	38.4	49.2	77.9	21.6	0.5	8.0	26.0	10.0	56.0
Joy stick	10.0	90.0	6.9	42.7	50.4	56.5	37.8	5.7	15.4	55.9	9.1	19.6
Round knob	16.4	83.6	11.3	32.4	56.3	74.2	24.2	1.6	25.0	39.0	11.5	24.5
Lever	37.7	62.3	37.6	44.2	18.2	21.9	53.4	24.7	45.6	42.4	6.1	5.9
Crank	2.8	97.2	47.5	36.7	15.8	21.2	43.3	35.5	14.3	20.0	15.7	50.0
Handwheel	2.9	97.1	31.6	49.7	18.7	18.2	47.6	34.2	15.0	35.0	17.5	32.5
Continuous adjustment thumbwheel	1.8	98.2	15.2	40.3	44.5	84.2	14.0	1.8	14.0	48.0	14.0	24.0

Range of adjustment/Continuous action				Control dimension		Setting time			
0° ≤ x ≤ 90°	0° ≤ x ≤ 180°	0° ≤ x ≤ 360°	360° < x	Uni.	Multi.	Slow	Int.	Quick	Very quick
100.0	0.0	0.0	0.0	94.2	5.8	20.3	19.0	32.7	28.0
98.8	3.2	0.0	0.0	83.2	16.8	21.9	36.3	33.9	7.9
75.0	25.0	0.0	0.0	60.5	39.5	5.6	15.2	19.1	60.1
31.2	22.5	23.7	22.5	82.0	18.0	37.5	42.0	17.2	3.3
28.5	21.9	28.4	21.2	78.3	21.7	23.5	32.7	35.8	8.0
55.0	36.2	7.5	1.3	98.7	1.3	6.2	18.1	32.9	42.8
100.0	0.0	0.0	0.0	92.5	7.5	4.2	15.4	30.9	49.5
76.6	6.7	6.7	10.0	88.2	11.8	6.7	19.1	48.6	25.5
29.4	13.1	13.1	49.4	11.6	88.4	14.9	48.3	33.3	3.5
67.1	12.1	7.4	13.4	16.7	82.3	7.5	35.5	43.3	13.7
20.0	22.9	26.7	30.4	85.3	14.7	16.9	51.0	24.3	7.8
80.6	11.2	2.1	6.1	86.7	13.3	12.0	54.0	26.5	7.5
2.0	4.2	9.5	84.3	82.7	17.3	56.2	33.3	7.5	3.0
3.7	10.8	20.5	65.0	82.1	17.9	41.8	38.7	14.7	4.8
19.0	19.7	26.8	34.5	83.0	17.0	35.0	39.5	22.0	3.5

Source: Reprinted from B. M. Pulat and M. A. Ayoub, 1984. "A Computer-Aided Display-Control Selection Procedure for Process Control Jobs." *IIE Transactions*, 16(4). pp. 377–378. © Institute of Industrial Engineers, 25 Technology Park, Norcross, Georgia 30092.

TABLE 10.10 IMPORTANCE RATINGS USED IN UNISER

	Importance ratings					
	Continuous settings	Adjustment precision	Force required	Range of adjustment	Control dimension	Setting time
Foot pushbutton	1.04	0.41	1.76	1.01	0.64	2.14
Foot pedal	0.75	1.26	2.18	1.01	0.43	1.37
Keyboard	1.03	1.30	1.01	0.98	1.12	1.56
Detent thumbwheel	1.32	1.15	0.97	2.02	0.47	1.06
Rotary selector switch	1.45	1.26	0.89	1.98	0.33	1.09
Hand pushbutton	1.08	0.94	1.00	1.40	0.50	2.08
Legend switch	1.08	0.70	0.84	1.54	0.56	2.28
Toggle switch	1.18	1.06	0.84	1.79	0.56	1.57
Ball controller	1.04	1.67	0.70	1.84	1.01	0.74
Joystick	0.84	1.91	0.75	1.45	0.91	1.14
Round knob	1.28	1.23	0.74	2.61	0.34	0.80
Lever	1.05	1.04	1.20	1.11	0.95	0.95
Crank	0.75	1.68	1.48	1.80	0.54	0.75
Handwheel	1.05	1.58	1.35	1.51	0.45	1.06
Continuous adjustment thumbwheel	1.04	2.20	0.92	1.59	0.35	0.90

Source: Reprinted from B. M. Pulat and M. A. Ayoub, 1984. "A Computer-Aided Display-Control Selection Procedure for Process Control Jobs." *IIE Transactions*, 16(4), pp. 377–378. © Institute of Industrial Engineers, 25 Technology Park. Norcross, Georgia 30092.

For selection of the best display for a given task, no new database was developed. Instead, a classification of displays was done in terms of the type of information displayed (static, historical, predictive, etc.) and the characteristics of specific displays (counters, legend and indicator lights, cathode ray tubes, horns, buzzers, legend plates, etc.) were then matched with the classification obtained to develop clusters of displays for different purposes. Then it is a trivial task to match task requirements with the individual displays for a composite that minimizes injuries and maximizes productivity and comfort in use.

10.12 SUMMARY

The problem of best design, selection, and arrangement of displays and controls has lured ergonomists for many years. The reason for this is that such equipment constitutes the human interface points of machines, equipment, and systems. For best human performance, displays and controls must be designed, selected, and installed with ergonomics in mind.

QUESTIONS

1. What is meant by ergonomic design of controls and displays?
2. Define *display*.
3. Discuss briefly the principal types of displays.
4. List and discuss three conditions that positively affect detectability of signals.
5. What is a quantitative display?
6. What are the advantages of a counter or digital display over other quantitative displays?
7. What are the recommended numerical progressions for display scales?
8. What is the recommended character width/height ratio for numerals?
9. Discuss the difference between legibility and readability.
10. What do qualitative displays show?
11. How do gestalt principles help display/control arrangement?
12. Discuss the basics of auditory displays.
13. List four considerations in the design of warning signals.
14. How does one measure speech intelligibility?
15. What is signal-to-noise ratio?
16. What guidelines govern S/N ratio with respect to auditory display design?
17. When is it advantageous to use tactual displays?
18. What is a control? How does it differ from a control task?
19. What are the most important three variables in the design and operation of controls?
20. List the different methods of control coding.
21. What is elastic resistance as applied to control design?
22. Discuss the C/R ratio.
23. Discuss the three types of compatibility between displays and controls.

24. What are the disadvantages of the QWERTY keyboard layout?

25. What are the six major task specific dimensions of control selection considered in UNISER?

EXERCISES

1. Go to your power plant and observe two displays: one quantitative and one qualitative. Evaluate each. Propose improvements.

2. Given that a display will present important information that is expected to be read from a distance of 2 m (50.8 in.) under an illumination level of 1.5 fc, develop the numeral and letter characteristics that must be designed into the display.

3. Select a video display terminal in your environment and evaluate its physical characteristics. Propose improvements.

4. Using a frequency analyzer, evaluate the characteristics of an auditory display. Any improvements necessary?

5. Select a small panel in your environment that contains a minimum of five displays and controls. Evaluate it with respect to compatibility issues. Are there any improvements to be made?

6. Evaluate the effectiveness of the QWERTY layout for keyboards using a paragraph of written material. Discuss your observations.

7. Apply UNISER for a control selection task. Discuss improvements.

REFERENCES

1. Sanders, M. S., and McCormick, E. J. 1993. *Human Factors in Engineering and Design*, 7th ed. McGraw-Hill, New York.
2. Patterson, R. D. 1982. Guidelines for Auditory Warning Systems on Civil Aircraft. *CAA 82017*. Civil Aviation Authority, London.
3. Munson, W. A. 1947. The Growth of Auditory Sensitivity. *Journal of the Acoustical Society of America*, 19, p. 584.
4. Deatherage, B. H. 1972. Auditory and Other Sensory Forms of Information Presentation. In *Human Engineering Guide to Equipment Design*. Van Cott, H. P., and Kinkade, R. G. (Eds.). U.S. Government Printing Office, Washington, D.C.
5. Eastman Kodak Co. 1983. *Ergonomic Design for People at Work*. Lifetime Learning, Belmont, CA.
6. Grether, W. F., and Baker, C. A. 1972. Visual Presentation of Information. In *Human Engineering Guide to Equipment Design*. Van Cott, H. P., and Kinkade, R. G. (Eds.). U.S. Government Printing Office, Washington, DC.
7. Sinclair, H. J. 1971. Digital versus Conventional Clocks: A Review. *Applied Ergonomics*. 2(3), pp. 178–181.
8. Whitehurst, H. O. 1982. Screening Designs used to Estimate the Relation Effects of Display Factors on Dial Reading. *Human Factors*, 24(3), pp. 301–310.
9. Peters, G. A., and Adams, B. B. 1959. These Three Criteria for Readable Panel Markings. *Product Engineering*, 30(21), pp. 55–57.
10. Berger, C. 1944. I., Stroke-Width, Form, and Horizontal Spacing on Numerals as Determinants of the Threshold of Recognition. *Journal of Applied Psychology*, 28, pp. 208–231.

11. Berger, C. 1944. II., Stroke-Width, Form, and Horizontal Spacing on Numerals as Determinants of the Threshold of Recognition. *Journal of Applied Psychology*, 28, pp. 336–340.

12. Helander, M. G. 1987. Design of Visual Displays. In *Handbook of Human Factors Engineering*. Salvendy, G. (Ed.). Wiley, New York.

13. Kahneman, D. 1973. *Attention and Effort*. Prentice Hall, Englewood Cliffs, NJ.

14. Wertheimer, M. 1923. Untersuchungen zur Lehre von der Gestalt, II. *Psychologische Forschung*, 4, pp. 301–350.

15. Reynolds, R. F., White, R. M., Jr., and Hilgendorf, R. I. 1972. Detection and Recognition of Colored Signal Lights. *Human Factors*, 14(3), pp. 227–236.

16. Heglin, H. J. 1973. NAVSHIPS Display Illumination Design Guide, II. Human Factors. *NELC/TD223*. Naval Electronics Laboratory Center, San Diego, CA.

17. Theeuwes, J., and Alferdinck, J. W. A. M. 1995. Rear Light Arrangements for Cars Equipped with a Center High-Mounted Stop Lamp. *Human Factors*, 37(2), pp. 371–380.

18. Theeuwes, J. 1991. *Center High Mounted Stop Light: An Evaluation* (Report I2F 1991 C-3). Soesterberg, Netherlands: TNO Human Factors Research Institute.

19. Kline, D. W., and Fuchs, P. 1993. The Visibility of Symbolic Highway Signs Can Be Increased among Drivers of All Ages. *Human Factors*, 35(1), pp. 25–34.

20. Zwaga, H., and Easterby, R. 1982. Developing Effective Symbols for Public Information: The ISO Testing Procedure. *Proceedings of the International Ergonomics Association Conference*, Tokyo.

21. Hinsley, D. A., and Hanes, L. F. 1977. Human Factors Design Considerations for Graphic Displays. *Westinghouse R&D Center Report 77-IC57-GRAFC-R*.

22. Snyder, H. L., and Maddox, M. E. 1978. Information Transfer from Computer Generated Dot-Matrix Displays. Reports DAFC04-74-G-0200 and DAAG 29-77-G-0067. U.S. Army Research Office, Research Triangle Park, NC.

23. Cakir, A., Hart, D. J., and Stewart, T. F. M. 1979. *The VDT Manual*. IFRA, Darmstadt, Germany.

24. Bailey, R. W. 1982. *Human Performance Engineering*. Prentice Hall, Englewood Cliffs, NJ.

25. Gould, J. D. 1968. Visual Factors in the Design of Computer Controlled CRT Displays. *Human Factors*, 10(4), pp. 359–376.

26. Korge, A., and Krueger, H. 1984. Influence of Edge Sharpness on the Accommodation of the Human Eye. *Graefe's Archive of Clinical Experimental Opthalmology*, pp. 222, pp. 26–28.

27. Gomer, F. E., and Bish, K. G. 1978. Evoked Potential Correlates of Display Image Quality. *Human Factors*, 20, pp. 589–596.

28. Plath, D. W. 1970. The Readability of Segmented and Conventional Numerals. *Human Factors*, 12(5), pp. 493–497.

29. Stammerjohn, L. W., Smith, M. J., and Cohen, B. G. F. 1981. Evaluation of Workstation Design Factors in VDT Operations. *Human Factors*, 23, pp. 401–412.

30. Smith, A. B., Tanaka, S., and Halperin, W. 1984. Correlates of Ocular and Somatic Symptoms among VDT Users. *Human Factors*, 26, pp. 143–156.

31. Dainoff, M., Happ, A., and Crane, P. 1981. Visual Fatigue and Occupational Stress in VDT Operators. *Human Factors*, 23, pp. 421–438.

32. Laubli, T., Hunting, W., and Grandjean, E. 1981. Postural and Visual Loads at VDT Workplaces. 2. Lighting Conditions and Visual Impairments. *Ergonomics*, 24, pp. 933–944.

33. Hedman, L. R., and Briem, V. 1984. Short Term Changes in Eye Strain of VDU Users as a Function of Age. *Human Factors*, 26, pp. 357–370.

34. Gyr, S., Nishiyama, K., Grierer, R., Laubli, T., and Grandjean, E. 1984. The Effect of Various Refresh Rates in Positive and Negative Displays. In *Ergonomics and Health in Modern Offices*. Grandjean, E. (Ed.). Taylor & Francis, London.

35. Gould, J. D., and Grischkowski, N. 1984. Doing the Same Work with Hard Copy and with Cathode-Ray-Tube (CRT) Computer Terminals. *Human Factors, 26*, pp. 323–337.

36. Kruk, R., and Mutter, P. 1984. Reading of Continuous Text on Video Screens. *Human Factors, 26*, pp. 339–345.

37. Gould, J. D., Alfaro, L., Barnes, V., Finn, R., Grischkowski, N., and Minuto, A. 1987. Reading Is Slower from CRT Displays Than from Paper: Attempts to Isolate a Single Variable Explanation. *Human Factors, 29*(3), pp. 269–299.

38. Hollands, J. G., and Spence, I. 1992. Judgements of Change and Proportion in Graphical Perception. *Human Factors*, 34(3), pp. 313–334.

39. Wickens, C. D., and Carswell, C. M. 1995. The Proximity Compatibility Principle: Its Psychological Foundation and Relevance to Display Design. *Human Factors*, 37(3), pp. 473–494.

40. Kvalseth, T. O. 1985. Work Station Design. In *Industrial Ergonomics: A Practitioner's Guide.* Alexander, D. C., and Pulat, B. M. (Eds.). Industrial Engineering and Management Press, Atlanta, GA.

41. Sorkin, R. D. 1987. Design of Auditory and Tactile Displays. In *Handbook of Human Factors Engineering.* Salvendy, G. (Ed.), Wiley, New York.

42. Fidell, S. 1978. Effectiveness of Audible Warning Signals for Emergency Vehicles. *Human Factors, 20*, pp. 19–26.

43. Cooper, G. E. 1977. A Survey of the Status and the Philosophies Relating to Cockpit Warning Systems. *NASA-CR-152071.* NASA Ames Research Center, CA.

44. Liberman, A. M., Cooper, F. S., Shankweiler, D. P. and Studdert-Kennedy, M. 1967. Perception of the Speech Code. *Psychological Review*, 74, pp. 431–461.

45. French, N. R., and Steinberg, J. C. 1947. Factors Governing the Intelligibility of Speech Sounds. *Journal of the Acoustical Society of America*, 19, pp. 90–119.

46. Chapanis, A., and Kinkade, R. G. 1972. Design of Controls. In *Human Engineering Guide to Equipment Design.* Van Cott, H. P. and Kinkade, R. G. (Eds.). U.S. Government Printing Office, Washington, DC.

47. Caldwell, L. S. 1959. The Effect of the Special Position of a Control on the Strength of Six Linear Hand Movements. *Report 411.* U.S. Army Medical Research Laboratory, Fort Knox, KY.

48. Rohmert, W. 1966. *Maximal krafte von Mannern im Bewegungsraum der Arme und Beine.* Westdeutscher Verlag, Koln, Germany.

49. Katchmer, L. T. 1957. Physical Force Problems: 1. Hand Crank Performance for Various Crank Radii and Torque Load Combinations. *Tech. Memo 3-57.* Human Engineering Laboratory, Aberdeen Proving Ground, Aberdeen, MD.

50. Conrad, R., and Hull, A. J. 1968. The Preferred Layout for Numeral Data-Entry Keysets. *Ergonomics*, 11, pp. 165–173.

51. Bradley, J. V. 1969. Desirable Dimensions for Concentric Controls. *Human Factors,* 11(3), pp. 213–226.

52. Dvorak, A. 1943. There Is a Better Typewriter Keyboard. *National Business Education Quarterly,* 12, pp. 51–58.

53. Ray, R. D., and Ray, W. D. 1979. An Analysis of Domestic Cooker Control Design. *Ergonomics,* 22, pp. 1243–1248.

54. Weiss, B. 1953. Building "Feel" into Controls: The Effect on Motor Performance of Different Kinds and Amounts of Feedback. *USN/ONR. TR 241-6-11.*

55. Dempster, W. T. 1955. The Anthropometry of Body Action. Annals of the New York Academy of Science, 63, pp. 559–567.

56. Chapanis, A., and Lindenbaum, L. A. 1959. A Reaction Time Study of Four Control-Display Linkages. *Human Factors,* 1(4), pp. 1–7.

57. Kanis, H. 1993. Operation of Controls on Consumer Products by Physically Impaired Users. *Human Factors*, 35(2), pp. 305–328.
58. Payne, S. J. 1995. Naive Judgements of Stimulus-Response Compatibility. *Human Factors* 37(3), pp. 495–506.
59. Bradley, J. V. 1954. Desirable Control-Display Relationships for Moving-Scale Instrument. *Technical Report 54-423.* Air Force, Wright Air Development Center.
60. Warrick, M. J. 1947. Direction of Movement in the Use of Control Knobs to Position Visual Indicators. In *Psychological Research on Equipment Design*. Fitts, P. M. (Ed.). Army, Airforce Aviation Psychology Program.
61. Andre, A. D., and Wickens, C. D. 1992. Compatibility and Consistency in Display-Control Systems: Implications for Aircraft Decision Aid Design. *Human Factors*, 34(6), pp. 639–653.
62. Courtney, A. J. 1994. Hong Kong Chinese Direction-of-Motion Stereotypes. *Ergonomics*, 37(3), pp. 417–426.
63. Bullinger, H.-J., Kern, R. and Muntzinger, W. F. 1987. Design of Controls. In *Handbook of Human Factors Engineering.* Salvendy, G. (Ed.). Wiley, New York.
64. Noyes, J. 1983. The QWERTY Keyboard: A Review. *International Journal of Man-Machine Studies,* 18, pp. 265–281.
65. Yamada, H. 1980. A Historical Study of Typewriters and Typing Methods: From the Position of Planning Japanese Parallels. *Journal of Information Processing,* 2, pp. 175–202.
66. Kinkead, R. 1975. Typing Speed, Keying Rates, and Optimal Keyboard Layouts. In *Proceedings of the Human Factors Society 19th Annual Meeting,* Santa Monica, CA, pp. 159–161.
67. Buesen, J. 1984. Product Development of an Ergonomic Keyboard. *Behavior and Information Technology,* 3, pp. 387–390.
68. Ilg, R. 1987. Ergonomic Keyboard Design. *Behavior and Information Technology,* 6, pp. 303–309.
69. Kroemer, K. H. E. 1965. Zur Verbesserung der Schreibmaschinen-Tastanter, *Arbeitswissenschaft*, 4(1), pp. 11–16.
70. Cakir, A. 1995. Acceptance of the Adjustable Keyboard. *Ergonomics*, 38(9), pp. 1728–1744.
71. Gerard, M. J., Jones, S. K., Smith, L. A., Thomas, R. E., and Wang, T. 1994. An Ergonomic Evaluation of the Kinesis Ergonomic Computer Keyboard. *Ergonomics*, 37(10), pp. 1661–1668.
72. Lutz, M. C., and Chapanis, A. 1955. Expected Locations of Digits and Letters on Ten-Button Keysets. *Journal of Applied Psychology,* 39, pp. 314–317.
73. Greenstein, J. S., and Arnaut, L. Y. 1987. Human Factors Aspects of Manual Computer Input Devices. *Handbook of Human Factors.* Salvendy, G. (Ed.). Wiley-Interscience, New York.
74. Hoffmann, E. R., Tsang, K. K., and Mu, Andrew. 1995. Data Entry Keyboard Geometry and Keying Movement Times. *Ergonomics*, 38(5), pp. 940–950.
75. Pulat, B. M., and Ayoub, M. A. 1984. A Computer Aided Display-Control Selection Procedure for Process Control Jobs. *IIE Transactions,* 16(4), pp. 371–378.

_____CHAPTER 11

Design of Products and Information Aids

11.1 INTRODUCTION

In this chapter we focus on the ergonomic bases of product, hand tool and information aids design. Most of the underlying theory has been presented in Chapters 3 and 4. Here we concentrate on use factors, legal issues, and provide practical design recommendations. Product design with use factors in mind assures effective use of the product with minimal injury potential even in misuse conditions. Such practices also provide protection against product liability law suits. Maintainability design aims at extending the life cycle of products while maintaining their functional effectiveness.

Hand tools are used very frequently in industrial environments. Poorly designed hand tools and their use in awkward postures may lead to cumulative trauma disorders, in addition to causing a drop in work efficiency. Ergonomic design of forms, signs, labels, and computer screens assures an effective display of information for enhanced processing. Fast, error-reduced use of such information aids is important in overall job design. We first discuss general topics in product design supported by maintainability design. Then special topics of hand tools and information aids are discussed.

11.2 PRODUCT DESIGN

One of the definitions of ergonomics is *design for human use*. More often than not, products are designed for repeated use by human beings. Although this book is oriented toward industrial environments, there too, many products are used by employees, whether blue collar or white collar. Hence it is important to consider product design in addition to workplace, work methods, and equipment design. In this section we first provide a step-by-step generic approach to design for human use. Then, a list of human-focused specific product design guidelines is given. Finally, a review of the *product liability* concept is provided along with its application to product design.

11.2.1 Design for Human Use

The six-step approach to designing for human use is as follows:

1. *Define user's needs.* Focusing on the user's needs and defining the purpose of use is the first step. Unless the user's needs are clearly understood, design cannot be guided toward the correct development paths. User needs can be defined based on a market review, interviews with potential users, and personal experiences. A clear definition of the use environment is also necessary.

2. *Detail functional requirements.* The specific functions of the product that satisfy user's needs must then be detailed. These can be listed in an itemized format with each function clearly defined. At this step, make sure that the functions listed span the purpose(s) of the product.

3. *Perform task analysis.* For each function, a detailed task analysis will show use patterns, hardware needs and possible misuse of the product. The latter is important in safeguarding the product against liability suits. All possible uses of the product, including maintenance and setup, must be considered.

4. *Design user interfaces.* Step 3 showed all user interface points with the design. Considering the user's information and functional needs, these points must be detailed and all hardware and software elements developed.

5. *Develop the product.* At this stage we have all functional requirements established, all hardware needs detailed, and user-interface points evaluated. The next step is the development of several product prototypes for testing.

6. *Allow user testing.* Before the product is manufactured in mass, a comprehensive test is recommended. The most important element here is testing of the product by the potential users. Several actual use conditions can be created in both the laboratory and the field. Feedback obtained from the users and the personal observations of the testers should be critically evaluated. Redesign and redevelopment may follow depending on user experiences with the product.

Even though at this point a marketable product has emerged, it is highly advised that the product be evaluated continually during field use. Many revision ideas to better serve customer needs are developed through such field evaluations. A function that is already an integral part of the product and that is not used should be deleted. The product may also be redesigned to include additional functionality. Such revised designs may be introduced under different model numbers.

11.2.2 Product Design Guidelines

In this section we provide several product design guidelines to support use efficiency, human factors, and safety in use. The first section covers the general concepts [1]. In the second section we provide specific guidelines [2].

General design concepts.   Major areas of product design focus are:

1. *Design for reliability.* The product should display reliable performance during use. Otherwise, the user will be frustrated with the product. A product that does not fulfill its intended purpose and one that breaks down often does not get user praise.

2. *Design for comfort.* The product should not give discomfort to the user during use. *Comfort in use* can be evaluated via subjective ratings by users. Several variables affect such ratings, including the weight of the product, amount of forces required during use, stress imposed on the soft tissue, hand grip size, and clearances.

3. *Design for durability.* A product should function well during its designed life. Naturally, this assumes normal use. Unexpected product failure may frustrate users and sometimes lead to injuries and deaths. Users will prefer a product that serves its purpose well without repeated failures.

4. *Design for usability.* Usability is a function of the degree to which a product fulfills its intended purpose and comfort in use. A product is more usable if it fully achieves the intended design purposes and is comfortable to use.

5. *Design for efficiency in use.* A product is more effective and more accepted by users if it is simple to operate. A product that is unnecessarily complex to operate will frustrate users.

In addition to the above, product manuals should be easy to follow and readable. They should be supported by accurate illustrations. Where a product is to be used in different countries, manuals developed in several languages are recommended.

Specific guidelines.   These draw from the general theory of the human–machine system and the limitations of the human element in this system. The product designer must understand the user and his or her performance reliability in different environments. A designer who fails to do so exposes the manufacturer to substantial legal liability and negative publicity that will adversely affect the success of the product in the market.

Significant injury risk develops when people are exposed to:

· Environmental agents such as extreme noise, air contaminants, and chemical substances over time (months, years)
· Physical impact with an object

· Localized overstress due to task demands, such as repetitive use of joints in awkward postures

Specific product design guidelines aim at limiting people's exposure to such situations during product use. Table 11.1 gives a representative set of such guidelines.

11.2.3 Product Liability

Many Western countries are now experiencing a change in the relationship between the manufacturer and the consumer. The old philosophy of *caveat emptor* (let the buyer beware) is changing to *caveat venditor* (let the seller beware). The old design atmosphere of strict technical emphasis, such as manufacturability, marketability, design of machine elements and circuits, strength of materials, and mechanics, is being challenged. Backed by legislation and court rulings, product users and consumers are demanding that in addition to sound engineering, products must meet society's design expectation. This expectation manifests itself in a much more comprehensive design process, including failure modes analysis, fail-safe designs, fault-tolerant designs, product safety audits, and hazard analysis. In the United States, over 20 million people are being injured each year as a result of product-related incidents. Of these, it is estimated that 30,000 die and 100,000 get permanently disabling injuries [3].

The Consumer Product Safety Act of 1972. The most important law to consider when studying product liability is the Consumer Product Safety Act (CPSAct). It was passed in October 1972 and was amended several times after that. This act is intended to

TABLE 11.1 PRODUCT DESIGN GUIDELINES

1. Compact products are more appealing and usable than others.
2. Handles should have adequate hand and finger clearance.
3. Vibrating tools should be dampened to minimize effect on arms.
4. Products should be convenient to carry.
5. The weight-holding and force application and control functions should be separated. Then one hand may hold and the other may apply forces or control.
6. Do not use glossy paint or highly polished surfaces.
7. Color coding may be applied for functional separation.
8. Provide slip-resistant grip surfaces.
9. Minimize rotational movements on the hand. Keep center of gravity of the product aligned with the center of grasping hand and close to it.
10. The product should be designed such that the hand and the forearm are aligned during use.
11. Force applications should be held to a minimum.
12. Grasping surfaces should be free of protrusions.
13. Setscrews should not protrude from product surface.
14. Round edges and corners.
15. All moving parts should be guarded.
16. Use low-voltage circuits for electrically powered products.
17. Electrically powered products should be double insulated.

Source: Reprinted with permission. L. Greenberg and D. B. Chaffin, 1977. *Workers and Their Tools: A Guide.* Pendell Publishing, Midland, MI.

provide protection for the public from unreasonably dangerous products. The emphasis is on the nonworkplace environment, such as the home, school, and recreation areas.

Similar to OSHAct and OSHA, CPSAct created the Consumer Product Safety Commission (CPSC). It was activated in May 1973 to administer and enforce the act. One of its primary responsibilities is to develop safety standards for consumer products sold in the United States. It does this by requesting standard developing organizations to develop and submit proposals. Proposals may become standards after internal reviews and public hearings. The commission also enforces the standards developed. It has legislative authority to do so in the courts with civil and criminal penalties, including jail terms and fines. Other responsibilities of the commission include:

1. Assist consumers in evaluating the safety of consumer products.
2. Support research, development, and investigation into the causes and prevention of consumer-product-related injuries and fatalities.
3. Assist states in developing uniform related local rules.

Certain products, such as aircraft, boats, drugs, foods, alcohol, firearms, motor vehicles, medical devices, tobacco, and pesticides, are not covered by the CPSA. Other agencies have jurisdiction over these products. CPSC also maintains injury/fatality data related to consumer products. It collects the relevant data through hot lines and an electronic system called NEISS (National Electronic Injury Surveillance System). This system is operated under contract with close to 120 hospital emergency rooms. If CPSC determines that a particular product is unreasonably dangerous (from the data collected), it can require the manufacturer to:

1. Recall and repair the product.
2. Replace the product.
3. Refund the purchase price.

It is a good business practice to review the commission's existing product safety standards before designing a new product. During the product design stage, the design engineer can communicate with the commission to make the product safe for its use.

Product liability insurance. A manufacturer may do everything possible and apply all the available information to make a product safe for use. Despite this, liability suits may be encountered. These require defense and possibly remedy for damages. The internal design review may be complemented with product liability insurance to minimize claims risk. Product liability insurance is ordinarily provided by an independent carrier. It is not uncommon to experience significant increases in annual premiums. Typically, the manufacturer is covered to a certain limit (assume $150,000) per plaintiff (person or agency claiming damage compensation). Additional insurance can be purchased for umbrella coverage.

In addition to damage compensation, product liability insurers provide other services to the manufacturer or the seller. These services include advice for minimizing product liability risk, general loss control, claims investigation, and defense help. Insurers are experienced in these areas and provide vital help to the manufacturer.

Product liability concepts and causes for claims. The two legal bases that underlie product liability are related to the principles of contract and tort law. A *contract* establishes a binding relationship between two parties. In relation to product liability, a contract relates to the sale of a product. A contract is a legally binding relationship, and if breached, a remedy must be provided. The concepts of contract law that are most relevant to product liability are breach of warranty and privity of contract. *Privity of contract* establishes direct relationships between two parties. In the domain of a sale, a *warranty* may be express or implied. They both have the same legal effects. They both imply that the seller has made an express guarantee, governed by the contract law, when a sale was made. *Express warranty* establishes the basis for a sale. It assumes that the buyer purchased the product on the assumption that whatever has been claimed by the seller (words, printed or written statements) is true. These become parts of the contract between the parties. However, in terms of *breach of warranty,* the important one is the *implied warranty.* Implied warranty is automatically activated by law at the time of sale and guarantees various product attributes. The most important attribute is fitness for use, which implies that the product is safe for intended use.

A *tort* is a civil wrong independent of contract. It is a wrongful act or failure to exercise due care. Such an act may result in an injury and civil legal action may follow. Tort law establishes acceptable human act standards. Breach of these standards requires a remedy. One of the tort theories is *negligence*. Negligence relates to the failure to exercise due care. The standard act is what a "reasonable and prudent" person would have done under the circumstances. The second and most important element of tort with respect to product liability is *strict liability in tort*. Under this doctrine, care of the manufacturer is irrelevant. In court, the plaintiff no longer has to prove the manufacturer's negligence. To recover damages, all that the plaintiff has to prove is that the product was defective, unreasonably dangerous, and the proximate cause of harm. Hence the focus is on fault in the product and not on fault of the manufacturer.

Strict liability in tort is not the same as absolute liability. The plaintiff still has to prove that the product was defective (unreasonably dangerous). The concepts of "unreasonably dangerous" and "unreasonable risk" are subjective concepts. What attributes constitute unreasonable risk and danger are subject to much debate which the lawyers argue during a court case.

As discussed earlier, in the United States, there are three theories under which liability is imposed on the manufacturer and the seller:

1. Negligence
2. Breach of express or implied warranty
3. Strict liability in tort

Generally, the circumstances involved dictate which theory to use in a claim. However, strict liability in tort is the one generally used since under this doctrine, the manufacturer cannot use the contributory negligence and statue of limitations defenses.

Minimizing liability risk. Modern product design requires a comprehensive view of the design stage. The entire life cycle of the product must be studied. Every planned use and foreseeable misuse must be considered. The designer must review injury potential to the user as well as to anyone nearby. Both typical and atypical users must be considered. If it is not feasible to guard against all hazards that make the product dangerous; warnings must be included in labels on products, in instruction manuals, and in other material associated with the product. Ergonomists often serve as expert witnesses in the claims courts, sometimes for the plaintiff, sometimes for the defendant. Lawyers are also available who specialize in product liability cases.

11.3 DESIGN FOR MAINTAINABILITY

Any presentation on equipment and product design is not complete without a section on maintainability design. Simply stated, the functional status of equipment must be maintained during the operational phase. Maintainability design is concerned with a review of the design cycle of equipment with the purpose of making it available to use for a maximum amount of time. No product can be designed for 100% availability. However, proper care during all stages of the development cycle may assure that the product will be operational most of the time. Such care may also result in minimum downtime for repair.

With the background above, we can state the objectives of design for maintainability as follows:

1. Minimize:
   · Down time
   · Maintenance time
   · Damage to equipment
   · Damage to personnel
   · Overall maintenance costs
2. Maximize:
   · Availability

11.3.1 Maintainability Design Features

Maintainability design features are developed during the conceptual stage of product development. During initial stages of the design, factors such as accessibility, visibility, and provisions for troubleshooting, lubrication, and the like are considered. All alternative designs may be subjected to maintainability demonstrations. As discussed before, the purpose is to have the end product maintainable by properly trained workers.

There is a direct link between equipment and product reliability and maintainability. Naturally, a product that is fully reliable is available. Such a product meets all the objectives of design for maintainability. However, not all products are fully reliable. This is where maintainability comes into play. Before discussing several specific maintainability features, it may be beneficial to review two types of maintenance:

1. *Preventive maintenance.* This is to help keep the product in good working condition. Even though preventive maintenance (PM) often results in product shutdown for short

periods of time, overall it may be argued that this type of maintenance will be more cost-effective. Naturally, PM requires a planned maintenance schedule.

2. *Corrective maintenance.* Corrective procedures are applied when a product or equipment breaks down and needs maintenance. The objective is to return malfunctioning equipment to operational condition. Corrective maintenance does not require a schedule. It requires maintenance resources available at all times.

In addition to the two types of maintenance described above, a product may be overhauled (completely rebuilt) periodically to restore the original condition. All three cases require ergonomics to be applied for effective handling of the product and efficient execution of the maintenance process. In each case, specific features enhance maintainability. Table 11.2 presents such a set of features [4,5].

Maintenance procedures should be well documented. In addition to instructions, schematics and flow diagrams are recommended. Also, user characteristics and use environment(s) must be considered. Additional items to consider when developing maintenance procedures are the following:

1. Assign an optimal number of decisions to the maintenance worker, neither too few nor too many.

2. Maintenance procedures should be brief but not sacrifice important elements. Provide systematic, clear, and step-by-step procedures.

3. Follow the equipment manufacturer's suggestions as much as possible.

TABLE 11.2 MAINTAINABILITY DESIGN GUIDELINES

1. Locate hand, arm, and body access points such that the potential for accidental contact with energy sources (electricity, heat, moving machine elements) is minimum.
2. Provide guards for moving machine elements and nip points.
3. Allow for safety interlocks on accesses leading to equipment or parts carrying high voltages.
4. Locate controls away from high voltages.
5. Design holding devices for parts that may tip over or slide and injure personnel.
6. Provide safety locks on hydraulic equipment to prevent accidental dropping of a load in case of hydraulic failure.
7. Locate accesses on exposed surfaces.
8. Locate accesses on the same side of equipment as related displays and controls.
9. Equipment that require more frequent maintenance should be located in more accessible locations.
10. Color-code wires for easy troubleshooting.
11. Provide warning labels on all accesses leading to high voltages and other hazards.
12. Consider availability of replacement parts.
13. Arrange modules such that each unit can be adjusted, checked, and lubricated independently.
14. Solder connections must be adequately spaced for the operation expected.
15. Hand-operated plug-in contacts are better suited for maintenance.
16. Alignment pins, keyways, keys, and the like facilitate proper alignment of connectors.
17. Allow for at least 9 cm (3.5 in.) opening for bare one-handed access.
18. Quick-connect handles may help the handling operation where permanent handles are not desirable.
19. Locate handles such that the center of gravity of the unit handled is below the grip.
20. Designate maintenance components uniquely for quick identification.
21. Use coding to distinguish between test and service points.
22. Use minimum number of fasteners consistent with stress demands.

Source: Adapted from Refs. [4] and [5].

11.3.2 Troubleshooting Performance

Maintenance effectiveness depends not only on maintainability design features, but also on the effectiveness of human troubleshooting performance. Furthermore, the characteristics of troubleshooting performance help identify maintainability design features that otherwise may not have been thought of. The remainder of this section focuses on the background theory in human troubleshooting performance.

Su [6] suggests that there is considerable correlation between real-life troubleshooting performance and results obtained through simulator-based research. Such studies evaluate troubleshooting performance in terms of several overall time scores, and subscores on access, correction, and special test procedures. Furthermore, specific designs can be evaluated in terms of number of parts removed and number of checks required.

Highland et al. [7] and Chase and Simon [8] showed that technical knowledge and understanding of the underlying logical relationships between subsystems positively affect troubleshooting performance. Whereas novices focus on surface features of problems, experts orient their efforts around the underlying theory and the resulting meaningful relationships. This has important troubleshooting training implications. For example, system charts may be developed using pictures showing hierarchical relationships between subsystems.

Research also shows that the fault-locating behavior is not complete or optimal. Rouse [9] showed that human beings tend to focus on "positive" indications of abnormality (an expected signal is absent or a value is in significant deviation from normal) and seldom on "negative" (all normal) indications. Along the same lines, Mehle [10] claims that troubleshooters use incomplete hypothesis sets in looking for faults. A quick solution to this "inefficient" fault-locating behavior may be to provide, for each problem case, a complete list of possible reasons for the observed fault and let the maintenance operator eliminate candidates one at a time.

McDonald et al. [11] observed that complexity of equipment to be troubleshot also affects fault-locating behavior. Although there are still issues with respect to the definition of complexity, it may be a good training exercise to develop troubleshooting flowcharts that contain only the necessary information.

There are marked individual differences in troubleshooting performance. This can be observed even in groups with similar training. Hence some performance variance can be expected between workers. A managerial follow-up may be necessary to assign those who are not proficient on the assigned job to other tasks.

11.4 HAND TOOL DESIGN

After a general discussion of product design, we now turn our attention to a more specific case, hand tool design. A hand tool extends the capabilities of human beings to do tasks that are otherwise not possible with mere use of the body. A power screwdriver, a jackhammer, a handsaw, and a soldering iron are examples. Hand tools are used in machine shops and research and development laboratories, on production floors and assembly lines, and in the home. Hence the design guidelines reviewed in this section apply to all use environments, not only to industrial work. There is more emphasis on tool use in the industrial environment, since in these situations hand tools are used in a repetitive manner, which is a major factor in evaluating potential impact on body tissue.

The primary consideration in the design of hand tools is their potential use. "What is the purpose of this tool?" is the first question to be asked and answered by the designer. A tool is of no value unless it can fulfill the design purposes. Misuse is also possible. However, injuries and damage caused by misuse are always questioned in the court. The best safeguard against misuse is to review, at the design stage, all possible uses of the tool along the lines of intended purpose. More information on this subject will be provided later.

In addition to the primary "use purpose" consideration, the hand tool must also not pose any safety and health hazard to the user. This is true in all use situations for which the tool has been designed. In general, engineers are so concerned with the design of the production process and the associated workstations that proper acquisition and design of needed tools are of last priority. However, workers use these tools day in and day out. In the long run, poorly designed hand tools cause significant tissue damage, in addition to lowered productivity.

11.4.1 Musculoskeletal Impact

One of the major occupational risk factors is associated with the use of hand tools. While many tools do not cause problems, some may lead to significant damage to the body tissue over time. The damage is usually to the musculoskeletal tissue as discussed in Chapter 3. Many CTD cases can be traced directly to the design of hand tools and the methods of use. A review of hand injury statistics in the United States [12] showed that hand tools were involved in 9% of all reported disabling injuries. However, this ratio is probably low since detailed records with respect to CTD cases have been available only since the early 1980s. About three-fourths of these injuries are related to manual tool use. In 1994, tool use accounted for 18% of repetitive-motion injuries. Armstrong et al. [13] observed that in the thigh-skinning department of a poultry processing plant, the cumulative trauma incidence rate was 130 cases per 200,000 hours of work. An incidence rate (explained in Chapter 13) of 10 or above normally attracts much management attention.

Armstrong [14] claims that muscle activity level in the hand and the resultant tendon tension depends on the grip configuration, and to a lesser extent, hand and wrist anthropometry. Furthermore, the angle of the wrist during grip exertion affects the amount of normal forces acting on the tendons and their synovia. Large intrawrist forces affect the integrity of the synovia, resulting in inflammation, swelling, pain, and at later stages, entrapment of the median nerve in the carpal tunnel, leading to carpal tunnel syndrome (CTS). The design guidelines presented next are based on these concepts.

11.4.2 Guidelines Based on Hand Tool Design

Adverse effects of hand tool use may originate from two major causes: poor design and improper use. Although they are frequently related, we will distinguish the design guidelines based on these concepts [15,16].

1. *Shape handles for grip assist.* The primary emphasis here is to distribute forces across the hand and not concentrate them on one small area. Although the center of the

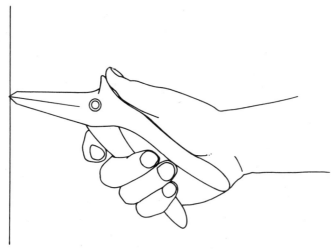

Figure 11.1 A thumb stop adds extra stability in tool use.

palm is designed to have a good grip on small objects, it is not suitable for stress concentration. The arteries, the median nerve, and the synovium of the finger flexor tendons in the region make the palm highly vulnerable to repeated force exertions. The length of the tool handle should be sufficient to distribute load on either side of the palm. Furthermore, wider and cushioned handles allow for a wider area of the palm to bear the resultant forces. The handle should also have a small curvature, depending on the application, which helps in the force distribution. Stops on the handle to prevent tool slippage, especially in sweaty hands, are also recommended (Figure 11.1).

2. *Consider handle clearance for optimal grip span.* The size of the object grasped has a direct impact on the extent of grip strength. Tichauer [17] showed that if an object is very small, sufficient grip forces cannot be applied. If an object is rather large, again, not enough force can be applied. Similarly, a force that acts on the distal segments of the fingers leads to more damage than the same force acting on more proximal segments. Fitzhugh [18] conducted studies on 50 male and 50 female employees of a large electronics component manufacturing plant with respect to linear grip span. Figure 11.2 gives the relationship between grip strength and grip span measured at the center of the hand [2]. Grip span is the distance between the open handles that needs to be grasped with force. As can be observed, maximum grip forces can be exerted at about 7.5 cm (2.95 in.) handle opening. Women demonstrate one-half of the grip strength of men, and there is a large variability in strength values between population percentages within the same gender. If the tool handle is round, optimal diameter for maximum grip strength seems to be 4 cm (1.57 in.) [19].

3. *Consider finger clearance.* Tools and manual handling containers must consider clearance for fingers for safe operation. Clinched, crushed fingers are frequently the result of inadequate clearances. A use analysis coupled with data on finger anthropometry should be sufficient for designing with finger clearance in mind.

4. *Minimize concentrated stress over soft tissue.* Stress concentrated over small areas of the palm may lead to obstructed blood flow and nerve function. Small, narrow, and short handles, such as a paint scraper with the handle running into the middle of the palm during

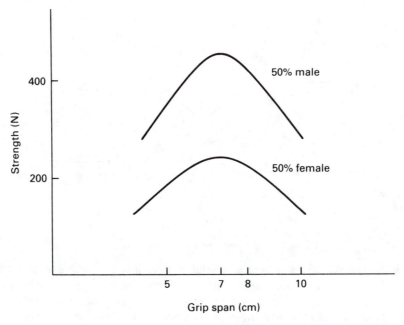

Figure 11.2 Optimal grip span. (Adapted from Ref. [2].)

use, will cause such ailments. Furthermore, finger grooves may produce focused stress, since the hand of a user will seldom conform with the molded pattern [20].

5. *Protect hand from external energy.* Tool handles must protect hands from possible ill effects of external sources, such as heat and cold, vibration, and electric shock. If not insulated, air-powered tools may expose hands to cold. Cold hands cannot exert forces. This may lead to reduced productivity if not to loss of grip and possible accidents. The handle must also be insulated against heat or electric shock potential. Compressible rubber handles will help reduce transmission of vibration to the hand as well as help with the grip without hand grooves. Bouenzi et al. [21] suggest that chipping hammer use should not exceed 4 hours per person per shift if the acceleration of the tool is 20*g* or more.

11.4.3 Guidelines Based on Hand Tool Use

The following guidelines focus on tool use factors:

1. *Maintain straight wrist.* Maintenance of wrist in a straight position (neutral posture) during tool use is the primary design consideration. This may require a bent handle, as illustrated in Figures 11.3 and 11.4. Use of a tool with the wrist in a deviated posture (ulnar, palmar, dorsal, and radial) may lead to CTD cases. Furthermore, the hand may not be able to exert sufficient forces under wrist deviation. In a study with industrial workers, 60% of those using straight-handled pliers developed wrist disorders (tenosynovitis, carpal tunnel, etc.) versus 10% of those using the bent-handle design [22]. Krohn and Konz [23] suggest that the idea of handle bending must not be abused. Not every job will get help due to this modification. The critical parameter is the use pattern, which depends on factors

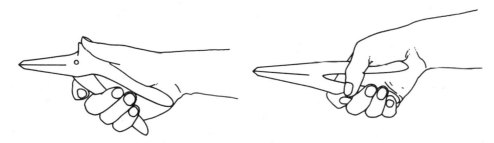

Figure 11.3 Bent pliers allow user to maintain stright wrist.

such as job and workplace design. The idea of bending the handle has also been applied to consumer products, such as the Reach toothbrush, which is claimed to have a larger contact area with the teeth and gums than do its competitors.

2. *Use gloves with caution.* Gloves may sometimes be worn during tool use. There are several points to be considered when using gloves. They require extra clearances for fingers and hands. Gloves reduce finger and wrist dexterity. Forces that can be applied with gloves are less than forces that can be applied with bare hands. Loose or bulky gloves may get caught in rotating tool elements, possibly causing injury. Gloves should be free of skin irritants. Tichauer [22] proposed that tight gloves between fingers may give a false sensation of a secure grip. This may lead to tool slippage and possibly accidents. On the other hand, Mital et al. [24] tested several types of gloves on force/torque exertion capability, using seven different nonpowered hand tools. The results indicated no change in muscle activity between glove and no-glove conditions, and peak torque exertion capability of individuals increased with glove use.

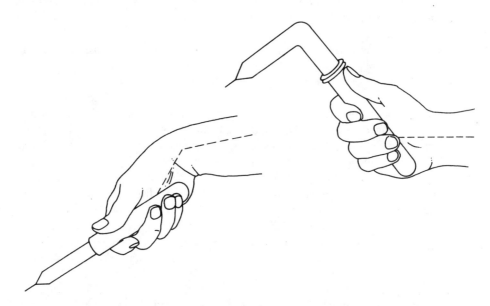

Figure 11.4 Straight wrist with bent soldering iron handle. (Adapted from Ref. [16].)

3. *Prevent tool slippage during use.* Flat metal and smooth plastic tool handles can easily slip during use, especially in hot and humid environments. Such tools will also require higher grip forces. The best alternative is compressible rubber handles, which do not absorb liquids and oils.

4. *Avoid repeated finger actions.* A finger must not be required to open or close tool handles during use. This should be accomplished by a spring mechanism. In addition, trigger mechanisms must not be operated with extended fingers, which may lead to *trigger finger* syndrome. If force needs to be applied, the preference is for triggers that are operated by several fingers [25].

5. *Minimize static loading on the hands and the arms.* The workplace and work methods must not call for the use of hand tools, especially heavy tools, for extended periods of time. This will result in significant muscle fatigue and stiffness in joints. The weight of the tool and its distribution are major variables here. For extended period use, tool weight should not exceed 0.5 kg. The center of gravity should be close to the hand to minimize tool slippage and turning during use. Additional weight may be created by power cords and air hoses attached to the tool. A tool balancer suspending the tool during or after use may be a consideration for weight management. Counterbalancing of weight is another alternative. In this respect, the workplace should be arranged such that the tool is used with elbows close to the body and not elevated.

11.5 FORM DESIGN

Ergonomic design of information aids is important in effective information transfer between people and equipment or systems. We now turn our attention to presenting specific information on the design of three most-used information aids: forms, computer screens, and signs/labels.

Forms are used in many facets of life. There are tax forms, insurance forms, job application blanks, and so on. It is no different in the industrial environment. Resource forecasts, budget forecasts, and annual performance reviews are some functions that require the use of forms. Although with information automation many paper documents are now being replaced by electronic files, the basics of form (or means of data collection) design do not change. Only the communication medium changes. There are still questions to be answered and sequential relationships and other characteristics associated with each.

The ergonomic aspects of form design relate to the speed of use, accuracy of responses, and whether or not a form is used in the first place. It could be that the form is so complex and time consuming that the user may be frustrated and may not provide the necessary data. Although the organization expecting the data to be filled in may penalize the user in such a case, it should make every effort to human engineer the form so that it can be used effectively. It is humane to do so and it costs less to society to do so.

A well-designed form collects the needed data with the fewest errors in the least possible time with the least user frustration. A form must be self-sufficient in many respects for the purpose intended. Bailey [26] suggests that most data collection errors are related to the design of forms. Furthermore, Lieberman et al. [27] claim that poor printing quality and poor organization of information on the form may be the reasons that people do not use forms effectively.

Research shows that forms must satisfy the need to be informed quickly and accurately [28]. However, they should impose only a light reading burden. Thomas [29] showed that when people read, they tend to skip words or lines. People often try not to read, but instead try to do things. Kammann [30] claims that what is read is misunderstood one-third of the time. A major reason for this may be that many documents are written at an expert level. Pakin [31] proposes that a good form is clear, legible, scannable, applicable to the audience, complete, and organized. Broadbent [32] demonstrated that positive and active sentences are easier to understand than negative and passive sentences.

All of the research on form and documentation design, a sample of which has been provided above, leads to a number of suggestions for form design. Table 11.3 gives a collection of general recommendations. Table 11.4 gives several recommended physical characteristics of forms.

TABLE 11.3 GENERAL RECOMMENDATIONS FOR FORM DESIGN

1. State the purpose of the form.
2. Provide a sample of a completed form.
3. Provide general instructions on the form.
4. In addition, provide specific instructions for certain items.
5. Specify error correction procedures.
6. General instructions may be grouped at the top of the form or at the back.
7. Place specific instructions just prior to the specific items to which they relate.
8. Always test your instructions with a group of subjects and modify them if necessary.
9. Horizontal headings are recommended.
10. Use short sentences.
11. Use active sentences. They are easier to read.
12. Avoid negatives or double negatives. Positive sentences are better responded to than negative sentences by human beings.
13. Avoid double questions.

Source: Adapted from Ref. [26].

TABLE 11.4 SPECIFIC RECOMMENDATIONS ON PHYSICAL CHARACTERISTICS OF DATA-ENTRY FORMS

1. Use standard form dimensions. Odd dimensions cannot be processed by automated equipment.
2. If machines will print entries in answer spaces, consider their print characteristics.
3. For entries that are to be handwritten, allow 6 to 8.5 mm (0.236 to 0.3346 in.) vertical line space.
4. For horizontal spacing of handwritten characters, allow 3 to 4.5 mm (0.118 to 0.177 in.) per character.
5. Answer spaces should be as close to the questions as possible.
6. Writing on lines where subdivisions are denoted by small vertical marks (ticks) may be better than boxed delimiters. However, free format is the best.
7. Recommended minimum between-line spacing is 0.7 mm (0.02756 in.).
8. At least 13 mm (0.512 in.) of margin space is recommended.
9. Space groups of information by at least 2 mm (0.0787 in.).
10. Recommended type size is 3.5 mm (0.1378 in.).

Source: Adapted from Ref. [26].

11.6 COMPUTER SCREEN DESIGN

There are many similarities between form design and computer screen design. The major difference is that a form is mostly a data collection medium. A computer screen such as a CRT may be used to display and/or collect data. Similar to form design considerations, a computer screen must be simple to use and clean (i.e., free from irrelevancies). The messages presented must be understandable. Visual clarity is affected by numerous factors including information organization, legibility, relevancy, and clutter. The last item deserves special mention. To improve screen clarity and readability, Dunsmore [33] listed separate items on different lines on a screen. His original design had included the separate items listed on the same display line to conserve space. However subjects were 20% more productive with the separate-line version, indicating the importance of minimum clutter on the screen. Similarly, there are questions concerning the amount of information to be included on computer screens. A measure of the amount of information could be information density (percent of available screen area used). Danchak's [34] studies lead one to conclude that a density loading of 40% or less is desirable. Other studies found that the search time on inquiry screens increased rapidly when screen density exceeded 60%. Questions with respect to information density per line have been answered by Rehe [35], who suggested that a text line should contain no more than 40 to 60 characters.

Eye fixation studies show that initially the human eye usually moves to the upper-left center of the display and follows a top–bottom, clockwise search pattern [28]. This suggests that information ordered in this fashion on a computer screen should be compatible with human expectations.

The above ideas provide a lead-in to specific computer screen design guidelines. Many such recommendations have been offered over the years by various researchers. Table 11.5 gives a sample set based on Galitz [28] and Eastman Kodak [36].

11.7 LABEL AND SIGN DESIGN

Labels and signs convey human-to-human information. Hence the designers must make sure that the messages get conveyed properly with minimum confusion. In addition to human recall, interpretation, and discrimination skills, the designer must consider the environment in which labels and signs function. Since labels and signs generally contain alphanumeric information, ergonomic recommendations with respect to their design (such as character height, stroke width, etc.) as given in Chapter 10 apply here. In this section we concern ourselves primarily with formatting, location, and use conditions.

With the objective of enhancing the efficiency of information transfer, the following list summarizes design recommendations [36]:

1. Each label or sign must have a clear and useful purpose of design.
2. Use brief and concise messages. Avoid ambiguous wording. Use active phrases.
3. If possible, use standardized labels and signs.
4. Engraved labels should not be used in areas where dirt is likely to accumulate. Protect labels and signs from heat and corrosive chemicals if necessary.

TABLE 11.5 COMPUTER SCREEN DESIGN RECOMMENDATIONS

1. Recommended location for error messages is the bottom of the screen. Various highlighting methods may be used to attract attention. If a flash rate is used, the suggested rate of flash is 3 Hz.
2. Reserve the top-right of the screen for status information. Maintain location consistency between screens.
3. Information of critical nature may be presented near the center of the screen.
4. Keep the information density (amount of space filled with characters) to less than 50%. This may be accomplished via:
   · Using coded information.
   · Using graphics.
   · Displaying only high-priority items. Other items may be accessed when need arises.
5. Provide groupings of screen elements by using blank spaces, surrounding lines, different intensity levels, and so on. However, do not break the screen into too many small windows.
6. Instructions on how to use a screen should be at the top of the screen.
7. Disposition instructions on a completed screen should be at the bottom of the screen.
8. Provide only information that is required for making a decision or taking an action. Do not crowd the screen with unnecessary information.
9. Group all data related to one task on one screen. Do not require users to recall data from one screen to the next.
10. Fonts:
    · For text, use lowercase with the initial sentence letter capitalized.
    · For captions, labels, and headings, use uppercase (capitals).
11. Text:
    · Use short, active sentences.
    · Include no more than 50 to 60 characters on each line.
    · Place a period at the end of each sentence.
12. Use line drawings to illustrate or supplement text wherever possible.
13. Present information in a directly usable form. Do not require reference to documentation, interpolations, and so on.
14. Visual appearance and procedural usage should be consistent.

Source: Adapted from Refs. [28] and [36].

5. Locate signs and labels such that reflections and shades will not make them difficult to read.

6. Labels on curved surfaces (pipes, etc.) must be readable from at least one comfortable point.

7. Headings and short messages may be developed using capital letters.

8. Readability may be improved by a border around the label. Consider accessibility and maintainability in locating labels and signs. Place borders around critical labels if several are clustered in the same area.

9. Warning labels on consumer products should include signal words such as *danger, warning,* and *caution,* as well as an icon, such as an exclamation point surrounded by a triangle, to communicate various degrees of potential hazards [37,38].

10. Mixed modality (word plus symbol) signs most effectively communicate intended action [39].

11. Perceived urgency of warning labels increases with enhanced text size and border width around the signal word. Red signal words generate greater urgency than black words [40].

11.8 SUMMARY

Ergonomic design of products, hand tools, and information aids help in effective and injury-reduced use. In addition, such designs provide protection against product liability suits. The designer rests in ease, and the user enjoys comfort in use.

It is the ergonomist's or engineer's responsibility to make sure that the equipment selected for use does not violate the concepts discussed in this chapter. Alternatively, equipment design considering these concepts minimizes injury potential and maximizes use effectiveness.

QUESTIONS

1. What is the reason behind ergonomic design of products?
2. Briefly discuss the six-step approach to design for human use.
3. What is the benefit of field evaluation of product performance?
4. Discuss design for comfort in use.
5. What is meant by the term *product liability*?
6. What is CPSA?
7. Provide a framework of the activities of CPSC.
8. Why do manufacturers buy product liability insurance if they take precautions during the design stage to make products safe for use?
9. Briefly discuss the two legal bases for product liability suits.
10. Define *tort*.
11. Discuss negligence under tort.
12. Who is an expert witness?
13. Briefly define *maintainability design*.
14. What is the relationship between maintainability and availability?
15. Discuss the difference between preventive and corrective maintenance.
16. What is the primary consideration in the design of hand tools?
17. How is optimal grip span determined in designing tool handles?
18. How does one reduce concentrated stress on the hand due to the handle of a tool?
19. How can one maintain a straight wrist during hand tool use?
20. Discuss four cautions in using gloves with hand tools.
21. What is trigger-finger syndrome?
22. What are the objectives of ergonomic form design?
23. What are labels used for?
24. List and elaborate on two considerations in label design.

EXERCISES

1. Take a consumer product. Evaluate it from an ergonomics point of view. Suggest improvement alternatives.

2. Discuss the relationship between design/product cost and ergonomic design.
3. Evaluate a car engine for design for maintainability. Propose improvements.
4. Go to a local industry. Select three hand tools in production use and evaluate their design from an ergonomics view. Also talk with workers who are currently using them. Are the results of your evaluation correlated with user experiences? Discuss.
5. Select a tax form and evaluate. Are there any improvement suggestions?

REFERENCES

1. Zurwelle, D. W. 1985. Product Design. In *Industrial Ergonomics: A Practitioner's Guide.* Alexander, D. C., and Pulat, B. M. (Eds.). Industrial Engineering and Management Press, Atlanta, GA.
2. Greenberg, L., and Chaffin, D. B. 1977. *Workers and Their Tools: A Guide.* Pendell Publishing, Midland, MI.
3. Thorpe, J. F., and Middendorf, W. H. 1979. *What Every Engineer Should Know about Product Liability.* Marcel Dekker, New York.
4. Altman, J. 1990. Designing for Maintainability. In *Industrial Ergonomics: Case Studies.* Pulat, B. M., and Alexander, D. C. (Eds.). Industrial Engineering and Management Press, Atlanta, GA.
5. Crawford, B. M., and Altman, J. W.1972. Maintainability Design. In *Human Engineering Guide to Equipment Design.* Van Cott, H. P., and Kinkade, R. G. (Eds.). U.S. Government Printing Office, Washington, DC.
6. Su, Y.-L. D. 1984. A Review of the Literature on Training Simulators: Transfer of Training and Simulator Fidelity. *Report 84-1.* Georgia Tech, School of Industrial and Systems Engineering, Atlanta, GA.
7. Highland, R. W., Newman, S. E., and Waller, H. S. 1956. A Descriptive Study of Electronic Troubleshooting. In *Air Force Human Engineering, Personnel, and Training Research TR 56-8.* U.S. Air Force Research and Development Command, Baltimore, MD.
8. Chase, W. G., and Simon, H. A. 1973. The Mind's Eye in Chess. In *Visual Information Processing.* Chase, W. G. (Ed.). Academic Press, New York.
9. Rouse, W. B. 1978. A Model of Human Decision Making in a Fault Diagnosis Task. *IEEE Transactions on Systems, Man, and Cybernetics,* SMC-8, pp. 357–361.
10. Mehle, T. 1980. Hypothesis Generation in an Automobile Malfunction Inference Task. *TR 25-2-80.* Decision Processes Laboratory, University of Oklahoma, Norman, OK.
11. McDonald, L. D., Waldrop, G. R. and White, V. T. 1983. Analysis of Fidelity Requirements for Electronic Equipment Maintenance. *TR-81-C-0065-1.* Naval Training Equipment Center, Orlando, FL.
12. Ayoub, M. M., Purswell, J., and Hoag, L. 1975. Research Requirements on Hand Tools. *NIOSH Technical Report.* University of Oklahoma, Norman, OK.
13. Armstrong, T., Foulke, J., Joseph, B., and Goldstein, S. 1982. Investigation of Cumulative Trauma Disorders in a Poultry Processing Plant. *American Industrial Hygiene Association Journal,* 43(2), pp. 103–116.
14. Armstrong, T. J. 1983. An Ergonomics Guide to Carpal Tunnel Syndrome. *AIHA Ergonomic Guide Series.* American Industrial Hygiene Association, Akron, OH.
15. Nemeth, S. E. 1985. Handtool Design. In *Industrial Ergonomics: A Practitioner's Guide.* Alexander, D. C., and Pulat, B. M. (Eds.). Industrial Engineering and Management Press, Atlanta, GA.

16. Chaffin, D. B., and Andersson, G. 1984. *Occupational Biomechanics*. Wiley-Interscience, New York.

17. Tichauer, E. R. 1966. Some Aspects of Stress on the Forearm and Hand in Industry. *Journal of Occupational Medicine*, 8(2), pp. 63–71.

18. Fitzhugh, F. E. 1973. Grip Strength Performance in Dynamic Gripping Tasks. *Occupational Health and Safety Engineering Technical Report*, Department of Industrial and Operations Engineering, The University of Michigan, Ann Arbor, MI.

19. Ayoub, M. M., and Lo Presti, P. 1971. The Determination of an Optimum Size Cylindrical Handle by Use of Electromyography. *Ergonomics*, 4(4), pp. 503–518.

20. Konz, S. 1979. *Work Design*. Grid Publishing, Columbus, OH.

21. Bouenzi, M., Petroniv, L., and DiMarino, E 1980. Epidemiological Survey of Shipyard Workers Exposed to Hand-Arm Vibration. *International Archives of Occupational and Environmental Health*, 46, pp. 251–266.

22. Tichauer, E. R. 1978. *The Biomechanical Basis of Ergonomics*. Wiley-Interscience, New York.

23. Krohn, R., and Konz, S. 1979. Bent Hammer Handles. In *Proceedings of the Human Factors Society, 26th Annual Meeting*, Santa Monica, CA. Human Factors Society, pp. 413–417.

24. Mital, A., Kuo, T. and Faard, H. F. 1994. A Quantitative Evaluation of Gloves Used with Non-Powered Hand Tools in Routine Maintenance Tasks. *Ergonomics*, 37(2), pp. 333–343.

25. Sanders, M. S. and McCormick, E. J. 1993. *Human Factors in Engineering and Design*, 7th Ed. McGraw Hill, New York.

26. Bailey, R. W. 1982. *Human Performance Engineering*. Prentice Hall, Englewood Cliffs, NJ.

27. Lieberman, M., Selig, G., and Walsh, J. 1982. *Office Automation: A Manager's Guide to Improved Productivity*. Wiley, New York.

28. Galitz, W. O. 1984. *Humanizing Office Automation*. QED Information Sciences, Wellesley, MA.

29. Thomas, J. C. 1982. Ergonomics Takes Many Types of Experts: IBMer. *Computerworld*, 7, p. 22.

30. Kammann, R. 1975. The Comprehensibility of Printed Instructions. *Human Factors*, 17(2), pp. 183–191.

31. Pakin, S. 1980. Evaluate User Documentation before You Buy the Software. *Infosystems*, October, pp. 91–96.

32. Broadbent, D. E. 1977. Language and Ergonomics. *Applied Ergonomics*, 8(1), pp. 15–18.

33. Dunsmore, H. E. 1982. Using Formal Grammars to Predict the Most Useful Characteristics of Interactive Systems. *Office Automation Conference Digest*, San Francisco, April 5–7. pp. 53–56.

34. Danchak, M. M. 1966. CRT Displays for Power Plants. *Instrumentation Technology*, 23(10), pp. 29–36.

35. Rehe, R. F. 1974. Typography: How to Make It More Legible. In *Design Research International*. Carmel, IN.

36. Eastman Kodak Co., 1983. *Ergonomic Design for People at Work*. Lifetime Learning, Belmont, CA.

37. American National Standards Institute. 1991. *American National Standard on Product Safety Signs*: Z535. 1–5. New York.

38. FMC Corporation. 1985. *Product Safety Sign and Label System*. Santa Clara, CA.

39. Kline, T. J. B., and Beitel, J. A. 1994. Assessment of Push/Pull Door Signs: A Laboratory and Field Study. *Human Factors*, 36(4), pp. 684–699.

40. Adams, A. S., and Edworthy, J. 1995. Quantifying and Predicting the Effects of Basic Text Display Variables on the Perceived Urgency of Warning Labels: Tradeoffs Involving Font Size, Border Weight and Colour. *Ergonomics*, 38(11), pp. 2221–2237.

_____CHAPTER 12

Special Topics

12.1 INTRODUCTION

In this chapter we focus on several special topics that are of interest to an ergonomist. The topics were selected on the basis of relative importance to an industrial environment. Each subject discussed can be extended into a full chapter; however, it is not the purpose of this book to discuss them in detail. The reader may refer to the references for further reading.

We look first at *human error*. As discussed before, a major element of human performance is accuracy. It is known that humans are not perfect with respect to performance reliability. If we can predict where and when humans will be highly vulnerable to making errors, then, through system or procedure redesign, we can minimize the effects on system performance. In this section, various other topics, such as human error rate banks and system reliability analysis procedures, are also discussed.

The next topic discussed is *selection and training*. The objective here is to provide the best match between job requirements and skills and abilities and other background possessed. Those who fall somewhat short of the requirements may be trained. Hence, after this process, all positions for system performance are staffed by fully capable employees.

Next, we discuss *job performance aids*. These help a worker perform his or her job efficiently and in a timely manner. Any process, hardware, software, memory aids, or other "things" that help the worker in his or her performance qualifies as a job performance aid. Designers must be aware of the fact that human performance can be supported by such aids for best input to the functioning of the system.

Shift work is discussed next. For economic reasons, many industries operate in a shift-work environment. Various types of shift schedules are available. In this section we discuss these along with the effects of shift work on human well-being and social life. Ergonomists have much input that they can provide in this area to make shift work have as few distractions as possible.

The next topic discussed is *design for the elderly*. American population is getting older. The trend is likely to continue. Designers must consider the differences between the younger and older population when developing products, work centers, and work methods. Such differences are discussed in this section together with several relevant human characteristics.

Finally, *ergonomics of advanced manufacturing* is discussed. The objective here is to note new areas of involvement for ergonomics in advanced manufacturing environments. Such operations are increasing at a rapid rate, much too rapid for ergonomics to catch. In this section we provide insights on the application of previous experiences and directions for future work in such environments.

12.2 HUMAN ERROR

Meister [1] defines *human reliability* as the probability of completing a task successfully within a given time period, if such a requirement exists. Swain and Guttmann [2] define it as the probability of correct performance of a system-required activity within a required time period (if such a requirement exists) while not degrading the system in any other manner. Therefore, any behavior or action that falls outside the limits of acceptability is *human error.*

Not all human errors result in system malfunction. Many errors are caught during subsequent inspection processes. Furthermore, if multiple channels are involved in performance (two or more people carrying out the task simultaneously and at the same physical location), one may catch the error of the other. Swain [3] calls these the *recovery factors*.

The criticality of an error depends on the expected impact on system performance. If an operator counts and places 101 parts in a container destined to an assembly process, even if only 100 were required by the process, this error will not create any ill effects. On the other hand, if the operator sent only 90 and no excess buffer of this part was present at the process, several assemblies will wait for the remainder. Depending on the duration of supply period, significant costs and inefficiencies will be incurred. Another example can be given from the air transportation industry. A baggage wrongly routed will probably not cost much money and human suffering; however, an airplane crash due to an inappropriate or forgotten emergency handling process will.

The ergonomist's role in system design with respect to human errors is to anticipate such errors. Then the human interface points can be designed in a manner that minimizes error potential. Another alternative or complementary procedure could be the development of *fault-resistant systems*. Such systems anticipate human error and have internal checks or correction procedures before a final input is accepted.

12.2.1 The Reasons for Human Error

There may be many reasons for human error. The following probably span 95% of the root causes:

    1. *Malevolent behavior.* Errors made on purpose are normally outside the limits of human error definition. The intention is important. If the operator intentionally committed an error, this act is not considered an error that requires ergonomic investigations. Although this is the accepted standard, one of the causes of human error remains to be malicious behavior.

    2. *Human capability less than the job requirement.* Whenever the job requirements exceed human capability, operators may have to make estimates, not perform within the time period allotted, or do an inferior job. In all cases, errors will be registered. Since one of the purposes of ergonomics is to develop the best match between the two, sound ergonomics applications will minimize such errors.

    3. *Insufficient or overstress.* As discussed before, too little stress will lead to boredom, lassitude, and drowsiness. These are situations where people will be most likely to make errors. By the same token, too much stress causes exhaustion. These too will result in many errors. Human performance is best when there is some stress, but not too much or too little.

    4. *Work not human-engineered.* Fitting the task to the person is the essence of ergonomics. When workplaces do not allow for adequate clearances, reach geometry, and adjustability for a large proportion of the working population; and work methods do not consider expected human behavior, motion economy, quality of working life, and desired work postures, then human efficiency and effectiveness will be degraded. This will lead to errors.

    5. *Insufficient/incorrect training.* The workplace and the work methods may be ergonomically designed within optimal environmental conditions. However, if the worker is lacking the skills necessary for acceptable performance, he or she cannot perform effectively. Among the inefficiencies will be errors committed. It is the designer's responsibility to specify the skills necessary for acceptable performance on the job and the personnel specialist's responsibility for staffing the jobs with properly trained operators.

    It is obvious from the discussion above that ergonomics has a lot to offer to minimize human-initiated errors.

12.2.2 Classification of Human Errors

Human errors may be classified in several ways. The two most widely accepted ways are:

    1. Classification based on the underlying behavior [4]
    2. Classification based on the type of operation [5,6]

Swain's [4] behavioral classification divides errors into errors of *omission* and errors of *commission*:

1. *Errors of omission.* This is the case where the operator does not do something that he or she should. A step in a task may be omitted or the entire task may be omitted. An example is the failure to print the part number on a document where it is needed and required. Errors of omission may arise from inadequate training or too much or too little stress, but the most likely reason is the former.

2. *Errors of commission.* Here the operator does the task but does it incorrectly. Possible reasons include incorrect selection of the applied behavior (read instead of write), incorrect task sequence application (fire and aim instead of aim and then fire), failure to complete task in time, or insufficient application (too little force applied on the torque-wrench). All of the possible reasons for making errors are indicated here.

Meister [5,6] classified errors based on the type of activity that led to them. In this respect, the following types can be listed:

1. *Operating errors.* These are errors made by the operating personnel in a field-use environment. Any type of error could be made by such personnel during equipment use.

2. *Assembly errors.* Errors made by assembly workers while assembling something. These are workmanship errors, which may be found during in-plant inspection or after experiencing product failures in the field.

3. *Design errors.* These are due to inadequate or insufficient design by designers. Causes may include insufficient design time and inadequate design background.

4. *Inspection errors.* Inspectors are not 100% accurate. They may reject items/assemblies that are good or miss items that are bad.

5. *Installation errors.* These are errors made during equipment installation. Causes are insufficient installation experience and failure to follow instructions.

6. *Maintenance errors.* Maintenance workers also make mistakes. Incorrect repair of the equipment and incorrect calibration are examples.

Errors are usually at their peak during the initial stages of equipment operation. Another peak occurs toward the end of the life of the product. Design errors, installation errors, and maintenance errors are at their maximum during and a little after product delivery. As the product gets used, operating errors first rise and then fall to stay at a flat rate. All other error rates fall. During the middle of the life of a product, the smallest error sum is observed. Toward the end of the product life, maintenance errors start rising. This is the result of more frequent maintenance due to wear-out. The total errors also rise due to relatively flat operator, design, and installation errors and increasing maintenance errors.

Meister [7] and Willis [8] estimated human-initiated failure percentages on different types of tasks. These estimates attributed a significant proportion of system failures to human errors. An average error rate between 40 and 50% is not uncommon. The 1985 aircraft accident percentage share involving either the pilot (63.6) or the personnel (36.4) has been given as 100 [9]. This means that in each accident, human error was one of the contributory reasons.

Several data banks have also been established for human error rates in industrial tasks. Table 12.1 gives a representative set of values [10–12] Unfortunately, the data banks established have not registered extensive human reliability data. This is probably the most serious problem in this field.

12.2.3 Human Reliability Analysis (HRA) Techniques

Since the early 1950s, scientists have been looking into developing new techniques or modifying existing ones for human reliability analysis in systems where multiple human and equipment functions exist. We will not present all of the techniques developed for this purpose. It will suffice to mention a few and dwell on one a little deeper.

Meister [13] provides a critical review of the available techniques as of early 1970. Using the same terminology, several available techniques can be classified into two categories:

1. *Operability methods.* These focus on estimating performance reliability on tasks that relate to equipment operation. Some of the better known ones are the AIR Data Store, THERP (Technique for Human Error Rate Prediction), TEPPS (Technique for Establishing Personnel Performance Standards), the throughput ratio technique, the Pontecorvo technique, HOS (Human Operator Simulator), ORACLE (Operations Research and Critical Link Evaluator), and SAINT (Systems Analysis of Integrated Networks of Tasks).

2. *Maintainability methods.* These techniques have been developed to predict performance reliability on maintenance tasks. Several of the better known ones are ERUPT

TABLE 12.1 HUMAN RELIABILITY ESTIMATES

Task	Reliability
Turn a control in the incorrect direction under high stress	0.5000
Read technical instructions	0.9918
Read electrical or flowmeter	0.9945
Install gasket	0.9962
Read a graph	0.9900
Record reading	0.9966
Mend hole in solder	0.9300
Read pressure gauge	0.9969
Connect electrical cable	0.9972
Position hand valves	0.9979
Read time from watch	0.9983
Remove nuts, plugs, and bolts	0.9988
Remove pressure cap	0.9988
Remove drain tube	0.9993
Mate a connector	0.9900
Use a checklist correctly	0.5000
Keypunch (per entry)	0.9997
Print (per character)	0.9995
Inspection performance	
Conforming item accepted	0.9–0.99
Nonconforming item rejected	0.8–0.9

Source: From Refs. [10], [11], and [12].

(Elementary Reliability Unit Parameter Technique), the Personnel Reliability Index, and the MIL-HDBK 472 prediction methods.

Some of the more recent human reliability analysis techniques are MAPPS (Maintenance Personnel Performance Simulator) [14], OAT (Operator-Action Tree Method) [15], SLIM (Success Likelihood Index Methodology) [16], and STAHR (Sociotechnical Approach to Assessing Human Reliability) [17].

THERP is a well-known technique for human error rate prediction and evaluating the degradation to the system likely to be caused by human errors (along with equipment and procedural reliability). It was developed in the early 1960s. It makes use of human error rate data from any reasonable set, including expert judgments. THERP requires that system missions be segregated into several functionalities, and that for each functionality, a detailed task analysis be carried out indicating all sequential relationships. Then human tasks are assessed as *basic error rates* and equipment tasks are assessed as failure rates. Joint probabilities are then developed for tasks composed of units of behavior that are concatenated in time (such as entering a nine-digit part number). Considering recovery factors and task dependencies (ZD: zero dependence, LD: low dependence, MD: medium dependence, HD: high dependence, CD: complete dependence; see Table 12.2), conditional probabilities are calculated between tasks. Then for each operational function, an event tree diagram is constructed (Figure 12.1). From there on it is a trivial task to calculate functional success probabilities and finally, system success probability.

12.2.4 Minimizing Human Errors

The most effective method of controlling human errors is the *ergonomic design* of work. As explained earlier, most of the causes of human error are related to user-hostile work design. Human beings make fewer errors in environments that are compatible with their

TABLE 12.2 EQUATIONS FOR CONDITIONAL PROBABILITIES OF SUCCESS ON TASK *N* GIVEN SUCCESS PROBABILITY (*n*) ON TASK *N* − 1

Dependence level	Success equation
ZD	$\Pr[S_N \mid S_{N-1} \mid \text{ZD}] = n$
LD	$\Pr[S_N \mid S_{N-1} \mid \text{LD}] = \dfrac{1 + 19n}{20}$
MD	$\Pr[S_N \mid S_{N-1} \mid \text{MD}] = \dfrac{1 + 6n}{7}$
HD	$\Pr[S_N \mid S_{N-1} \mid \text{HD}] = \dfrac{1 + n}{2}$
CD	$\Pr[S_N \mid S_{N-1} \mid \text{CD}] = 1.0$

Source: A. D. Swain and H. E. Guttman, *Handbook of Human Reliability Analysis with Emphasis on Nuclear Power Plant Applications*, U.S. Nuclear Regulatory Commission, Washington, DC, NUREG/CR-1278, 1983.

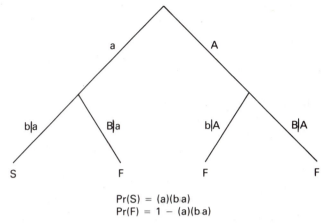

$$Pr(S) = (a)(b \cdot a)$$
$$Pr(F) = 1 - (a)(b \cdot a)$$

a: Probability of successful performance of task A
A: Probability of unsuccessful performance of task A

Figure 12.1 Event tree Diagram for two units connected in series. (After Ref. [19].)

expectations and in which they function effectively. Another approach in the past has been *replacing the operator*. The assumption of this approach is that poor performance is due to personnel factors such as deficient dexterity, poor vision, partial deafness, or inadequate skills [18]. A more proficient operator will make fewer errors. Finally, systems can be *designed with human error in mind*. It is not possible to eliminate human errors totally. Hence if these can be anticipated and the system designed with either internal checks of input data or with fault-tolerant mechanisms, errors will be of no significance.

12.3 CASE ILLUSTRATION 7

Pulat [19] applied THERP at a facility in order to estimate a material management and handling system reliability, including multiple human and equipment tasks. For class A MRP II user status, a minimum of 95% inventory count accuracy is required. The objective of this application was to estimate the handling process capability, compare it with the present count accuracy, and make the necessary changes if capability and current accuracy fall short of the minimum.

Several system functions were determined:

1. Receiving
2. Stocking
3. Pick and delivery
4. Accumulation
5. Bin link
6. Bin unlink

For each function, all requirements of THERP were developed. Figure 12.2 gives a section of the operational flowchart for the accumulator function. Figure 12.3 gives a portion of the event tree diagram for the same function.

Calculated success probabilities for the above-mentioned functionalities ranged between 95.14% for pick and delivery and 99.8% for bin linking. These results were comparable with the relative complexities of the two functions. The system success probability or the process capability was calculated at 94.7%. Several improvement ideas were tested and implemented in order to bring this percentage up. The improvements focused on giving extra ownership to the workers (modify behavior), and ergonomic modifications of the user-computer interfaces. As a result, the system is now functioning at 97 to 98% accuracy level.

12.4 SELECTION AND TRAINING

Personnel selection and training are related concepts. They both aim at staffing positions with qualified employees. We will investigate these concepts separately while elaborating on their interrelationships.

12.4.1 Personnel Selection

Personnel selection is concerned with the assignment of the right person to the right job. Its methodology spans the total process of development, use, and validation of employee selection techniques under specific job requirements [20]. The general steps of personnel selection are (Figure 12.4):

1. Develop job requirements (mental, physical, psychomotor, education, etc).
2. Develop or select testing materials.
3. Carry out the testing and interview process.
4. Select personnel on the grounds of best match between job requirements and background, skills, and abilities possessed.
5. Assign people to jobs.
6. Follow-up in the operation phase for possible reassignment and/or dropouts.

Reassignment needs arise from overlooked factors during assignment or insufficient job requirement specifications. It is also possible that a certain percentage of the population assigned will not continue the job prospect and drop out of the program.

Whichever methodology is exercised during selection, it is important that the following minimum factors related to the personnel screening process be kept in mind:

1. *Criterion-related validity.* Another term for this is *predictive validity.* Simply stated, the testing process must possess good predictive validity for proficiency on the job. This depends to a large extent on proper definition of the job requirements, selection of predictors, and statistical analysis of the data collected. This is the most important validity that a screening tool must have.

2. *Content validity.* This refers to the extent to which selection procedures actually sample relevant job behaviors. An applicant screening tool must sample from a domain of important job behaviors.

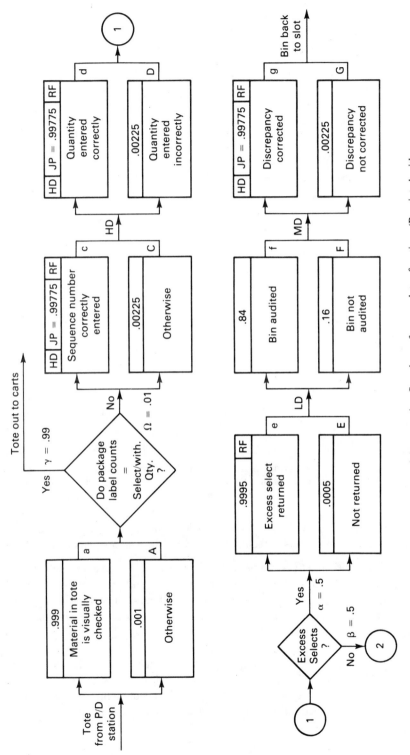

Figure 12.2 A portion of the operation flowchart of accumulator function. (Reprinted with permission, Elsevier Science Publishers. B. M. Pulat, 1988. "Human Reliability: A THERP Application at ASRS," *Trends in Ergonomics/Human Factors V*, Aghazadeh, F. (Ed.), p. 41.)

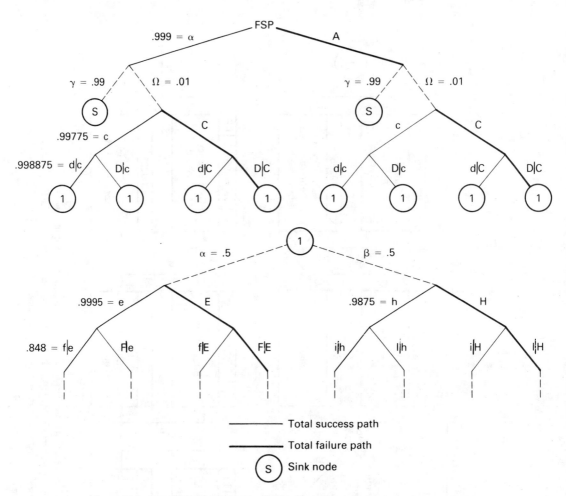

Figure 12.3 A portion of the HRA Event Tree Diagram for accumulator function. (Reprinted with permission, Elsevier Science Publishers. B. M. Pulat, 1988. "Human Reliability: A THERP Application at ASRS," *Trends in Ergonomics/Human Factors V*, Aghazadeh, F. (Ed.), p. 41.)

3. *Construct validity:* A construct is a trait or a hypothesized attribute of people. Construct validity refers to the interpretation of the test or screening tool scores.

In addition to the validity questions, one must also be concerned with reliability, and practicality [21]. *Reliability* refers to the accuracy and precision of the measurement procedure. *Practicality* contains a wide range of factors, such as administration time, economy, and interpretability. We will investigate the selection process with special emphasis on selection for physically demanding jobs.

Selection for physical jobs.   Due to many reasons, including variability in the demonstrated capability and many existing work methods and workplaces that have not been designed with ergonomics in mind, worker selection for physically demanding jobs is an acceptable procedure. The objective is to enhance work capacity to meet job demands.

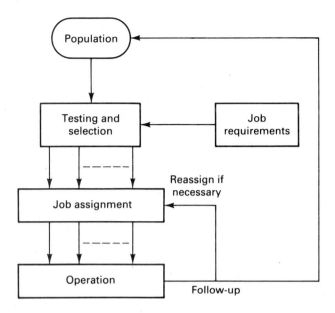

Figure 12.4 Elements of personnel selection.

The risk of musculoskeletal injury increases with age. Women may be at more risk than men. Positive history of back pain or other musculoskeletal problems is a sign of increased risk. For these reasons, taking a medical history and performing a physical examination on the job applicant are recommended procedures. A preemployment x-ray of the spine (especially the lumbar region) is counterindicated unless *spondylolisthesis* (anterior displacement of one vertebra over another) or degenerative changes are suspected [22]. Range of motion and anthropometric measurements may be taken if these are important for performance. Chaffin et al. [23,24] claim that static and dynamic muscle strength affect one's ability to perform on physical tasks that require force application. The general fitness of the person is also important relative to acceptable performance on such jobs.

Selection for other jobs. Jobs that do not specifically call for physical parameters may get help from screening procedures that consider cognitive abilities, background checks (biographical data), reference checks, and motor ability. Hunter and Hunter [25] showed that cognitive ability had the highest adjusted validity for entry-level jobs. Finally, legal aspects of personnel selection must be considered. Many countries have passed laws that have had far-reaching impact on personnel selection practices.

12.4.2 Training

The objective of training is to achieve acceptable performance or improve one's skills to qualify for jobs. In addition to skills, training also adds to one's knowledge and attitude. In relation to staffing, training has two major implications:

1. Preemployment training
2. Continuous training

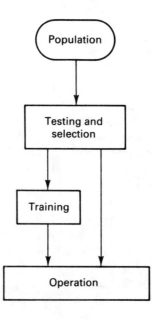

Figure 12.5 Training in the overall selection process.

Preemployment training takes place with job applicants who do not quite possess the needed attributes, but in the opinion of the testers may become proficient employees if trained. Figure 12.5 portrays this concept with the major elements of selection displayed.

Continuous training is highly visible in those industries that experience severe competition and hence change. These companies expose their employees to multiple skills for utmost flexibility in personnel utilization. New technology application also necessitates ongoing training. There are several training methods available. The most important of these are:

1. On-the-job training
2. Job rotation
3. In-class training

On-the-job (hands-on) training is probably the most effective. Here the trainee is actually performing the task. Simulator training is an extension of on-the-job training. However, Ellis [26] showed that on simple tasks hands-on-training may not benefit one significantly. In this research, half the people had training sessions on a task requiring little moving skill that also included an opportunity for practice. The other half had no such practice. Ellis found that for this task there was no improvement in performance when people were provided the opportunity to practice during training.

Job rotation is another form of on-the-job training; however, there is no trainer in this case. The subjected employee learns the task by doing it. In-class training is easy to set up and a very cost-effective training method. One or several trainers may be exposed to many trainees in a classroom environment with training films and other equipment [27].

Several points that should be kept in mind during the design of a training program are [28]:

1. Keep trainees active.
2. Make use of repetition.
3. Make use of reinforcement.
4. Promote understanding.
5. Encourage creative thinking.
6. Train under different conditions, if possible.

Training for physical jobs.   Through training and education, injury and fatality rates may be reduced. Exposing workers to information relevant to injury-causing situations may develop awareness of unsafe acts and conditions. Instruction on the correct lifting, carrying, pushing, pulling, and lowering methods may significantly reduce injury rates due to manual material handling. Highly repetitive tasks should be worked into pace. Starting with several hours per day, rather than full 8-hour shift work, is highly recommended. When a person has been away from these kinds of jobs for more than several weeks, the same warm-up period is recommended.

Jobs that require heavy load handling must be paced by the worker during training. An untrained person should not be exposed to the same loads as a trained person. Carefully select workers for jobs that require handling of heavy loads. Acclimatization to heat or cold is a special training method. Trainees should be allowed frequent breaks in environments that display significant heat or cold. In addition, general fitness training for all physical jobs is highly recommended.

Training for other jobs.   Jobs that do not place physical demands on the operator can benefit from general training concepts and methodologies. Several additional considerations include [28]:

· Provide feedback on the quality and quantity of output.
· Train in discrete tasks first and then on total task sequence.
· Review training progress with the trainee after several days.
· Maintain the job content and challenge during training.
· Sequence training with respect to task similarity if the trainee is being exposed to several tasks.

12.5 JOB PERFORMANCE AIDS

As systems and machinery become more complex, human skills and abilities become less and less reliable in coping with the operational requirements. It is a well-known fact that people cannot be expected to function effectively under sensory and mental overload characteristic of the information age. In many situations, human performance must be supported by job performance aids, tools, equipment, and other means for adequate system performance.

Rifkin and Everhart [29] define a job performance aid (JPA) as a device or document that provides the needed information to carry out a task effectively. A JPA can also be any other mechanism that assists people in the performance of work activities. In many instances, JPAs also help extend capacity to store and maintain information and make it available on the job. Hence they extend human capabilities to do work. They reduce the amount of decision making necessary to perform a task by providing step-by-step guidance to a probable solution/cause. According to Swezey [30], the tasks that are most helped by JPAs are those that involve calculation, stringent memory requirements, accuracy, difficult decisions, and multiple judgments. JPAs are also appropriate for repetitive and boring tasks.

It is evident from the discussion above that whereas training materials have an effect on future performance, JPAs affect present performance on the job (Figure 12.6). Training materials may be modified to become JPAs. There are several benefits of JPAs:

1. Reduction in errors
2. Increase in speed of performance
3. Reduction in training requirements

An effective JPA reduces training requirements since it will be available at work for reference. Many errors can be reduced via JPA use. Overall speed of performance also increases since the operator does not have to redo many tasks that were done incorrectly. Examples of JPAs are many. A machine manual is a JPA. Labels next to controls and displays, a short list of telephone numbers of maintenance crew, and a calculator on a part picker are all examples of JPAs.

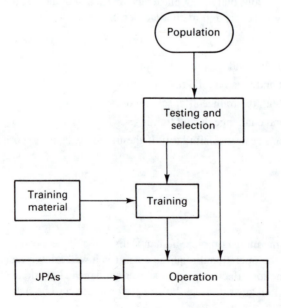

Figure 12.6 JPAs relation to selection and training.

JPA design. Several suggestions in JPA development are [28]:

1. Develop the JPA type (manual, short note, software) consistent with the expected use and an established purpose.
2. Be specific in wording. Do not require interpolations. Use short words and sentences. Use the active voice in the present tense.
3. Use symbols and abbreviations that are familiar to users.
4. Check the effectiveness before general use.
5. Only include facts, not opinions or assumptions. Combine related information.
6. Important sections may be highlighted via color, space, borders, and so on.
7. Use cross-referencing sparingly. Do not repeat information.
8. Design the JPA such that it does not interfere with the user tasks.
9. Clearly state the title, version number (possibly needed for software help), and the purpose.
10. If the JPA is multipart, include a contents section.
11. Use uniform notation throughout.
12. Place examples close to the text to which they refer.
13. The text and diagrams must be clearly legible and readable.
14. Allow adequate spacing for margins and between lines and words.
15. Edit before final printing.
16. Ensure that the users targeted get the JPA.

12.6 CASE ILLUSTRATION 8

In this section we describe the main functions of a job performance aid in the form of software and associated hardware, developed for the purpose of helping a maintenance crew to perform its task more effectively. The task is error recovery in a multimillion-dollar miniload AS/RS at an electronic assembly facility. Everyday, thousands of parts are moved to the manufacturing shops from the AS/RS. Thus it is extremely important that the throughput capacity of the system be maintained in order not to affect the downstream manufacturing operations adversely.

In terms of maintenance attention, the most critical element in the AS/RS is the crane. Its moving elements, interfaces with other hardware, such as the rack structure and power and guide rails, and communication interfaces with host computers and manned stations make it highly vulnerable to encountering downtime (error condition). Once a crane is down, stocking, retrieval, and audit activities in that aisle cease.

12.6.1 Crane Fault Conditions

Each crane's work is downloaded from the MRP (material requirements planning) system throughout the day. Furthermore, system administrators may schedule retrieval orders. The crane is under computer control during stocking or retrieval as long as it encounters no electromechanical problems. Once a problem occurs (pins jammed, bad limit switch, etc.), the crane control is automatically switched to manual mode and the crane stops.

Prior to the development of the job performance aid, the stocking operator followed a manual "quick-fix" procedure, through job knowledge, when an error condition was detected (Figure 12.7). Due to the complexity of most error conditions and the associated mental load (recall requirements), the attempt usually failed. Then the AS/RS maintenance group was informed. However, by that time, significant downtime had been encountered, and in some cases, the original fault condition was propagated into a complex problem. The original fault diagnosis and correction process also did not allow for recording any historical data of errors by the crane. Such data may be invaluable in pinpointing causes of faults and may aid in quicker error recovery.

Thus it was obvious that the existing process was incapable of consistently maintaining high levels of crane uptime. A job performance aid was necessary which would allocate the function of fault interpretation to a mechanized process and present simple suggestions to the maintenance crew as to the best error recovery scheme for each of over 70 possible fault conditions. Furthermore, the process would automate the communication between the crane and the maintenance crew in addition to storing error condition data.

12.6.2 A Software Tool As JPA

A number of people took part in the development, including material-handling engineers, the ergonomist, information systems developers, the AS/RS software developers, and relevant personnel from the crane vendor. Protocols were established for passing information between the AS/RS software and the crane logic unit software. Shared memory locations were designed for instant on-line reporting. Special software was developed to translate the

Figure 12.7

data into informative reports. The crane field engineer was interviewed to develop accurate rules for error recovery. These rules were then mapped into a rule-based system. Hardware was provided to the maintenance crew and other relevant personnel using the system.

A main menu guides the user to the appropriate function of the JPA. The five major functions are as follows:

1. *Interactive error recovery.* This is an expert system and the most important module (Figure 12.8). It interactively steps the user through a fault condition to suggest the most effective means of recovery. This may require adjustments on any combination of over 400 crane parameters in the crane logic unit software. The driver here is the rule-based knowledge system. For the more experienced users, the JPA allows for graphical fast step-through.

2. *Preventive maintenance schedules.* The software maintains and issues all required preventive maintenance schedules upon user request. Everyday, maintenance personnel check this function to carry out maintenance work on vital components before problems arise.

3. *Request maintenance.* In this function a special maintenance request is documented. A section chief may request maintenance work on any part of the AS/RS using this option. The request is automatically printed on the maintenance printer, logged into the computer records, and a copy forwarded to the engineer. The request is assigned a number that must be canceled by the engineer after work completion.

4. *Print special reports.* A number of reports can be obtained using this function to allow an investigator to carry out analyses using past performance records on any crane (Figure 12.9). In many cases this capability proved to be invaluable in pinpointing the

Figure 12.8 Reprinted from *Industrial Ergonomics: Case Studies,* © Institute of Industrial Engineers, 25 Technology Park, Norcross, GA 30092

Figure 12.9

causes of crane problems. Some examples of crane performance statistics are downtime, error codes, and error frequencies by the minute, by day, and by crane. A special instruction list is also available to perform certain duties, such as putting the crane in random mode for testing. Higher-level charts may also be generated to review overall crane performance.

5. *Monitor bar-code scans.* Using this function, the maintenance personnel monitor the performance, such as misreads, of all bar-code scanners for potential preventive maintenance.

The manufacturer is achieving notable crane downtime reduction and direct labor-hour savings at the AS/RS through supporting the maintenance crew with a job performance aid. The crane downtime percentage has been reduced fourfold with well-controlled variability. Furthermore, a 50% reduction in labor hours has been observed with 99% reduction in the engineer's attention time.

The last point to be made is that the aid was not imposed on the operators. Thus they did not feel that they are "trained robots" who can only perform the simplest tasks. A valid business need triggered this development as voiced by the operating personnel. Thus the aid was accepted readily by everyone.

12.7 SHIFT WORK

The fifth special topic that we will review is shift work. Shift work and overtime work are usually exercised in order to make better use of the existing resources, increase production, and prolong the period over which a service is provided. This may include work over weekends. In some industries it is almost a mandatory exercise. Process industries that engage in continuous manufacturing, such as the chemical industry, brick, cement, glass, and petroleum processing operations, must employ shift work to avoid heavy penalties of disrupting the process. Shift work is a way of life in about 20% of industries in Europe and the United States [31,32]. Furthermore, compared to the 1950s and 1960s, the proportion of industries employing shift work is increasing.

12.7.1 Circadian Rhythms

The main human variable that is related to shift work is the circadian rhythm. Many bodily functions show a rhythmic behavior in a 24-hour cycle. This is called the *circadian rhythm*. The body functions that increase by day and decrease by night include body temperature [33] (Figure 12.10), heart rate, blood pressure, mental abilities, adrenalin production, and physical capacity.

As can be observed in Figure 12.10, the body temperature peaks at around 7:30 P.M. and is at its lowest point at around 3:30 A.M. for day-shift work (sleep between 10 P.M. and 7 A.M.; work between 8 A.M. and 5 P.M.). On the other hand, there is a flattening in this behavior for night-shift work (totally reversed sleep/work cycle). The important point here is that there is *no reversal* in body temperature fluctuation when one moves from day shift to night shift. Only a flattening or reduced variability is observed. According to Akerstedt [34], both laboratory and field studies indicate that it takes 10 days for circadian adjustment to be complete (flatten).

Other bodily functions do not reach their maxima or minima at the same time. In general, though, all functions are on-the-go during daytime. The nighttime is for recovery

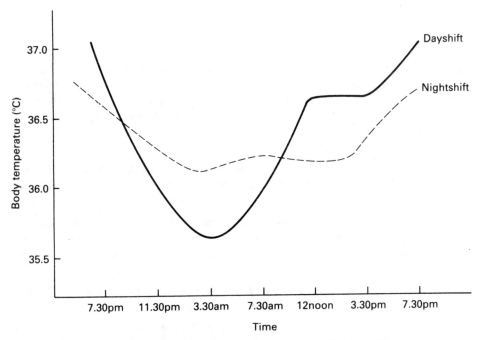

Figure 12.10   Average body temperature versus time. (After Ref. [31].)

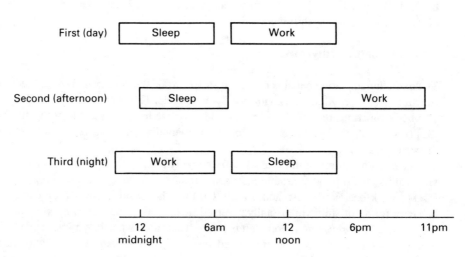

Shift	Activity

First (day) — Sleep — Work

Second (afternoon) — Sleep — Work

Third (night) — Work — Sleep

| | | | | |
| 12 midnight | 6am | 12 noon | 6pm | 11pm |

Figure 12.11 Relationship between shifts and activity pattern.

and renewal of resources. Figure 12.11 gives the protocol that will be the basis of the discussion that follows.

12.7.2 Physiological Response to Shift Work

Several effects of shift work on the body have been documented:

1. Quality of sleep is affected. Day sleep is not as effective as night sleep. There are more interruptions. It usually takes two rest days to pay the sleep debt collected due to night work. After a lengthy survey of shift workers, Tepas et al. [35] showed that the third-shift (night-shift) workers sleep the least, second-shift (afternoon-shift) workers sleep the most, and first-shift (day-shift) workers report a sleep length that is somewhere between the two groups.

2. Physical capacity to do work at night is less. Although circadian adjustment problems constitute the primary reason, part of the reason is the feeling of being sleepy and tired [35].

3. Mental capacity is also affected. Johnson [36] reports that reduced mental capacity affects vigilant behavior on tasks such as quality control and monitoring. Furthermore, Kelly and Schneider [37] claim that errors may increase significantly (80 to 180%) due to increased shift length.

4. Nervous disorders have been reported among night-shift workers. Sleep deprivation and social effects of shift work are the primary reasons.

5. Gastrointestinal problems are not uncommon. Thiis-Eversen [38] reported these problems with 6000 workers in Norwegian factories as given in Table 12.3.

TABLE 12.3 GASTROINTESTINAL PROBLEMS IN NORWEGIAN FACTORIES

Gastrointestinal problems	Percent of workers studied	
	Day work	Night work
Stomach troubles	10.8	35
Ulcers	7.7	13.4
Intestinal disorders	9	30

Source: Ref. [38].

12.7.3 Social Effects

In addition, shift work affects social life:

1. Family life is disrupted.
2. There are fewer opportunities to interact with friends.
3. Group activities are disrupted.

Although shift work seems to have many negative effects, there is usually more pay associated with such work, as well as more freedom at work. There are clearly fewer bosses around. Furthermore, shift work involving only morning and late afternoon shifts presents fewer problems.

12.7.4 Performance Effects

All the previous discussion may lead one to believe that as a result of physiological and social effects, performance should be degraded on the night shift. Wyatt and Marriott [39] confirm this. Browne [40] found that the delay of teleprinter switchboard operators in answering calls increased drastically on the night shift. Bjerner et al. [41] observed significantly higher errors made by meter readers in a gas works on the night shift than on other shifts. Monk and Embrey [42] claim that most of these effects are due to low alertness of workers on the night shift. Figure 12.12 supports this proposition.

On the other hand, Folkard and Monk [43] claim that some tasks may not be negatively affected by the third shift. They proposed a descriptive model in which on-shift performance is seen to be dependent on the type of task, type of shift system, and type of person, with these factors interacting through the worker's circadian rhythms. For example, Folkard and Monk claim that cognitive tasks that involve high storage load may be performed better during the night on a rapidly rotating shift system.

12.7.5 Recommendations

As indicated before, circadian rhythms do not reverse when work hours are altered drastically. Most of the flattening occurs in 5 to 7 days. Hence shift changes that occur once a week are particularly not recommended. Experts are divided as to whether a rapid or slow

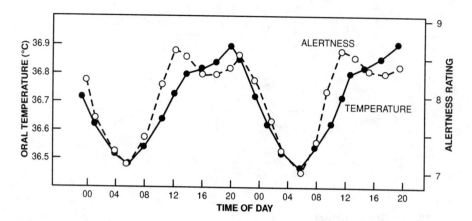

Figure 12.12 Circadian rhythms of oral temperature (solid dots) and self-rated alterness (open dots) in rotating shift workers. (Reprinted from T. H. Monk and D. E. Embrey, "A Field Study of Circadian Rhythms in Actual and Interpolated Task Performance," in *Night and Shift Work: Biological and Social Aspects,* Reinberg and Andlauer (Eds.), copyright 1981, with kind permission from Elsevier Science Ltd., The Boulevard, Langford Ln., Kidlington OX5 1GB, UK.)

rate of shift change should be implemented. Slowly rotating systems use lengthy shift rotation periods, such as a schedule maintained for 6 months to a year. These are not as undesired as weekly changes; however, the social problems that this pattern brings do not make it desirable.

Based on both social and physiological factors, modern views support a short-term shift rotation scheme. The 2–2–2 (metropolitan rota) and 2–2–3 (continental rota) systems in use in England seem to comply with these views. The main caveat in this recommendation is that the night shifts must be followed by at least a 24-hour rest. An example of the 2–2–2 system is given in Table 12.4. Although not shown in Table 12.4, this system compensates for losses in legitimate free days (2 days each 7-day period) by skipping one of the shifts and allowing extra off days based on a cycle (i.e., ...MMAANN–AANN–NNA...). The 2–2–3 system is demonstrated in Table 12.5.

TABLE 12.4 THE 2–2–2 SHIFT ROTATION SYSTEM[a]

First week	M	T	W	Th	F	Sa	Su
Shifts	M	M	A	A	N	N	—
Second week	M	T	W	Th	F	Sa	Su
Shifts	—	M	M	A	A	N	N
Third week	M	T	W	Th	F	Sa	Su
Shifts	—	—	M	M	A	A	N
Fourth week	M	T	W	Th	F	Sa	Su
Shifts	N	—	—	M	M	A	A

Source: Ref. [32].
[a] M, morning shift; A, afternoon shift; N, night shift.

TABLE 12.5 THE 2-2-3 SHIFT ROTATION SYSTEM[a]

	M	T	W	Th	F	Sa	Su
First week	M	T	W	Th	F	Sa	Su
Shifts	M	M	A	A	N	N	N
Second week	M	T	W	Th	F	Sa	Su
Shifts	—	—	M	M	A	A	A
Third week	M	T	W	Th	F	Sa	Su
Shifts	N	N	—	—	M	M	M
Fourth week	M	T	W	Th	F	Sa	Su
Shifts	A	A	N	N	—	—	—
Fifth week	M	T	W	Th	F	Sa	Su
Shifts	A	A	N	N	—	—	—

Source: Ref. [32].
[a]M, morning shift; A, afternoon shift; N, night shift.

Other Recommendations include [44]:

1. Extended workdays (9–12 hours) are acceptable if the nature of work and work-load are suitable.
2. Early start of the morning shift (4–5 A.M.) should be avoided.
3. Quick changeovers (same day) between shifts should be avoided.

12.8 DESIGN FOR THE ELDERLY

We now turn to the subject of designing for the elderly. There is no clear-cut definition of old age; however, many agree that ages 50 to 55 are probably the start of one's later portion of life. Many individuals beyond this range still work in industrial environments, and many products are in the market for use by the older population. Hence this section will briefly cover ergonomic concerns that relate to design for older people.

The older population in the United States is increasing. In 1900, 4% of the population was over 65. By 2020 this percentage is expected to exceed 16. A notable majority (80 to 85%) of this population is able bodied. The rest suffer major physical or mental losses. In this section, various characteristics of the older groups are summarized with respect to sensory, cognitive, physical, and work habit parameters [45,46]. Designers must allow for such considerations, especially if the expected user population is the older group.

Sensory characteristics. Many changes occur with respect to sensation as one gets older. Lower-order visual functions such as static visual acuity decrease after age 60. Higher-order functions such as dynamic visual acuity and motion perception exhibit deterioration after age 50. [47]. There is a linear loss in accommodation with age. Dark adaptation is not as effective between the ages of 60 to 89. Although the aged require much more light in the environment, they are also notably susceptible to glare. After the age 55, there is also a shrinkage of the visual field. Color perception is degraded, especially around 70 years of age and beyond.

As discussed before, hearing deteriorates with age. There is progressively more hearing loss at frequencies of 1000 Hz and above. Speech intelligibility also decreases progressively. The reduction for age 60 is 10% compared to ages 20 to 29.

Cognitive characteristics. Schaie [48] claims that cognitive abilities of people until age 60 are mainly intact. After that age, losses in intellectual ability, reasoning, memory, and speed of comprehension are common. Fozard [49] claims that the short-term memory store of older people does not hold as much as it did when they were younger. Old people often must be supported by memory aids for acceptable performance. Welford [50] suggests that increased task complexity is a definite slowing risk for older people compared with younger ones. However, if stimulus qualities and feedback mechanisms are enhanced, much of the slowing can be negated [51]. As people get older, they obtain more experience and judgment ability. These are advantages over younger people. They are also able to learn with a lot more selectivity.

Physical and psychomotor characteristics. Stoudt [52] and Annis [53] report that there is progressive decline in both standing and sitting height after midlife. Hip measurements and weight increase up to age 50, then stabilize until 60, with a small increase after that. Abdominal and body breadth measurements increase in midlife and show continuous decrease after that. Changes in bone, muscle, and nervous tissue also occur with older age. Degenerative processes in cartilage and muscle lead to decreased mobility and increased risk of injury. Hence it is critical that jobs which older people perform should not demand notable work capacity and muscular strength, endurance, speed, and flexibility.

General strength is maximum toward midlife and declines continuously afterward. Almost 50% of strength is lost by age 65. However, hand strength declines by only 16%. Elsayed et al. [54] suggest that older people can gain their isometric strength back after rehabilitation just as younger individuals can. Reduction in body strength inhibits certain movements and restricts coordinated activity. This may be the reason for higher incidence of slip-and-fall accidents by the elderly [55]. Psychomotor behavior slows by age. Reaction times show at least a 20% increase by age 60 compared to 20. The brain waves of the elderly are also slower than those of their younger counterparts.

Thermoregulatory responses. Drinkwater and Horvath [56] report that, in general, exercise-heat tolerance is reduced in older persons. Wagner et al. [57] reported higher heart rates, mean skin and core temperatures, and lower sweat rates for older men. However, Pandolf [58] matched older and younger people for pertinent physiological and/ or morphological variables and showed that there is little impairment of the thermoregulatory system during exercise-heat stress at least through the fifth decade of life. The most important parameter in this regard seems to be the aerobic fitness of the individual.

Young [59] reports that older men appear less able than the young to defend their core temperature during experimental cold exposures. However, older women appear to defend their core temperature as well or better than young women.

Work habits and attitudes. Old people elect to focus on one thing at a time. They should not be given multiple tasks to do. They should not be rushed. Older people are better suited to work at their own pace. Since older people have had longer to learn what to expect, compatibility is particularly important with the elderly. The relationships between control movements and objects being controlled must be the expected ones. Older people have increased concern for safety. However, when they are off the job for an injury, they take longer to return to work. They display greater awareness in product deficiencies.

12.9 ERGONOMICS OF ADVANCED MANUFACTURING

As the last special topic in this chapter, we will review the main elements of ergonomics of advanced manufacturing, a frequently overlooked subject. The main characteristic of advanced manufacturing is automation. It is also characterized by the application of state-of-the-art technology to fabrication, assembly, and material-handling operations. Automatic identification systems (transponders, barcodes, radio frequency, speech input), automated storage and retrieval systems, carousels, conveyor and automated guided-vehicle systems, machine assembly and test systems, robotic workstations, flexible manufacturing cells, and automated packaging are all examples of state-of-the-art technology. In addition, integration of computers into manufacturing (CIM), vendor relationships, and dock-to-dock material flow arrangements all support these operations.

CIM plays a particularly important role in advanced manufacturing. CIM emerged primarily in the late 1970s and early 1980s as an advanced manufacturing philosophy (in addition to MRP II: manufacturing resources planning and JIT: just-in-time) to support the various types of production, such as job shop, batch, and continuous manufacturing of the early and mid-nineteenth century. CIM is the automated movement and management of manufacturing information from product design, through order entry and manufacturing, to distribution. It proposes a layered architecture of business control from corporate levels to an equipment in a machining cell.

Many of the earlier applications of CIM and hard automation created insulated systems, hence islands of automation. A survey [60] indicates that up to 75% of manufacturing companies in the United States are still operating in this mode. In the late 1970s, information automation applications expanded. Risk-taking managers pushed for implementation of MRP II systems that required interfacing of several functions, including manufacturing planning, product definition, and inventory management. Thus interfaced systems were born with communication channels established between computers. True integrated systems came into being in the 1980s via interfacing many processors for functional effectiveness, such as cycle-time reduction, rework elimination, and WIP (work-in-process) reduction. It is estimated that true CIM implementations exist in less than 5% of manufacturing concerns in the United States.

In the 1990s, production systems will become more flexible and will facilitate the management of production in addition to supporting production. The value-adding functions, such as fabrication, assembly, and testing, will be the primary focus of information systems architecture. These systems will integrate planning, execution, and control throughout the manufacturing process.

12.9.1 Effects of Automation

There is no question that for many reasons, including economics and competition, automation will continue in the future. Effects of automation on human work are many. Ergonomists must be familiar with these to smooth the process while preparing the workers for its impact. The main effects include:

1. *Job fragmentation.* Not all jobs or functions can be automated. In general, jobs that lend themselves to automation are simple, routine, and do not require complex decision making. Furthermore, available proven technology in the market also limits the extent of automation applications. However, one thing is obvious. With automation, some jobs are taken over by machines and the remaining ones are left to the human workers. Hence automation leads to a degradation of human roles in the system. The *complete job* concept is lost. In automated functions, normally what remains to the operator are jobs such as loading, monitoring, intervention, and manual operation in case of malfunction. The last two are especially difficult since a human being gradually loses work skills in situations where the machine is the primary performer. Hence retraining of workers is important.

2. *Supervisory control.* Automation requires more time spent on supervising the process than otherwise. Supervisory skills are different from performer skills. This indicates that workers must be trained on supervisory skills in addition to skills required for manual control in case of malfunction. Human beings have distinct cognitive advantages compared to computers, which can also supervise the process. Sharit et al. [61] count flexibility, inductive logic, pattern recognition, diagnosis, and decision making as some of the advantages.

3. *Training.* As mentioned before, automation requires significant training and retraining. Training is also required for new skills with new machinery and equipment. Surveys show that the biggest problem that operators and maintenance workers voice due to automation is inadequate training.

4. *Changing role of labor.* Automation assigns many direct labor responsibilities to machines. Depending on the extent of mechanization, some direct jobs may or may not remain. However, indirect jobs are not affected much. Perhaps the area that expands under automation in terms of jobs is the maintenance area. Many workers who held direct labor positions may become indirect laborers or maintenance people.

5. *Decreased autonomy.* With some processes controlled by computers, including planning actions, there is less autonomy on the job. There are few production decisions to be made, if any.

6. *Increased dependence on others.* CIM crosses organizational boundaries. Automation introduces much new equipment and machinery into a process. As a result, there is increased dependence on maintenance and indirect workers. White-collar employees also collaborate in operations that are data driven.

7. *Safety.* There may or may not be more safety concerns, depending on the starting point. An already unsafe environment looks much cleaner and safer when automated. However, new equipment and machinery may introduce more safety concerns.

8. *Displacement of workers.* It is often said that automation displaces workers. This is generally true. Assuming that the facility is not increasing production, many workers may be displaced. The result is a morale problem with those who are left. If there is a labor union at the facility, it may object to layoff decisions.

9. *Boredom and stress.* Due to less autonomy, boredom may be a problem. However, the maintenance worker's task will be more challenging. Stress may come about due to more complex computer systems and machinery.

12.9.2 Robotic Workstations

A specific case for automation is robot applications. Since the early 1980s, many robotic workcells have been installed in the Western world. Their economy, consistency, and accuracy make robots more preferable than human workers, especially on monotonous, strenuous, and hazardous jobs. In certain cases, labor shortages warrant robot use. However, robots are considered potential causes for unemployment.

In terms of ergonomics of robot versus human assignment to a job, checklists such as those given by Nof [62] and Bullinger et al. [63] may be used. Such checklists compare the relative advantages/disadvantages and capabilities/limitations of the two for best assignment. Table 6.2 presented one such comparison.

Robot use leads to less physiological strain, but more psychological stress may accrue due to programming and teaching needs. Safety problems will arise if a person steps into the work envelope of a robot during operation. Workers must be trained in robot use and programming. Maintenance workers must enhance their robot-handling skills. Production workers may be assigned only to loading and unloading operations in a robotic workcell. Otherwise, they spend most of their time supervising its operations.

12.10 CASE ILLUSTRATION 9

This illustration presents a case for the need to human engineer automation projects from the start for people integration into the development. It also details specific ergonomic applications in a buffer zone automation project, a unit-load AS/RS (automated storage and retrieval system) at an electronic equipment manufacturer.

12.10.1 The AS/RS Unit Loader

The unit-load AS/RS functions between factory receiving and the assembly shops. The system is composed of 10 aisles with aisle-captive man-on-board cranes. Each aisle has two P/D (pick and delivery) stations with three staging positions each (Figure 12.13). Palletized loads must be positioned on "base" pallets ($\frac{3}{4}$ in. plywood) to enable the crane to move the material. Each rack opening has an associated base pallet.

Materials buffered in the unit-load AS/RS are mostly frames, fans, fan housings, brackets, backplanes, and the like. These are purchased and handled in bulk form, primarily on pallets. Material routing decisions are made at central receiving. An incoming part to be consumed in manufacturing may be routed to several destinations, including the assembly

Figure 12.13 Unit-loader P/D stations.

shops, miniload AS/RS, unit-load AS/RS, and the warehouse. If the item is to be buffered prior to manufacturing, it is stored in a bin or rack location. It is picked and dispatched to the point of use after receiving a pull signal or an MRP (material requirements planning–manufacturing planning software) request.

12.10.2 Ergonomic Applications

Ergonomics received a major consideration in planning for the unit loader. The aim was to make the system flexible, easy to use, and operation compatible. Specific applications may be presented in three major sections:

1. *Operational procedures.* For maximum operational flexibility, the system was designed around operator-directed stocking philosophy. Briefly, this means that stocking location for a receival is determined by the crane operator and not by any other supervisory mechanism, such as software. This gives the operators the feeling that they are in control of the process. Another benefit of this approach is that special conditions (e.g., error conditions) can be resolved very fast since there are no software constraints imposed on the stockers. After stocking, operators report the material location to the tracking software. This approach also gives flexibility to the storeroom operating management to modify stocking procedures. On the other hand, picking action is under software control in order to speed the pick process, which averages to 80% of total system activity in such systems. The software directs the operator to the specific material location, also indicating the quantity to be picked. An output conveyor along the middle of the pick face in each aisle is used for fast material discharge.

An interesting application of "coding" allowed a very simple material transfer process from the receiving dock to the unit loader. The rack structure displays four different opening heights to allow for stocking loads with different heights. The problem was to match available stocking locations in the unit-loader with a given load delivered from the receiving dock. This problem required the forklift operator delivering the material to know the load height, as well as the unit-loader aisle that has a matching empty stocking location, without any software guidance. Furthermore, all this activity had to be performed in minimum time. A simple color-coding scheme brought an effective solution to both problems. The load height determination by the forklift operator was solved by coding several steel support structures with different colors: green from floor to 22 in., yellow between 22 and 36 in., blue between 36 and 44 in. and red from 44 to 60 in. Thus, as the forklift is passing by any one of these reference points (Figure 12.14), the operator notes the color with which the load lines up. If the load height falls within the yellow zone, the operator knows that the load requires a 36-in. rack for storage. It is not even necessary for the forklift operator to remember the numeric load height. All unit-load base pallets have been color coded with matching colors. They are also bar coded to indicate the rack opening to which they belong. The crane operator frequently moves base pallets from rack locations to P/D stations for load staging. This serves as an indication to the forklift operators regarding what size rack slots are available at each aisle for stocking. As soon as a base pallet is spotted with a matching color code, the load is staged on it. Bar coding on the base pallet allows the crane operator to take the load to the corresponding rack opening.

2. *Hardware design.* The labor interface points have also been human engineered for comfortable and efficient use of the system. The crane cab received most of the attention in this respect. A detailed analysis of operator tasks revealed equipment needs (CRT terminal, printers, weigh scale, intercom, etc) as well as task relationships for material-handling activity. In addition, task elements that relate to crane movement indicated needs with respect to crane control. These data were used to finalize the cab geometry and equipment placement, with human anthropometry in mind. An actual-size wood mock-up at the crane vendor's facility was used to test the resulting design and make minor modifications. The vendor then custom built each cab according to specifications (Figure 12.15).

The resulting design was well accepted by the operators, except for a pushbutton control, which was later converted to a trigger-type control. Operators alternate between

Figure 12.14   Color-coded height gauge.

Figure 12.15 Crane cab.

seated (during crane move) and standing (during select or location audit activity) postures. Seat pan height is adjustable for comfortable operation by 90% of the adult population. All controls are within easy reach, and visual and other sensory design requirements have been considered. In addition to general lighting in the cab, there is a pivoted floodlight for illuminating any desired point in the rack structure. Operators stand on a nonslip platform with adequate room for carrying out daily activity.

Other hardware parameters built into the design are:

· Instead of a plexiglass cab guard, which dulls, pits, and may be a glare source, a black wire mesh was used.
· The cab instrument panel was split into two. Those displays that monitor crane movement were grouped and located in close proximity to the normal line of sight at seated posture. All error and maintenance displays were grouped and located within the peripheral vision.
· Adjustable fans provide improved climate control.
· A hardwired communication link that operates in a U channel to minimize "crosstalk" and other interference was preferred over remote link technology.
· A full-length pad behind the seat protects and supports the back and the head.
· A locking drawer under the seat provides storage space for personal items.
· A pull cord in the aisle acts as a kill switch for an operator caught in the aisle during crane operation.
· An emergency dead-man switch is on the crane.
· Flashing red lights function during crane moves.
· Sensors detect unstable and overheight loads.

3. *Software design.* For picking and counting functions, the system is under complete software control. For these cases the software directs the operator to the material location, displays the quantity to be picked (for picking only), and expects the operator to report the results of picking and counting operations. For stocking, the operator determines the stocking location; however, the stocked location and the quantity stocked must be reported to the software. The primary reason for such tracking of material in the buffer zones is that the production planning systems need an accurate view of on-hand balances. Production activities for existing orders and purchasing decisions for planned orders rely on such vital information.

The software controls were designed after a two-pronged approach [64] to the problem. The *value-added* approach concentrated on supporting the required handling functions in the system. For this purpose a detailed mission/function/task analysis of the operator duty was carried out, including system interfaces. Those functions determined to add value to the process were supported. Furthermore, the *user-friendliness* approach aimed at engineering the human interface points for use parameters consistent and compatible with human expectations and characteristics. The objectives were to allow for more speedy and error-free data entry/retrieval performance with minimal operator frustration. Several examples are:

- At the storeroom receival, the MRP (material requirements planning) system shortage requirement is checked to fill shortages. This allows for automated access to two different systems with the use of only one screen.
- JIT (just-in-time) material handling is performed the same way. There is no requirement to access separate MRP screens for this function.
- One screen allows for several types of receivals to be made, such as new item, return to stock, restock, and so on. No menu selection is necessary for this purpose.
- Function keys are used extensively for fast access between screens. In general, such moves between screens also retain information to fill relevant fields on the following screen.
- Many internal checks assure data integrity and negate human-initiated errors.
- Electronic mail allows operators to communicate with the control unit instantaneously.
- Reverse video is used extensively to highlight desired data elements.
- Work is generated at a central location and distributed to the cranes.

A new release of the material tracking software will use bar-coded serial numbers to identify material and/or related information. This system further enhances human and system performance by faster and error-free transaction processing.

12.10.3 Conclusions

The ergonomic approach to people integration into a major development project such as the unit-load AS/RS at this facility was beneficial to say the least. The system has been accepted with minimal use problems. Operators feel that they are not overwhelmed by task demands. The simplicity of many of the operating procedures also contributed to rapid movement of the system from construction to production mode.

12.11 RESEARCH TECHNIQUES

Ergonomists frequently find themselves in situations which involve research of some kind. Such situations may arise largely in system design or system evaluation. It is important that the ergonomist be thoroughly familiar with major research techniques that involve humans. This may require a formal course in research techniques along with formal treatment of experimental design techniques. In this section, we will only provide a brief review of some of the most common research approaches and issues. Interested reader may refer to relevant publications in this area for more information [65,66].

There are two major types of research that the ergonomist has to grasp: basic research and applied research. Each has its own proper use as described below.

12.11.1 Basic Research

Basic research is aimed at developing or investigating a relationship between one or more variables. This relationship may be in the form of one or more variables having or not having an effect on one or more other variables. Variables can be categorized in two sets: the first is comprised of independent variables; the second is made up of dependent variables. For example, if one suspects a relationship between blood alcohol level and reaction time, reaction time may be treated as the dependent variable in a formal investigation, and the blood alcohol level may be treated as the independent variable.

Two types of basic research can be distinguished; *experimental* and *ex-post-facto* research.

Experimental research. This type of research involves a designed experiment with human subjects. Usually there is a hypothesis to be tested. As a result of the experiment, the researcher either states that there is evidence to support the hypothesis or that there is not. Such statements carry a confidence level, usually at 95%. The experimental hypothesis is tested via data collected during the experiment and subsequently analyzed. For example, in the reaction-time experiment, if the researcher measures human reaction time corresponding to several percentages of blood alcohol level and conducts a variance analysis with the resulting data, this may be termed experimental research provided that certain precautions are taken before, during, and after the research as described below.

Many factors affect the experimental results; hence, the researcher must exercise caution in conducting experiments. True designed experiments must pass four tests: randomization, replication, controls, and variable manipulation.

Randomization requires both random assignment of experimental units (human subjects in most cases) to experimental conditions and the randomization of the order of experimentation. Proper randomization helps with the validity of the model constructed by promoting normal and independent distribution of errors, and it also helps in averaging the effect of extraneous variables not controlled in the experiment. With simpler experiments, such as those involving only two experimental conditions (e.g., drug administered, no-drug taken-control condition; 10% blood alcohol level, no-blood-alcohol control; etc.), counterbalancing helps the experimenter to achieve the second objective. A simple method of counterbalancing may involve performing half the trials using one sequence of experimental conditions and the other half with the opposite sequence. For example, subject one may be exposed to drug and then no-drug sequence, and subject two may be exposed to the reverse sequence, i.e. no-drug and then drug condition.

Replication is simply repetition of the basic experiment—having more than one subject perform in the experiment, thereby collecting multiple dependent variable data and independent variable observations. By replication, one achieves two objectives. First, a measure of random error is obtained. In designed experiments, statistical significance of factor effects is tested by ratios of various variance estimates. Variance calculation is only possible when more than one data point exists. In a factorial experiment one achieves replication by assigning more than one experimental unit to each factor level combination. The second objective of replication is to increase the power of the experiment to an acceptable level, usually 90% and above. The power of an experiment indicates the probability that the experimental hypothesis is accepted when it is truly indicated by the data.

Whenever possible, *controls* on independent variables are necessary to minimize their effects on the dependent variable(s). It may be impossible or uneconomical to control some variables. Several methods of control are possible, including blocking, stratifying, leveling and covarying. Control of variables also helps in minimizing the possibility of confounding. Two variables are confounded when their effects cannot be isolated. For example, after a salary survey in small and large companies, one may erroneously conclude that small companies in general pay less for the same work, whereas this may be entirely due to larger companies hiring degreed professionals who often hold high academic achievements. In this case there is a possibility that preparedness and salary are confounded. A better comparison may be possible via a stratified sample attempting to equalize observations on educational preparation.

Variable manipulation is the last major characteristic of designed experiments. In experimental design, the experimenter varies certain factor levels on purpose and measures the effect of this manipulation on the dependent variable(s). This is a prerequisite of declaring cause-effect relationship between independent and dependent variables. The variable that is manipulated is the one of interest to the experimenter. For example, the blood alcohol level and reaction-time example previously cited displayed purposeful manipulation of alcohol level at several percentages.

In any experiment, the experimenter has a hypothesis to be tested or wants to draw certain inferences about a "hunch" concerning a process being studied. Most everyday events involve decision making under uncertainty. It is not any different in an experiment. Hence, the experimenter uses probabilities and applies a statistical approach to the decision-making process to reject or accept the hypothesis or the "hunch." Hypothesis testing and related decision making involve one or several samples (replication in designed experiments) to be drawn from a population; sample statistics to be calculated and compared with expected performance corresponding to a confidence level; and making "accept" or "reject" decisions about the hypothesis, based on whether sample statistics fall in a "critical" region.

For examining a single independent variable at only two levels, the *t*- or *z-test* comparing two means is appropriate. For example, in the reaction-time experiment the independent variable—blood alcohol level—may have been selected for investigation at only two levels; 0% and 10%. Such designs are extremely popular and powerful. Another term for these designs is *controlled* experimentation, in which one group of experimental units is exposed to nothing (placebo) and the other to an experimental condition. In the above example, one group of subjects is receiving no alcohol and the other enough to create a blood alcohol content of 10%.

On the other hand, for examining a single independent variable at more than two levels or more than one independent variable each at multiple levels, the t-test or the z-test are not appropriate. The problem is with type-1 error probability (risk). With one test, usually the risk is set at 5%, which corresponds to a confidence level of 95%. At four levels of the independent variable, the number of tests on the means increases to six, elevating the risk to 26%. Hence, there is only 74% confidence in results. As the number of levels of the independent variable increases, confidence in results gets progressively worse. An alternative method of testing is needed. The answer is analysis of variance (ANOVA), perhaps the most useful technique in the field of statistical inference. ANOVA is used in design of experiments most of the time, except in special and nonparametric designs.

Some of the most common experimental designs ergonomics researchers use are single-factor ANOVA and multi-factor ANOVA (or factorial) designs. The latter design is used when more than one independent variable, each with multiple levels, are suspected of affecting the dependent variable. In this case, it is also possible to investigate the effects of interaction between factors. Sensitivity of the design can be improved by blocking on one or more variables resulting in randomized complete block designs, Latin-square designs (blocking on two variables) and Graeco-Latin square designs (three-variable blocking). Repeated-measures design—where each subject serves as a block to control variables such as background, training, abilities, skill, etc.—is an extremely popular design in ergonomics. Nested designs and analysis of covariance are other appropriate designs. It is also possible to combine some designs, such as nested-factorial designs. Lately, new designs to economize on number of replications—and hence subjects required for the design—have been developed. These designs, such as fractional factorials and Taguchi designs, can handle more variables than what is practically possible via full factorial designs; however, not all interaction effects can be tested.

Ex-post-facto research. In some cases full experimental control can not be established via complete randomization, replication, variable control and purposeful variable/level manipulation, or it may not be economical to do so. In these cases some researchers opt to use ex-post-facto research which is not experimental. *Historical surveys*, *incident analyses*, and *accident analyses* are examples. Here the variables have already acted and the researcher only measures results, for example, linking certain behaviors to on-the-job accidents. The researcher may look at employee records and try to infer a relationship between recorded behaviors and accidents. However, since cause-and-effect relationship cannot be established, it may be dangerous to infer such relationships. The experimenter has to rule out other possible hypotheses as to what may have caused the accident history displayed.

12.11.2 Applied Research

Applied research in ergonomics is characterized by use of existing knowledge instead of developing new information, a characteristic of basic research. *Literature review* is an example. Most basic human capability/limitation/behavior information is published in research journals and handbooks. An ergonomist facing a design or evaluation decision may refer to such sources to develop the best strategy. One caution: much human-characteristic information is situation specific. Research journals provide information on the context as part of the study. On the other hand, design handbooks usually do not. Hence, care must be exercised in terms of context similarity before applying literature data.

Another type of applied research manifests itself in terms of *heuristics*, *techniques*, *models*, *procedures*, *algorithms*, and *software*. All these are decision aids to a designer. The developers of these decision aids have already researched the needed basic human-characteristic information and made decisions on input/output requirements, taken procedural steps, and provided documentation on proper use and training. Does this mean no care should be exercised in their use? Not at all. Still the ergonomist must make sure of the appropriateness of the aid to the specific problem being considered. Many times, such aids are interactive—i.e., the user must provide data or evaluation at certain steps of the process. Such interaction provides flexibility to the decision aid to improve its validity. Models cited in sections 7.8 and 7.9 are examples.

Ergonomists also confer with their partners/peers and sometimes visit other facilities in order to gain knowledge of how others have solved or approached a similar problem. This process may loosely be termed *technical exchange*; or, if it becomes a very structured process, it may be called *benchmarking* [67]. The purpose behind this idea exchange is cross-fertilization and application of best practices.

12.11.3 Relationships

Although both basic and applied research can be performed independent of each other, there is a precedence relationship between the two: basic research results are necessary for applied research to occur. There are people in corporations, government, and academia who primarily conduct/fund basic research. Results of such studies may lead to direct application or further research. An ergonomist facing a complex design may already possess the necessary information for appropriate decisions. However, there may be some vague areas which may have to be investigated with respect to how people may behave in particular situations. Hence, basic and applied research complement each other for best decisions.

Some specific applied and basic research techniques in several areas such as telecommunications, the automotive industry, and consumer products are given in Weimer's *Research Techniques in Human Engineering* [68].

12.12 SUMMARY

Many other subjects interest industrial ergonomists. In this chapter we presented a few of these topics. Of primary interest is human error. Understanding the reasons for human error and developing systems that minimize their effects are major responsibilities of the ergonomist. Personnel selection and training focus on the assignment of best personnel to do the job. Job performance aids aim at supporting on-the-job performance. Design of the most acceptable shift schedules is another task for the ergonomist. With the baby boom bubble approaching the older ages, the ratio of the older working population to the younger is increasing. The specific characteristics of older people must be considered in job and workplace design. The ergonomist must also be familiar with the advanced manufacturing concepts of today's industry to function effectively in that environment.

QUESTIONS

1. Define *human error*.
2. What are recovery factors?
3. Define *fault-resistant systems*.
4. Is malevolent behavior included in the definition of human error?
5. How does insufficient/incorrect training lead to human error?
6. Define *errors of omission*. What can be done to minimize such errors?
7. What is HRA? Give examples.
8. How can the designer minimize the effects of human initiated errors?
9. What is the objective of personnel selection?
10. Define the *predictive validity* of a personnel selection process.
11. What are the objectives of training?
12. What are the two elements of training in relation to staffing?
13. Compare and contrast the three methods of training.
14. How could employees be trained for physical jobs?
15. What is a JPA used for?
16. Compare the use of JPAs and training materials on the job.
17. Define *circadian rhythms*. Give examples.
18. Compare the body temperature rhythm for prolonged day-shift and night-shift workers.
19. How does shift work affect social life?
20. What pattern of shift work does the modern view support?
21. What happens to the visual and auditory capabilities as one gets older?
22. Elaborate on the safety behavior and injury recovery time of the older workers.
23. Discuss job fragmentation as it relates to automation.
24. What changes in the role of labor may be expected by automation?

EXERCISES

1. Review the task sequence of your secretary for at least 10 tasks. Flowchart your observations. Calculate the probability of successful completion of the total sequence using THERP. What was your main problem in the process?
2. Discuss the relationship between content and predictive validity in a personnel selection test.
3. Go to a local industry and review their personnel selection processes. Do they have different procedures for different categories of jobs?

4. Select a job performance aid. Review its characteristics and propose improvements.

5. Review an automation project at a local industry. Evaluate its impact on the organization and the labor force. Discuss your observations.

REFERENCES

1. Meister, D. 1966. Human Factors in Reliability. In *Reliability Handbook*. Ireson, W. G. (Ed.). McGraw-Hill, New York.

2. Swain, A. D., and Guttmann, H. E. 1983. Handbook of Human Reliability Analysis with Emphasis on Nuclear Power Plant Applications. Sandia National Laboratories, *NUREG/ CR-1278*. U.S. Nuclear Regulatory Commission, Washington, DC.

3. Swain, A. D. 1967. Some Limitations in Using the Simple Multiplicative Model in Behavior Quantification. In *Symposium* on *Reliability of Human Performance Work Quantification. AMRL-TR-67-88*. Askren, W. B. (Ed.). Aerospace Medical Research Laboratory, Wright Patterson Air Force Base, OH, pp. 17–32.

4. Swain, A. D. 1963. A Method for Performing a Human Factors Reliability Analysis. *Monograph SCR-685*. Sandia National Laboratories, Albuquerque, NM.

5. Meister, D. 1962. The Problem of Human-Initiated Failures. In *Proceedings of the 8th National Symposium on Reliability and Quality Control.* IEEE, New York, pp. 234–239.

6. Meister, D. 1976. *Human Factors: Theory and Practice,* Wiley, New York.

7. Meister, D. 1962. The Problem of Human Initiated Failures. *Proceedings of the 8th National Symposium on Reliability and Quality Control,* IEEE, New York, pp. 234–239.

8. Willis, H. 1962. The Human Error Problem. *Report M-62-76.* Presented at the American Psychological Association Meeting. Martin-Denver Co., Denver.

9. *U.S. News and World Report,* May 16, 1988.

10. Dhillon, B. S. 1986. *Human Reliability with Human Factors.* Pergamon Press, Elmsford, NY.

11. Woodson, W. E. 1981. *Human Factors Design Handbook.* McGraw-Hill, New York.

12. Sinclair, N. A. 1970. The Use of Performance Measures on Individual Examiners in Inspection Schemes. *Applied Ergonomics*, 10(1), pp. 17–25.

13. Meister, D. 1973. A Critical Review of Human Performance Reliability Predictive Methods. *IEEE Transactions on Reliability,* R-22(3), pp. 116–120.

14. Siegel, A. I., Bartter, W. D., Wolf, J. J., Knee, H. E., and Haas, P. M. 1984. *Maintenance Personnel Performance Simulation (MAPPS) Model: Summary Description.* NUREG/CR-3626. Vol. 1, U.S. Nuclear Regulatory Commission, Washington, DC.

15. Hall, R. E., Fragola, J.. and Wreathall, J. 1982. *Post Event Human Decision Errors: Operator Action Tree/Time Reliability Correlation.* NUREG/CR-3010. U.S. Nuclear Regulatory Commission, Washington, DC.

16. Embrey, D. E., Humpreys, P., Rosa, E. A., Kirwan, B., and Rea, K. 1984. *SLIM-MAUD: An Approach to Assessing Human Error Probabilities Using Structural Expert Judgement,* Vol. 1, *Overview of SLIM-MAUD, March;* Vol. 2, *Detailed Analysis of the Technical Issues.* NUREG/CR-3518. U.S. Nuclear Regulatory Commission, Washington, D.C.

17. Phillips, L. D., Humpreys, R. and Embrey, D. 1988. A Sociotechnical Approach to Assessing Human Reliability (STAHR). In *A Pressurized Thermal Shock Evaluation* of *the Calvert Cliffs Unit 1 Nuclear Power Plant.* Oak Ridge National Laboratories, Oak Ridge, TN.

18. Miller, D. P., and Swain, A. D. 1987. Human Error and Human Reliability. In *Handbook of Human Factors,* Salvendy, G. (Ed.). Wiley-Interscience, New York.

19. Pulat, B. M. 1988. Human Reliability: A THERP Application for an Automated Storage and Retrieval System. In *Trends in Ergonomics/Human Factors V.* Aghazadeh, F. (Ed.). Elsevier, Amsterdam.

20. Osburn, H. G. 1987. Personnel Selection. In *Handbook of Human Factors.* Salvendy, G. (Ed.). Wiley, New York.

21. Thorndike, R. L ., and Hagen, E. 1969. *Measurement and Evaluation in Psychology and Education.* Wiley, New York.

22. Chaffin, D. B., and Andersson, G. 1984. *Occupational Biomechanics.* Wiley-Interscience, New York.

23. Chaffin, D. B., Herrin, G. D., Keyserling, W. M., and Foulke, J. A. 1977. Pre-employment Strength Testing. *NIOSH Technical Report.* NIOSH Physiology and Ergonomics Branch, Cincinnati, OH.

24. Chaffin, D. B., Herrin, G. D., and Keyserling, W. M. 1978. Pre-employment Strength Testing. *Journal of Occupational Medicine*, 20, pp. 403–408.

25. Hunter, J. E., and Hunter, R. F. 1984. Validity and Utility of Alternative Predictors of Job Performance. *Psychological Bulletin*, 96, pp. 72–98.

26. Ellis, S. H. 1977. An Investigation of Telephone User Training Methods for a Multiservice Electronic PBX. In *Proceedings of the 8th International Symposium on Human Factors in Telecommunication.* Standard Telecom Laboratories, Harlow, Essex, England.

27. Eastman Kodak Co. 1986. *Ergonomic Design for People at Work.* Van Nostrand Reinhold, New York.

28. Bailey, R. W. 1982. *Human Performance Engineering: A Guide for System Designers.* Prentice Hall, Englewood Cliffs, NJ.

29. Rifkin, K. I, and Everhart, M. C. 1971. *Position Performance Aid Development.* Applied Science Associates, Valencia, PA.

30. Swezey, R. W. 1987. Design of Job Aids and Procedure Writing. In *Handbook of Human Factors.* Salvendy, G. (Ed.). Wiley, New York.

31. Kabaj, M. 1968. Shiftwork and Employment Expansion. *International Labour Review*, 98(3).

32. Grandjean, E. 1980. *Fitting the Task to the Man.* Taylor & Francis, London.

33. van Loon, J. H. 1980. Diurnal Body Temperature Curves in Shift Workers. In *Studies of Shift Work.* Colquhoun, W. P., and Rutenfranz, J. (Eds.). Taylor & Francis, London.

34. Akerstedt, T. 1977. Inversion of the Sleep Wakefulness Pattern: Effects on Circadian Variations in Psychophysiological Activation. *Ergonomics*, 20, pp. 459–474.

35. Tepas, D. I., Armstrong, D. R., Carlson, M. L., Duchon, J. C., Gersten, A., and Lezotte, D. V. 1985. Changing Industry to Continuous Operations: Different Strokes for Different Plants. *Behavior Research Methods, Instruments and Computers*, 17, pp. 670–676.

36. Johnson, L. C. 1982. Sleep Deprivation and Performance. In *Biological Rhythms, Sleep and Performance.* Webb, W. B. (Ed.). Wiley, New York.

37. Kelly, R. J., and Schneider, R. F. 1982. The Twelve-Hour Shift Revisited: Recent Trends in the Electric Power Industry. *Journal of Human Ergology*, 11, pp. 369–384.

38. Thiis-Evensen, E. 1958. Shiftwork and Health. *Industrial Medicine*, 27, pp. 493–497.

39. Wyatt, S., and Marriott, R. 1953. Night Work and Shift Changes. *British Journal of Industrial Medicine*, 10, pp. 164–170.

40. Browne, R. C. 1949. The Day and Night Performance of Teleprinter Switchboard Operators. *Occupational Psychology*, 23, pp. 121–126.

41. Bjerner, B., Holm, A., and Swensson, A. 1985. Diurnal Variation in Mental Performance: A Study of Three-Shift Workers. *British Journal of Industrial Medicine*, 12, pp. 103–110.

42. Monk, T. H., and Embrey, D. E. 1981. A Field Study of Circadian Rhythms in Actual and Interpolated Task Performance. In *Night and Shift Work: Biological and Social Aspects.*

Reinberg, A., Vieux, N., and Andlauer, P. (Eds.). Pergamon Press, Oxford, England.

43. Folkard, S., and Monk, T. H. 1979. Shiftwork and Performance. *Human Factors*, 21, pp. 483–492.

44. Knauth, P. 1993. The Design of Shift Systems. *Ergonomics*, 36(1–3), pp. 15–28.

45. Small, A. M. 1987. Design for Older People. In *Handbook of Human Factors Engineering*. Salvendy, G. (Ed.). Wiley, New York.

46. Poulton, E. C. 1970. *Environment and Human Efficiency*. Charles C Thomas, Springfield, IL.

47. Shinar, D. and Schieber, F. 1991. Visual Requirements for Safety and Mobility of Older Drivers. *Human Factors*, 33(5), pp. 507–519.

48. Schaie, K. W. (Ed.) 1983. *Longitudinal Studies of Adult Psychological Development*. Guilford Press, New York.

49. Fozard, J. L. 1981. Person-Environment Relationships in Adulthood: Implications for Human Factors Engineering. *Human Factors,* 23(1), 7–27.

50. Welford, A. T. 1977. Motor Performance. In *Handbook of the Psychology of Aging*. Birren, J. E., and Schaie, K. W. (Eds.). Van Nostrand Reinhold, New York.

51. Lawton, M. P. 1978. Sensory Deprivation and the Effects of the Environment on Management of the Senile Dementia Patient. In *Clinical Aspects of Alzheimer's Disease Conference Proceedings*. National Institutes of Mental Health, Bethesda, MD.

52. Stoudt, H. W. 1981. The Anthropometry of the Elderly. *Human Factors,* 23(1), pp. 29–37.

53. Annis, J. F. 1995. Aging Effects on Anthropometric Dimensions Important to Workplace Design. In *Advances in Industrial Ergonomics and Safety VII*. Bittner, A. C., and Champney, P. C. (Eds.). Taylor & Francis, London, pp. 27–32.

54. Elsayed, A.-M., Khalil, T., Asfour, S., Goldberg, M., Rosomoff, R., and Rosomoff, H. 1990. On the Relationship between Age and Responsiveness to Rehabilitation. In *Advances in Industrial Ergonomics and Safety II*. Das, B. (Ed.). Taylor & Francis, London, pp. 49–56.

55. Wright, J. and Mital, A. 1995. Human Strengths and Aging: What Do We Know? What Should We Know? In *Advances in Industrial Ergonomics and Safety VII*. Bittner, A. C., and Champney, P. C. (Eds.). Taylor & Francis, London, pp. 21–25.

56. Drinkwater, B. L., and Horvath, S. M. 1979. Heat Tolerance and Aging. *Medicine and Science in Sports,* 11, pp. 49–55.

57. Wagner, J. A., Robinson, S., Tzankoff, S. P., and Marino, R. P. 1972. Heat Tolerance and Acclimatization to Work in Relation to Age. *Journal of Applied Physiology,* 33, pp. 616–622.

58. Pandolf, K. B. 1990. Heat Tolerance and Aging with Application to Industrial Jobs. In *Advances in Industrial Ergonomics and Safety II*. Das, B. (Ed.). Taylor & Francis, London, pp. 19–26.

59. Young, A. J. 1990. Effects of Aging on Human Thermoregulation in Cold Environments. In *Advances in Industrial Ergonomics and Safety II,* Das, B. (Ed.). Taylor & Francis, London, pp. 41–48.

60. Inglesby, T. 1988. We Have Met the Enemy and It Are Management. *Manufacturing Systems,* June, pp. 16–19.

61. Sharit, J., Chang, T.-C., and Salvendy, G. 1987. Technical and Human Aspects of Computer-Aided Manufacturing. In *Handbook of Human Factors*. Salvendy, G. (Ed.). Wiley-Interscience, New York.

62. Nof, Shimon. 1985. *Handbook on Industrial Robotics*. Wiley-Interscience, New York.

63. Bullinger, H.-J., Korndorfer, V., and Salvendy, G. 1987. Human Aspects of Robotic Systems. In *Handbook of Human Factors*. Salvendy, G. (Ed.). Wiley-Interscience, New York.

64. Didner, R. S. 1988. A Value-Added Approach to Information Systems Design. *Human*

Factors Society Bulletin, 31 (5), pp. 1–2.

65. Montgomery, D. C. 1991. *Design and Analysis of Experiments* (3rd Ed.), John Wiley and Sons, New York, NY.

66. Schmidt, S. R., and Launsby, R. G. 1994. *Understanding Industrial Designed Experiments* (4th Ed.) Air Academy Press, Colorado Springs, CO.

67. Camp, R. C. 1989. *Benchmarking: The Search for Industry Best Practices that Lead to Superior Performance.* ASQC Quality Press, Milwaukee, WI.

68. Weimer, J. 1995. *Research Techniques in Human Engineering.* Prentice Hall, Englewood Cliffs, NJ.

CHAPTER 13

Industrial Safety and Health

13.1 INTRODUCTION

A subject that is very closely related to ergonomics is *safety and health*. Industrial applications of safety and health concepts interest industrial ergonomists, since one of the objectives of ergonomics is maintenance of employee well-being (see Chapter 1). The focus on safety and health at the workplace extends back at least a century, although at those times there was less emphasis on human values. In 1893, when the Railway Safety Act was being discussed, a railroad executive claimed that burying a man would cost less than installing air brakes on a car [1]. Since then, many views have changed. The government stepped up its efforts in providing guidance for enhanced safety and health at the workplace, standards developing agencies have emerged, independent organizations have come into existence promoting these concepts, and in general, more safety and health awareness has been created in industry. However, one cannot claim that all jobs in the United States are now being performed in safe environments. Many realize that total safety is impossible. There are just too many variables that affect safety performance, and every day, new factors are emerging. Soon after the *Challenger* accident in 1987 there was a massive investigation with respect to the total design of the vehicle system. Hundreds of design flaws were found and possibly corrected. It is interesting to hear such findings when the space program values quality at all costs. Although one still cannot guarantee a hazard-free environment, research is continuing to solve unknowns. As long as serious effort along these lines is supported by the government and independent sectors, the gap between total safety and the existing state of knowledge and implementation will continue to shrink.

There are many topics common to ergonomics and safety and health focus, such as reducing physiological and psychological stress and designing safe work areas and work methods. On the other hand, many areas are unique to safety and health, such as toxicology and fire prevention and control. Also there are several topics that are not so much discussed in safety but are integral parts of ergonomics. Examples are information acquisition and processing, motor control, and anthropometry.

Industrial safety deals with hazard recognition and control with respect to acute or instantaneous cases, such as sudden release of energy and possible injury or fatality due to it. A fall, or getting a hand stuck between reciprocating machine parts, are examples. Industrial health deals with hazards that show cumulative effects. A natural result of health hazards is illness. Lung cancer due to exposure to an airborne carcinogen is an example. Industrial health specialists are also known as industrial hygienists. On the other hand, an industrial safety specialist usually deals with factory process safety. In this chapter we provide more information on safety and health concepts of interest to an industrial ergonomist. Various legislative issues, along with hazard recognition, factors contributing to accident causation and hazard control, are covered in a summary format.

13.2 SAFETY AND HEALTH LEGISLATION AND THE GOVERNING BODIES

It is impossible to detail all related legislation, organizations, and governing bodies in a section of this chapter. However, the most important ones are covered in a summary format. The ergonomist must be familiar with these in order to appreciate the legal aspects of the area. Furthermore, the ergonomist should become familiar with the forces behind applicable government and consensus standards and rules.

The most important legislation in this respect is the *Occupational Safety and Health Act (OSHAct)* of 1970, also known as the Williams-Steiger Act. OSHAct was signed into law in December 1970 and became effective on April 28, 1971. The main objective of this law is to provide safe and healthy working conditions to every working man and woman covered by it. Those exempted include workers already covered by another similar federal safety program.

The law created several new organizations. Occupational Safety and Health Administration (OSHA) under the Department of Labor is one of these. OSHA has been given the responsibility of setting and enforcing safety and health standards. It accomplishes these tasks with the help of standards-writing institutions, its own experiences, research carried out or supported by NIOSH (National Institute of Occupational Safety and Health) and its inspectors (OSHA Compliance Officers). NIOSH is another organization created by OSHAct. It has been charged with the responsibility of funding and carrying out research and education functions related to occupational safety and health. One of its major responsibilities is to develop and recommend to the Secretary of Labor new safety and health standards. It is also authorized to conduct research in order to establish tolerable levels of hazardous substances. Each year NIOSH publishes lists of known toxic substances and the levels at which toxic effects will develop.

A third new agency created by OSHAct is the OSHRC (Occupational Safety and Health Review Commission). This is a quasi-judicial board that hears and reviews alleged standards violations, and whenever necessary, assesses civil penalties and corrective orders. Its three members are appointed by the president.

OSHAct brought several standardized record-keeping requirements with it. OSHA Form 200, "Log of Occupational Injuries and Illnesses," is the basic form used to note all "recordable" cases, which include those that resulted in fatality, medical treatment, and one or more lost workdays. Other cases are also recordable as specifically defined by OSHA. A section of Form 200 is an annual summary of all recordable cases, which must be displayed at a prominent place at the facility no later than February 1 of each year and to remain posted until March 1. All records are to be saved for at least 5 years. Beside the log/summary, Form 101, "Supplementary Record of Occupational Injuries and Illnesses," provides additional details on each recordable case. Each page of the supplementary record corresponds to a single-line entry in the log/Form 200 (Figures 13.1 and 13.2).

Under OSHAct, only the employer can be penalized for failure to comply with the law. The employer has two major responsibilities as detailed by OSHAct:

1. *General duty.* Furnish a safe working environment free of recognized hazards that can cause death or serious physical harm.
2. *Specific duty.* Comply with specific OSHA standards; keep records of work-related injuries, illnesses, and deaths; and keep records of employee exposure to toxic substances and harmful agents.

In return, the employee must comply with all applicable rules and standards relevant to his or her own actions.

Until now, OSHA has developed many industry standards and adopted many others (consensus standards). Organizations such as ANSI (American National Standards Institute), NFPA (National Fire Protection Association), ASME (American Society of Mechanical Engineers), and others provided many of the initial (consensus) standards to OSHA. To ensure compliance with these standards, OSHA compliance officers may inspect work locations according to a priority rank order. Top priority is given to those cases where a fatality has occurred or where five or more employees got injured. Such cases must be reported to OSHA (or to one of its local offices) within 48 hours.

The compliance officers must issue citations for violations of OSHA standards. This must occur within 6 months of the observance of the violation and must include a reasonable time for elimination or abatement of the hazard. OSHA can also request that the employer stop immediately any operation where an imminent danger exists.

Individual states may take over all OSHA responsibilities if they can show that they have a program that is at least as effective as OSHA's. This must be done in writing, and the Secretary of Labor must approve. The state then pays for half the cost of the approved program and the federal government for the other half. The U.S. Department of Labor monitors the program for 3 years to ensure its effectiveness.

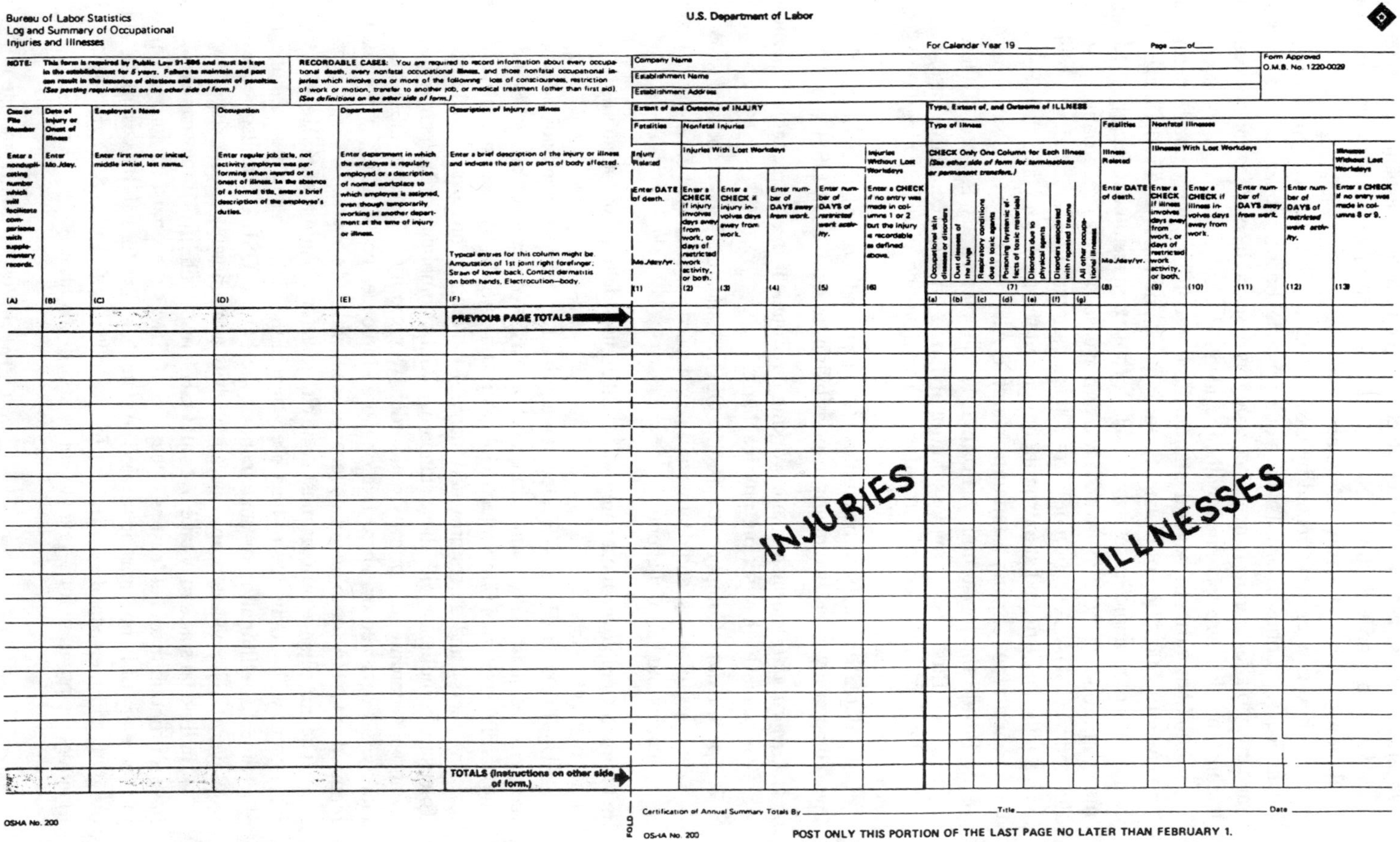

Figure 13.1 OSHA 200 log of daily injuries. (From U.S. Department of Labor, Bureau of Labor Statistics.)

Bureau of Labor Statistics
Supplementary Record of
Occupational Injuries and Illnesses

U.S. Department of Labor

◈

This form is required by Public Law 91-596 and must be kept in the establishment for 5 years. | Case or File No. | Form Approved
Failure to maintain can result in the issuance of citations and assessment of penalties. | | O.M.B. No. 1220-0029

Employer

1. Name

2. Mail address (No. and street, city or town, State, and zip code)

3. Location, if different from mail address

Injured or Ill Employee

4. Name (First, middle, and last) | Social Security No.

5. Home address (No. and street, city or town, State, and zip code)

6. Age | 7. Sex: (Check one) Male ☐ Female ☐

8. Occupation (Enter regular job title, not the specific activity he was performing at time of injury.)

9. Department (Enter name of department or division in which the injured person is regularly employed, even though he may have been temporarily working in another department at the time of injury.)

The Accident or Exposure to Occupational Illness

If accident or exposure occurred on employer's premises, give address of plant or establishment in which it occurred. Do not indicate department or division within the plant or establishment. If accident occurred outside employer's premises at an identifiable address, give that address. If it occurred on a public highway or at any other place which cannot be identified by number and street, please provide place references locating the place of injury as accurately as possible.

10. Place of accident or exposure (No. and street, city or town, State, and zip code)

11. Was place of accident or exposure on employer's premises? Yes ☐ No ☐

12. What was the employee doing when injured? (Be specific. If he was using tools or equipment or handling material, name them and tell what he was doing with them.)

13. How did the accident occur? (Describe fully the events which resulted in the injury or occupational illness. Tell what happened and how it happened. Name any objects or substances involved and tell how they were involved. Give full details on all factors which led or contributed to the accident. Use separate sheet for additional space.)

Occupational Injury or Occupational Illness

14. Describe the injury or illness in detail and indicate the part of body affected. (E.g., amputation of right index finger at second joint; fracture of ribs; lead poisoning; dermatitis of left hand; etc.)

15. Name the object or substance which directly injured the employee. (For example, the machine or thing he struck against or which struck him; the vapor or poison he inhaled or swallowed; the chemical or radiation which irritated his skin; or in cases of strains, hernia, etc., the thing he was lifting, pulling, etc.)

16. Date of injury or initial diagnosis of occupational illness | 17. Did employee die? (Check one) Yes ☐ No ☐

Other

18. Name and address of physician

19. If hospitalized, name and address of hospital

Date of report | Prepared by | Official position

OSHA No. 101 (Feb. 1981)

SUPPLEMENTARY RECORD OF OCCUPATIONAL INJURIES AND ILLNESSES

To supplement the Log and Summary of Occupational Injuries and Illnesses (OSHA No. 200), each establishment must maintain a record of each recordable occupational injury or illness. Worker's compensation, insurance, or other reports are acceptable as records if they contain all facts listed below or are supplemented to do so. If no suitable report is made for other purposes, this form (OSHA No. 101) may be used or the necessary facts can be listed on a separate plain sheet of paper. These records must also be available in the establishment without delay and at reasonable times for examination by representatives of the Department of Labor and the Department of Health and Human Services, and States accorded jurisdiction under the Act. The records must be maintained for a period of not less than five years following the end of the calendar year to which they relate.

Such records must contain at least the following facts:

1) *About the employer*—name, mail address, and location if different from mail address.

2) *About the injured or ill employee*—name, social security number, home address, age, sex, occupation, and department.

3) *About the accident or exposure to occupational illness*—place of accident or exposure, whether it was on employer's premises, what the employee was doing when injured, and how the accident occurred.

4) *About the occupational injury or illness*—description of the injury or illness, including part of body affected; name of the object or substance which directly injured the employee; and date of injury or diagnosis of illness.

5) *Other*—name and address of physician; if hospitalized, name and address of hospital; date of report; and name and position of person preparing the report.

SEE *DEFINITIONS* ON THE BACK OF OSHA FORM 200.

Figure 13.2 OSHA form 101. (From U.S. Department of Labor, Bureau of Labor Statistics.)

OSHA has several voluntary protection programs—the Star, Merit, and the Demonstration programs. Worksites that have effective, ongoing safety and health programs (measured by performance indicators such as injury incidence and lost workday injury rates, cooperation between labor and management, and annual self-evaluation) may apply for one of these programs. These are designed to augment OSHA's enforcement efforts with encouragement and recognition of performance beyond OSHA's expectations. Participation exempts a worksite from OSHA's programmed inspections. The highest level of recognition is the Star. Merit recognition is an effective stepping stone to the Star award. The Demonstration Program provides a basis for alternative and innovative approaches to safety and health management.

Many states have also passed *workers' compensation laws*. The fundamental principle of these laws is to compensate workers for any loss of income due to an injury or fatality. Many changes in workers' compensation laws made it possible to compensate workers for permanent injuries that do not cause loss of income.

Benefits, specifically outlined by the applicable law depending on the case involved, may have to be paid over a long period of time. To guarantee that the benefits will be paid when and as long as required, the workers' compensation laws require that the employer obtains insurance (workers' compensation insurance) or show that it is capable of supporting a foreseeable burden (self-insurance).

To receive benefits under workers' compensation, three fundamental conditions must be met:

1. The injury must have resulted from an accident.
2. It must have arisen from one's employment.
3. It must have occurred during the course of employment.

The *Federal Mine Safety and Health Act* was signed into law in November 1977. It became effective in March 1978. This is a very similar act to OSHAct. Its focus is on miners and mine safety and health. It created the MSHA (Mine Safety and Health Administration). Unlike OSHAct, miners can be sanctioned for violations of standards related to smoking [2].

The *Consumer Product Safety Act of 1972* was discussed in Chapter 11. This act's focus is the consumer product and its safe use for the purpose intended [3].

13.3 ACCIDENT MODELS

An accident may or may not lead to physical harm to an individual. No matter what the outcome is, the fundamental question to be answered is: How do accidents happen? Many suggestions have been offered for this purpose [4,5]. Denton [6] speaks of a two-tier cause chain with the first level focusing on unsafe acts and/or unsafe conditions. The second tier looks at what the author calls *situational factors*, including psychological and physical characteristics of the environment and the workers. For example, factors such as temperature [7], insufficient light [8], length of the working day [9], and layoff rate [10] have all been linked to causes of accidents. On the other hand, Smith et al. [11] showed that plants with well-designed work processes, lighting, ventilation, and clean workplaces had lower accident rates. Uusilato and Mattila [12] observed in Finland that active participation of management in safety programs and activities determined to a large extent which industries would display a low accident rate. In a recent study of Bangladesh industry, Khaleque and Karim [13] discovered that the managers and workers were more agreeable with respect to the causes of industrial accidents than in terms of effective prevention strategies. As common causes, they cited mechanical breakdown of equipment, faulty tool handling, poor working conditions, irresponsibility, inexperience, and uncaring attitude. Workers thought that incentives, improvement in working conditions, accident insurance, and safety training would be the most effective prevention methods. On the other hand, managers cited inspection, persuasion of workers to use safety equipment, enforcement of safety rules, and standardization of equipment as some of the effective safety techniques. All of the foregoing studies indicate that causes of accidents are complex and involve many agents and factors.

Focusing more on the human behavior, Ramsey [14] proposed a four-stage accident sequence model. In particular, the probability of accident is greater if the *perception, cognition, decision,* and *ability to avoid* stages of behavior result in negative consequences. The current view on accident causation is that there is a multiplicity of factors that affect a company's safety performance (Figure 13.3):

1. *The task.* Specific elements of the task being performed may affect the probability of an accident. The task may be paced. There may be both speed and load stress present. A task that does not seem to be paced may become stressful due to its tiring characteristics. Such task attributes as rewards and perceived value may influence the worker's motivation to perform well. Excessive physical requirements of the task will reduce worker capacity to perform effectively over time. Both mental and physical fatigue will develop. Probability of an accident increases notably with accumulating fatigue.

2. *The worker.* Many variables related to the worker may influence accident potential. If one portrays a worker as the controller of a three-chain event sequence, one realizes

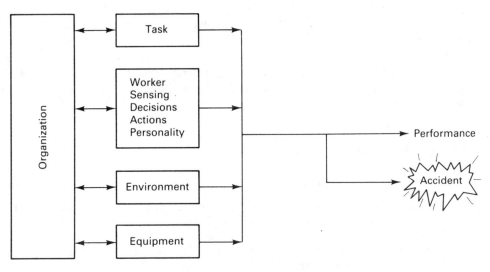

Figure 13.3 Major factors of safety performance.

that if anything goes wrong in any of the chains, the probability of an accident will increase. The human being is limited in terms of sensing. Wrong decisions may be made due to wrong input or improper training. Inappropriate actions may also lead to an incident. Physical capacities such as strength, endurance, and reach may also be important in determining whether or not an accident occurs. Physically conditioned workers may perform effectively on physically demanding tasks. Although personality variables have not been shown [15] to be major causes of accidents, they may nevertheless contribute to the elevation of accident risk.

3. *The environment.* Environmental conditions may significantly alter an otherwise safe work climate. Factors such as heat, glare, humidity, and vibration may affect human behavior negatively and lead to unsafe actions. Poor maintenance and housekeeping may be major determinants of accident causation. Poor lighting, excessive noise, and excessive cold may also affect worker behavior and the ability to sense important signals for safe performance.

4. *Equipment factors.* Machinery and specific equipment factors may contribute to accident risk. Inadequately guarded moving machine elements, high voltages, and excessive speeds are potential causes of significant trauma. Equipment factors present themselves among the unsafe conditions for accident causation.

5. *Organizational factors.* Management attitude toward safety is a critical variable in any organization's safety history. An organizational structure that promotes cooperation among workers to recognize and correct hazards will positively affect worker behavior. Such a structure will also motivate workers for extra care during work. Other variables that can be counted among organizational factors include supervisor–subordinate relationships and task–rest organization.

The worker/task/environment/equipment/organization system generally functions with no accidents. The reasons for accidents may be simple or complex. Sometimes a simple unsafe act such as not operating a control when it is required to do so may lead to an accident. Unsafe conditions that relate to the task, the environment, the equipment, and the organizational structure also may cause accidents. However, usually a combination of unsafe acts and unsafe conditions leads to incidents, which may cause a near accident or an accident. If an accident occurs, there may be only material damage or there may be physical harm to a worker or both. The next two sections dwell on unsafe conditions and unsafe acts.

13.4 UNSAFE CONDITIONS

The following is an excerpt from a company's accident records:

> *Employee's Activity at Time of Accident:* The employee was at his work position; he stood up and slipped on some wire on the floor. In an attempt to break his fall, he put out his right hand. His hand hit a machine screw protruding out from a wire guide on the front of a wire rack. The machine screw lacerated the palm of his right hand. The employee reported to medical and the laceration was closed with eleven sutures.

Reinstructions Given to Employee(s) Following Accident: To keep loose wire picked up off the floor.

Corrective Action Taken: The wire cart was modified with plastic caps being placed over the exposed screwheads to prevent employees from being cut or scratched.

The excerpt above exemplifies a classic accident. This employee had only one week experience on the job. The injury was caused by an exposed and protruding machine screw. There was no material damage. He was also instructed as to better housekeeping.

It can be predicted from the short description above that it was the responsibility of the employee to keep wire off the floor. This was probably a self-paced job. There were no equipment aids to catch falling wire. An exposed and protruding machine screw caused a nondisabling injury. The corrective action taken indicates that the management realized an unsafe condition and corrected it. However, the cause of this particular accident seems to be the loose wire on the floor. It can be argued that a simple device would probably have accumulated all loose wire that results from the operation in question. A personnel variable also played a role. With only a week's experience on the job, the operator probably did not expect loose wire on the floor to cause such a mishap.

Unsafe conditions lay the groundwork for accidents. Such conditions accumulate the necessary hazard potential for a subsequent trigger event. In this section we present several areas that focus on a particular process or hazard cluster to detail sources of hazards to be corrected. For further information, refer to [2] and [16–18].

13.4.1 Industrial Fire

From a fire hazard standpoint, the most dangerous industries seem to be chemical plants, grain elevators, mines, refineries, and warehousing operations. Principal causes of fires are hot processes, overheated parts, and hot equipment. Another major cause is poor housekeeping. Accumulated ordinary combustibles such as paper, wood, and cloth, paint residues, and airborne combustibles such as dusts and mists are good examples.

The best way to deal with industrial fires is to develop a good prevention program. If their occurrence can be prevented, their effects can be minimized. A preventive maintenance program for overheating machinery, good housekeeping, and isolating sources of heat and combustibles are some ways of preventing fire occurrence in industry.

13.4.2 Material Handling

Manual material handling, especially lifting, is the most dominant workplace injury category. Even mechanized handling (using industrial trucks, tractors, cranes, and conveyors) methods cause injuries and fatalities each year. Many material-handling injuries are caused by the material falling on or striking a person. In many cases the loads handled are massive. A bulky and heavy object striking a comparatively fragile person causes severe injuries. Manual handling activities may cause cumulative trauma as well as single-incident trauma. Relevant parameters have been discussed in Chapter 3 and will not be repeated here. Mechanized handling also displays unique hazards. A conveyor that is started by a person while another is performing preventive maintenance on it at a distance may inflict injuries.

Cranes, industrial trucks, and hoists have been designed to apply significant power. If abused or not properly mounted or maintained, these devices may cause notable damage and injuries. Refueling and recharging of industrial trucks must be done carefully. There are also hazards from gases and acids in battery-charging locations. Trucks should be operated in sufficient lighting; otherwise, they should be equipped with directional lights. Dock levelers are especially important in minimizing the occurrence of toppled or fallen fork trucks. Cranes should not be overloaded. Wind may create a hazard for overhead cranes working outdoors. Crane cabs should be easily accessible to workers. Cable hooks and ropes should be inspected periodically for cracks, metal fatigue, broken fibers, and wear. In-running nip points on conveyors may catch a body part. Emergency tripping devices alongside conveyors may save lives.

13.4.3 Machine Guarding

Over 30% of the citations issued by OSHA are due to inadequately guarded machine elements [18]. The largest number of injuries due to machines occur at the point of operation, where the machine tool engages the workpiece. Power transmission apparatus (belts, pulleys) also present a "caught in or between" type of hazard. Rotating or reciprocating machine parts present similar hazards. Raw material, belts, pulleys, and gears also create nip points at their contact with each other or other machine elements. All such power transmission elements, rotating parts, and nip points must be guarded.

An important area is guarding of points of operation. For maximum production flexibility, one would like to keep such points open as much as possible. On the other hand, due to the hazards involved, points of operation need to be guarded, especially during machine operation. There is obviously a trade-off here. Two-hand controls and trips, holdouts, pull-outs, and presence-sensing devices are used extensively in preventing injuries due to operation points.

Related areas include machine anchoring, tagouts and lockouts, interlocks, and location guarding. Machines that have reciprocating motions have a tendency to move around or "walk." Such machines must be anchored. A number of accidents occur when a machine is down for maintenance. A possible scenario is: When a maintenance man is close to the machine or inside the machine, another operator turns it back on. In the tagout system, the maintenance worker hangs a tag on the on–off switch warning of his presence. Better, a lockout device (a simple lock) provides positive protection to the maintenance worker since it can only be operated by the maintenance worker who has the key to it. An interlock stops a machine when the guard is not in place. As soon as the guard is removed, the interlock between the guard and the drive mechanism cuts the power to the machine and stops any rotation. Location guarding dictates that the hazard points (nip points, points of operation) be designed away from the operator.

There is an additional gain from machine guards. They protect flying parts, chips, and other particles from reaching the worker. Therefore, they serve as shields against foreign particles.

13.4.4 Electrical Hazards

The most helpful source of energy to human welfare, electrical energy, can cause severe injuries and fatalities. A unique characteristic of electricity is that it cannot be seen. One may hold an exposed power line charged with live electricity and get electrocuted. Electric current exceeding 70 milliamperes will usually be fatal.

Three effects of electricity on living tissue are evident:

- *Tissue burning.* Current in the range of 2000 milliamperes will severely burn living tissue.
- *Fibrillation.* The cardiac muscle (heart muscle) may fibrillate (i.e., convulse weakly and rapidly). This does not serve any purpose, and unless controlled electric shock is administered, death from fibrillation is almost a certainty.
- *Breathing difficulty.* Electric shock may also affect one's breathing. The muscles of the chest wall and the diaphragm may get cramps due to interference with their natural rhythmic activity. The person may die due to cramped breathing muscles. Artificial respiration may reverse the process.

Among the electrical hazards is possible fire. Overheating of wires due to too much current is a common cause of fires. Arcs and sparks due to conductors making physical contact may also cause fire. Many control devices, such as fuses and circuit breakers, protect against electrocution and especially fire.

13.4.5 Welding

Welding presents a multitude of hazards. Acetylene cylinders used in gas welding have a devastating destructive power. Acetylene gas is very unstable. It's pressurization in manifolds greater than 15 psig is prohibited. The acetylene cylinder must not be tilted more than 45 degrees from vertical to maintain the valve-end-up position.

Oxygen cylinders must not be banged around or dropped. Such acts may knock off the valve, which leads to the cylinder's flying around, ricocheting off the walls and killing anybody in the way. The cylinders must also not be rolled, for this may damage the cylinder and the valve. Oxygen and acetylene cylinders should not be stored together in the same room. If a leak occurs, the combination gas is highly dangerous. They must be separated by a 5-foot noncombustible barrier or moved apart at least 20 feet.

The major hazards of arc welding are health hazards, fires, explosions, eye hazards, and confined space hazards. The frame of the welding machine must be properly grounded. Resistance welding also presents electrical shock hazards. One of the principal causes of industrial fires is welding. Eye protection and face protection may be necessary in welding. Protective clothing is very important to a professional welder. The long-term effects of inhaled welding fumes should not be overlooked. The material and the coating of the surfaces to be welded may contaminate the atmosphere around the welder. The gases and fumes present may lead to severe health effects.

13.4.6 Heat Treating

Metal alloys are heat treated to improve the strength, impact resistance, hardness, durability, and heat and corrosion resistance [19]. These operations present hazards due to the furnace operations and handling nitrate and cyanide bath materials. Hot baths require exhaust ventillation. Canopies must be erected on oil, salt, and metal baths when sprinkler systems are used. Water entering such baths presents extremely hazardous conditions to nearby personnel. Hot conditions around furnaces require special clothing for the workers.

13.4.7 Chip Removal Operations

Metal machining operations present similar hazards, which are mentioned briefly here. Lathe, drill press, miller, shaper, planer, and surface grinder operations are examples of metal machining. Notable pressure and heat are generated at the interface between the cutting tool and the workpiece. An interface lubricant is provided to this point in a solid stream or mist to cool the interface and flush away the chips formed. Such fluids may lead to two health problems. First, skin contact with these fluids may cause contact dermititis. Second, inhalation of oil mist created may cause harmful effects [20]. Severe damage to the respiratory and the digestive systems may occur due to inhaled mists, depending on the conditions involved.

Machining operations with some metals, such as magnesium and titanium, may generate explosive concentrations of dust. These operations must be conducted with suitable ventilation. Exhaust hoods are commonly used to capture toxic metal dusts.

13.4.8 Cotton Operations

Cotton handling operations, from picking of cotton boll from the field to spinning, winding, and weaving, present unique hazards that must be controlled. Cotton dust causes a condition called *byssinosis,* a disease characterized by shortness of breath, coughing, and tightness in the chest. Symptoms are maximum upon return to work on Mondays and gradually subside during the week. Dust-producing machines must be equipped with local exhaust ventilation. Workplaces must also be supplied with continuously cleaned air. Deposited lint and cotton dust on machines and equipment also present a fire hazard.

13.4.9 Underground Mining

The hazards of underground mining are primarily exposure to the mineral being recovered, associated gases and vapors, fire, explosions, heat stress, and others. Physical contact with certain types of minerals may present problems. Mercury and asbestos are examples of such minerals. Dust hazards are also noteworthy. Coal dust and quartz dust are examples and must be controlled through exhaust ventilation. Gaseous emissions in mines also present poisoning and fire hazards. Methane gas in coal mines is a major problem due to its wide flammability range. Fire may also lead to dust explosions, releasing carbon monoxide and burying workers under fallen debris. Oxygen deficiency in the underground mines is another problem. Vibration exposure due to use of percussion drills may present specific hazards, such as Raynaud's syndrome. High noise due to mining equipment may also present hearing hazards.

13.4.10 Others

In this section we briefly mention hazards due to several other operations. The interested reader may refer to Refs. [2] and [16–19] for additional information. Many items are painted for protection against corrosion, for insulation, and for appearance. Painting presents hazards related to contact dermatitis and exposure to solvent or thinner mist.

Foundry operations are widely used for metal casting. Such operations present a variety of hazards, including exposure to air contaminants, noise, heat, vibration, and skin contact with resins.

Glass making involves the use of many elements in the periodic table. With the major ingredient being silica sand, there is a potentially severe silicosis hazard. Optical glass and certain decorative glasses require the use and handling of lead or lead oxide. This is another potentially serious hazard, which requires local exhaust ventilation and personal protective equipment. Heat stress may be a problem in furnaces and molten glass handling. Glass fiber must also be controlled for possible respiratory system effects.

Petroleum exploration, handling, and refining operations must be carried out with care. Exposure to heat and cold stress in the field, chemical exposure during handling, oxygen deficiency, and possible chemical exposure during tank cleaning are some of the problems.

Plastics production presents chemical exposure (contact and airborne) problems. Polyvinyl chloride (PVC) is a known carcinogen.

It is up to the designer to provide safe working conditions to workers. People design the production machines and equipment. People also put these components together to make a process. Futhermore, people design the job, the workplace, and its arrangement. It is the responsibility of the engineer, the ergonomist, and the unit manager that safe equipment are brought in, safe processes are established, and safe jobs are developed.

13.5 UNSAFE ACTS

Not only the working conditions, but also the personnel acts must be controlled for the total job to be safe. In general, a significant portion of the blame goes to the worker after an accident. One may argue that all accidents are due to unsafe acts. A person who designs and builds a machine that is not safe to operate is committing an unsafe act. However, the person who selects that equipment for use in the production process is also responsible. Equipment may be safe for use in certain conditions and unsafe in others. Hence situational factors also play a role in determining whether or not an accident will occur.

Assuming that a safe working environment has been created, a person's actions must also be safe for the total system to function in a safe manner. Several personnel attributes contribute to safe performance of the worker (Figure 13.4):

1. *Education.* The personnel must be educated in various facets of their jobs. The worker must understand thoroughly the job content and how it relates to other jobs. A primary building block for safe behavior is the worker's appreciation of the importance of his or her job.

2. *Training.* The overall background developed by education must be complemented by specific training on the job. The worker should develop regular and emergency job attendance skills. Contingencies must be reviewed with the worker and possible courses of

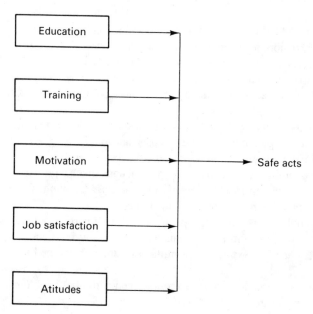

Figure 13.4 Employee factors that contribute to safe behavior on the job.

action must be detailed. Not only is normal job training necessary, but the worker must also be exposed to safety training. Probable unsafe acts and conditions may be reviewed with the worker and correction procedures may be discussed. The employee must also know who to contact in case an unsafe condition is detected. Specific emphasis may be placed on the potential unsafe acts and their consequences on the worker and others in the area.

3. *Motivation.* Employees must be motivated to perform in a safe manner. Positive consequences of safe behavior and negative consequences of unsafe behavior, when presented in a clear manner, may help employee motivation. Other methods may also be tried. Group prizes may be awarded for safe performance over a period of time such as 3 or 4 months. These may be small prizes, such as an umbrella or jacket with an attached safety emblem or slogan. At the end of the year, the group that had the fewest accidents and lost workday cases may be recognized by company management.

4. *Job satisfaction.* A worker who is satisfied with the job that he or she is performing, in general, will pay more attention to the events and display a safe working behavior. On the other hand, one who is not satisfied with the job will spend more time concentrating on things other than the job itself. It is this concentration on the job or lack of it that determines whether the employees' actions are also contributing causes of accidents or not. Motivation and job satisfaction have been covered in more detail elsewhere in this book.

5. *Attitudes.* Attitudes determine one's disposition to evaluate events, processes and other input [21,22]. A person's previous experiences partially determine his or her response to a stimulus. Development of correct attitudes and modification of incorrect ones are necessary for a worker's safe performance.

A primary element that affects one's safety motivation and attitude is management attitude toward safety. A safety program designed around management focus on the issue goes a long way toward affecting and changing worker behavior. Unsafe acts should be investigated carefully similar to investigations concerning unsafe conditions. The safety specialist must understand the reason behind such acts and correct them. The employee may lack education and training. He or she may not be motivated to do his or her best. The employee's attitudes may have to be changed. A well-planned monitoring system will spot the failure modes and attempt to eliminate root causes. Sometimes job rotation may help improve job satisfaction. A different job may also be beneficial, due to previous experiences on similar jobs. A thorough interview with the employee will provide many clues as to how unsafe acts may be prevented.

13.6 EVALUATING SAFETY PERFORMANCE

To compare one period's safety performance with another and also to evaluate the effects of specific safety programs across a business or a department, several safety indexes may be used. The traditional measures include the *frequency rate*, the *severity rate,* and *average days charged.* A more recent measure is the *incidence rate.* Such measures are not used only to evaluate one particular unit's safety performance over a period of time; they can also be used to make comparisons across industries. This is possible since the measures are not industry specific. The discussion below details these measures.

1. *Frequency rate.* This is a measure classified in the old ANSI Z16.1 standards system. It quantifies the frequency of accidents or disabling injuries over a fixed time interval, usually a year. In quantitative terms, the accident frequency rate is defined as follows:

$$\text{accident frequence rate} = \frac{\text{number of accidents} \times 1,000,000}{\text{person-hours of job exposure}}$$

The equation above calculates the frequency of accidents across a period per million person-hours of work. Hence if a company experienced 25 accidents in 1 year for 1,500,000 person-hours of work, the accident frequency rate comes out to:

$$\text{accident frequency rate} = \frac{25 \times 1,000,000}{1,500,000} = 16.67 \text{ per million worker-hours}$$

If the same company had experienced only 20 accidents during the previous year for 1,100,000 person-hours of work exposure, the rate for that year would be 18.18 per million worker-hours, which is worse. Although in this case the number of accidents is less, exposure hours are also fewer, and the accident rate comes out to be higher.

A similar measure, the disabling injury frequency rate, is defined as follows:

$$\text{disabling injury frequency rate} = \frac{\text{number of disabling injuries} \times 1,000,000}{\text{workers-hours of job exposure}}$$

This measure calculates the frequency of disabling injuries across a period on a per million person-hour exposure base. A business unit that experienced 30 disabling injuries in 1 year (including fatalities) for 1,250,000 worker-hours of exposure is charged with 24 disabling injuries per million worker-hours of work exposure.

2. *Disabling injury severity rate.* Some companies may experience many injuries, but they may be minor. Others may experience few injuries, each of which may be severe, including fatalities. To develop a common basis of comparison, ANSI established a severity measure across a fixed period using specific time charges for each incident. For example, ANSI Z16.1 defines 1800 lost workdays for loss of an eye or eye sight; complete loss of hearing in one ear carries loss charges of 600 days. When all such charges are accumulated over all injuries over a period (usually, a year), total number of days charged may be calculated. Then

$$\text{disabling injury severity rate} = \frac{\text{total days charged} \times 1,000,000}{\text{worker-hours of job exposure}}$$

Assume that a company experienced 38 injuries in one year. Time charges across these incidents come up to 19,000 days. Total work exposure has been 2,000,000 person-hours. Then,

$$\text{disabling injury severity rate} = \frac{19,000 \times 1,000,000}{2,000,000} = 9500 \text{ days}$$

If the same company had only 20 injuries with 12,000 days charged and 1,800,000 person-hours of work exposure in the next year, the corresponding severity rate value is 6666.67. Hence a notable drop in total days charged outweighs the drop in exposure hours and pulls the severity rate down by about a third.

3. *Average days charged.* This is a measure of severity for each injury. It is defined as:

$$\text{average days charged} = \frac{\text{total days charged}}{\text{number of disabling injuries}}$$

or

$$\text{average days charged} = \frac{\text{injury severity rate}}{\text{injury frequency rate}}$$

Assume that the total days charged for 20 injuries is 1000. Then average days charged is 50 per injury.

4. *Incidence rates.* OSHAct enlarged the Z16.1 system and defined a generic safety performance measure called the incidence rate. The measurement base is 200,000 worker-hours, equivalent to 100 full-time employees' work hours during a year. Specifically,

$$100 \text{ workers} \times 40 \text{ hours/week} \times 50 \text{ weeks/year} = 200,000 \text{ hours/year}$$

Under the incidence rate, not only disabling injuries are counted, but also those cases that resulted in an occupational illness, loss of consciousness, restriction of work or motion, transfer to another job, medical treatment, and termination of employment. Medical treatment does not include simple first aid, diagnostic procedures, or preventive medicine.

Incidence rate can be calculated separately for fatalities, illnesses, specific hazards, lost-workday cases, medical treatment cases, and the like. It is a more flexible measure than the disabling injury frequency rate. Assume that a business experienced 16 disabling injuries in a year in which a total of 800,000 employee work-exposure hours have been accumulated. Then the disabling injury frequency rate would be calculated as:

$$\frac{16 \times 200,000}{800,000} = 4 \text{ per 200,000 worker hours}$$

If during the same year, all other injuries (nondisabling) amounted to 150, then

$$\text{nondisabling-injury frequency rate} = \frac{150 \times 200,000}{800,000} = 37.5 \text{ per 200,000 worker-hours}$$

The reader should observe that the bases of the ANSI and the BLS (Bureau of Labor Statistics) frequency rate measures are different. Similarly, the severity rates would also have different bases. One is calculated on the basis of 1,000,000 person-hours of work, the other 200,000. Hence it is important to know the base in comparing the current safety experience with experience in the past.

The foregoing measures provide good measures of evaluating safety performance or effectiveness across different periods. The safety specialist and the ergonomist are encouraged to track such performance over multiple periods to evaluate trends.

13.7 INDUSTRIAL TOXICOLOGY

This section focuses on a different type of hazard, one that cannot be seen by the unaided eye, responsible for organic injuries. When such injuries arise out of and in the course of employment, they are called occupational diseases [23]. The primary cause of such injuries is toxic substances, which manifest themselves largely as chemicals. Scientists who deal with toxicology and related topics are frequently called *hygienists*.

A toxic substance may act on the body in a number of ways:

1. Cause neoplastic effects (cancerous tumors and other tumors)
2. Produce damage to an embryo
3. Irritate the skin, the eyes, and the respiratory tract
4. Diminish mental alertness
5. Alter behavior, general health, and sexual functioning
6. Cause short- or long-term illness [24]

There are over 50,000 known toxicants. The discipline that focuses on toxic effects of various substances is industrial toxicology, a branch of environmental toxicology. Specifically, toxicology examines human exposure to toxic agents. Not every chemical is toxic. By the same token, a toxic substance does not produce toxic effects in any concentration and duration of exposure. Chemical agents will produce undesirable effects, depending on several factors, including the type of agent, concentration, mode of entry (discussed in the next section), and duration of exposure. Every year, NIOSH publishes a list of known toxicants. Various other organizations also participate in this endeavor, including the American Conference of Governmental Industrial Hygienists, which publishes the *threshold limit values* (TLVs) for various chemicals, and the American National Institute, which publishes "Standards of Acceptable Concentrations of Toxic Dusts and Gases." OSHA also has a list of PELs (permissible exposure levels) for different substances given in terms of ppm (parts per million) or mg/m^3 (milligrams of particulate per meter cube of air).

13.7.1 Types of Toxicants

Toxic substances enter the body through three primary routes. In the order of significance in the industrial environment, they are:

- Inhalation
- Absorbtion
- Ingestion

Inhalation of the toxicants represents the most direct and rapid way of entry to the body. This is due to the fact that respiration and circulation are very closely linked in the lungs. Foreign elements can easily be carried to every corner of the body if they are inhaled and can diffuse into the bloodstream. Since breathing is an involuntary activity, airborne toxicants present the most serious health hazard in industry. Airborne toxicants are of two types:

- *Completely dispensed fluids:* gases and vapors
- *Suspended particles or liquid droplets:* aerosols, dusts, fumes, mists, smoke

Several factors determine the degree to which airborne contaminants enter the circulatory system:

- Duration of exposure
- Rate of respiration
- Solubility of contaminant in blood and other tissue
- Concentration of toxicant in the inhaled air
- Reactivity of the toxicant

As with other types, poisoning due to airborne substances occurs in two major ways: acute poisoning and chronic poisoning. The former is the case where a person gets exposed to a heavy concentration in a single instant. The latter is the result of repeated exposures to smaller concentrations. If the doses are not lethal, the chances of recovery from acute poisoning are greater than recovery from chronic poisoning. Since the latter occurs over a long period, the effects are not very apparent until too late and are more damaging. Airborne particulate toxicants may also accumulate in the lungs and do damage there. Their presence also leads to fibrous tissue development in the lungs, reducing their capacity. Byssinosis, cotton dust disease, and silicosis experienced by miners are examples.

Toxicants may also enter the body through absorbtion and ingestion. The epidermis, the outer layer of skin, resists absorption. The thicker the epidermis (e.g., in the palms and soles), the more resistant it is to foreign agents. The gastrointestinal (GI) tract is also a very effective mode of entry. However, the human being is more selective as to what he or she eats or drinks. Upon entry to the body, toxicants exert harmful effects, including irritation, tumor development, tissue damage or destruction, and immunological effects. Some toxicants may be excreted through exhaling, kidney function, and liver function.

13.7.2 Monitoring Methods

Environmental and biological monitoring are two methods by which exposure to contaminants may be controlled. Monitoring of substances in workroom air may be accomplished by periodic sampling or continuous monitoring. In the former, air is sampled periodically using simple (pumps and detector tubes) or sophisticated (atomic absorption spectrometers, gas chromatographs) equipment. In the latter method, ambient air is continuously (24 hours per day) monitored. Results of such measurements may then be compared with recommended safe concentrations (TLVs or PELs) of the monitored substance and decisions made on the safety of working in the air that is in question [25].

Biological monitoring involves routine analysis of human tissue or excreta (excreted substances, air, fluids) for evidence of exposure to toxicants [26]. Specimens may be collected from the body and analyzed for traces of toxicant presence. Urine is a frequently analyzed specimen. Other specimens that have been investigated include breath, blood, saliva, hair, nails, breast milk, and biopsy fat. Routine availability of the specimen is an important factor to be considered in deciding on the type to be used for biological monitoring. Other factors include route of exposure, availability of the published reference limits, and the metabolic profile of the specific chemical [27].

13.8 HAZARD AVOIDANCE

Hazard avoidance is the primary task of a safety and health specialist. It makes a lot of sense to try to prevent accidents rather than correct the situation after the fact. Hence avoidance and prevention are synonymous. There are no specific guidelines for hazard avoidance. This is due to the fact that one cannot be sure of the existence or accumulation of hazard potential. Although much information has been gathered and data banks developed with respect to hazard potential of many industrial processes, there still exists a lot more to be discovered and researched. It is this "dealing with quasi-known things" that makes the task of the safety and health specialists so hard. In addition to step-by-step prevention procedures, the specialist must also rely on general avoidance concepts and their implementations at the workplace.

Figure 13.5 gives the major elements of a hazard avoidance program. These elements must be applied diligently with feedback from actual safety performance. The elements described in Figure 13.5 are rooted in Figures 13.3 and 13.4. The basic concepts are derived from the accident model described in Section 13.3.

Enforcement. The element of enforcement implies that certain safety rules are set and those who do not meet the requirements are penalized one way or the other. There may be as many as several warnings before any punitive action is taken. Punishment may not be severe at first. Enforcement is very effective in correcting unsafe acts if it is not overdone. Not only unsafe acts but also unsafe conditions may be corrected through

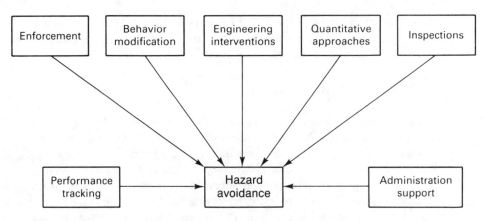

Figure 13.5 Methods of hazard avoidance.

enforcement. Engineers and managers who select and install unsafe equipment may be warned. Employees whose actions are not safe may also be subjected to enforcement procedures. As stated earlier, enforcement requires that mandatory standards or rules be developed and set. These must be very clear rules, not subject to interpretation, and everybody involved in the safety program must be educated in their implementation. One caution: Enforcement may not work if it is used as a means of planned punishment.

Behavior modification. If enforcement is coupled with behavior modification, better results may be obtained. Behavior modification includes all approaches to channel behavior into safe execution of assigned jobs. This may include rewarding safe behavior and attitudes, training in safe behavior, education, and enhancing safety awareness through publicity, such as posters, signs, and short articles in the company newsletter. It is always a good practice to have top management give related talks to the employees. Safety meetings may be conducted where near incidents and potential hazards may be discussed and corrective actions formulated and implemented. New and young workers may be especially vulnerable to behavior modification. It is a good practice particularly to influence their behavior when they are potential causes of unsafe acts.

Engineering interventions. As discussed in Chapters 3 and 9, engineering controls are much preferred over other types. This is so because engineering controls focus on unsafe conditions and aim at correcting them. Safe working environment with local exhaust ventilation, properly labeled displays, coded controls, and safe processes are examples of engineering interventions to make the overall job safe for the workers.

Quantitative approaches. Such approaches aim at developing models of accident causation, and with the use of single-event statistics, calculating the probability of an accident before even a job is developed. Furthermore, all other statistical analyses, including hypothesesis testing and behavioral modeling, are included here. There are unique advantages of quantitative approaches to safety and health evaluation. First and foremost is that a job need not be existent for such investigations to take place. A simple job plan along with task sequences and their probabilities of safe or unsafe performance and paper/

pencil tools are sufficient. "What if" analyses may also be performed to decide on such things as number of process checkers to use, size of a crew for firefighting, and resource size and type for in-plant safety and health inspections. The second advantage is that this approach usually complements engineering approaches to improve working conditions. Hence better engineering judgments can be made in the light of quantitative accident potential analyses. Cost and benefit analyses of alternative hazard prevention methods may also be classified under the quantitative methods.

Inspections. An excellent method of self-evaluation is inspection. These could be performed by plant personnel or outside consultants. The most important point is that inspections should be performed unannounced. Random inspections reveal potential hazards to be corrected. In-plant personnel may be biased in inspections. On the other hand, such methods cost less. Outside consultants may be very expensive. However, they will, in general, provide a true picture of the plant's safety performance. A very well planned inspection process is the major prerequisite of a successful inspection. Resources (personnel, equipment) must be adequate. A summary of the inspection process and documented results and suggestions are extremely helpful to management.

Administration support. Management backing of the safety and health program, including hazard control, cannot be overlooked. One of the most effective means of modifying behavior is to display the fact that top management is truly concerned with safety. Managers must also respect all safety rules, to set a model for others to follow.

Performance tracking. It is also important to monitor safety performance from period to period in order to evaluate the effectiveness of various hazard avoidance programs. It is with this feedback that adjustments can be made in specific areas for extra effectiveness. Factory accident and incidence records are valuable in providing such feedback to the safety professional. Furthermore, this item closes the loop between hazard avoidance programs and the actual safety experience.

13.9 PERSONAL PROTECTIVE EQUIPMENT

Protective equipment is worn by a worker to protect the body from foreign objects and energy. The objects in question may be gases, fumes, dust, flying particles, falling objects, or chemicals. Furthermore, such gear protects the body from unwanted energy, such as combustion, electricity, and extremes of light. Most of these devices are manufactured to comply with an existing standard.

Unwanted effects of energy, gases, and other particles can be controlled at three points:

- At the source
- Between the source and the receiver
- At the receiver

Engineering controls are necessary to block the effects at the source and within the transmission medium. Personal protective equipment may be used for controls at the receiver level. However, this is the last resort; it makes much more sense to correct the problem at the source. Three other issues confound the picture here. First, personal protective equipment very seldom provides complete protection against significant stress. Hence they are not 100% effective. Second, there are use problems. Employees frequently resist the discomfort of wearing such equipment. They may tamper with these devices to make them more comfortable while reducing their effectiveness. Employees may also have to be reminded (signs, audible messages) to wear the appropriate devices before entering an area that requires their use. Third, employee morale and job satisfaction will be higher if there is a permanent solution to the problem at the source. Although the primary correction of a hazard should focus on the source, for many reasons including technical impossibility, impracticality, and cost, personal protective equipment are used in industry. Hence the safety specialist and the ergonomist must be aware of their availability and use. The balance of this section summarizes these devices and the protection they offer. For further information, the reader is referred to Refs. [2], [17] and [28]. The discussion will proceed in head-to-toe order:

1. *Head protection.* Industrial head protection devices are categorized based on the exposure that may be encountered. There are many different types of helmets that provide protection from sound, heat, electricity, and impact. A wide variety of attachments are also available with helmets such as face shields and head lamps. It is important that the inner suspension system be independently adjustable so that the helmet shell does not rest on the head.

2. *Eye protection.* Devices that protect the eye are also available in a variety of forms, including safety glasses and goggles. It is important that safety glasses be equipped with side shields. Also, the protective lens must not shatter during impact.

3. *Hearing protection.* OSHA standards establish maximum noise levels permitted without hearing protection. Two types of hearing protection devices are those that are used in the ear canal (earplugs) and those that cover the ear (earmuffs).

4. *Face protection.* Devices that protect the face are primarily face shields, welder's helmets, and hoods. Face shields may be worn with a helmet. Some face shields also protect the ears. Welder's helmets provide protection against heat, electricity, and splashing molten metal.

5. *Respiratory protection.* The most complicated personal protective devices are those that focus on respiration. Three types of respiratory protection equipment are available. First, there are air-purifying devices that use the ambient air and provide it to the wearer after purification. Second, there are atmosphere-supplying respirators. These supply air to the wearer from an external source such as a blower. An air hose and a totally enclosed head cover are the basic equipment. There may or may not be a blower or compressor at the other end of the hose. Third, there are types of self-contained breathing equipment, which are basically compressed air tanks and associated equipment that the user carries.

6. *Body protection.* Exposure to fire, extreme heat, extreme cold, impact, and radioactive and otherwise hazardous substances require special clothing. Asbestos clothing, leather, and aluminized garments are examples. Specialized aprons are also used in less dangerous situations.

7. *Finger and hand protection.* Industrial finger and hand protection devices protect against chemicals, heat, cold, cuts, abrasions, electricity, and punctures. This is probably the only category of protective devices for which there are insufficient applicable standards. It should be borne in mind that whenever such equipment is worn, the user gives up some finger and hand dexterity and grip. Gloves are the dominant category of such equipment.

8. *Foot protection.* Safety shoes and metatarsal protectors that are worn over regular shoes are examples. They provide protection against a variety of hazards including electricity, impact, and fire.

9. *Fall protection.* Life lines, safety belts, harnesses, and safety nets are examples of fall protection devices. Maintenance and replacement of such devices are extremely important since they are the only lifesaving gear when the circumstances dictate.

Many of the devices cited above can be worn at the same time, offering combination protection. Furthermore, one device may protect multiple regions of the body. An example is protective clothing that covers the entire body.

13.10 SUMMARY

Industrial ergonomists or industrial engineers who are concerned with worker health and safety must be aware of occupational hazards and the means of avoiding them. Before any process is designed or implemented, a thorough hazard analysis should be carried out in addition to task and use analysis. It is also important that applicable safety and health standards be considered.

QUESTIONS

1. What is meant by *total safety?* Is this possible?
2. Discuss the differences between industrial safety and industrial hygiene.
3. Why was OSHAct of 1970 passed?
4. What is the responsibility of NIOSH under OSHAct?
5. Discuss the responsibilities of OSHRC.
6. Discuss the documentation requirements under OSHAct.
7. What are consensus standards?
8. Which event receives the top priority for a safety and health inspection?
9. What is the fundamental principle of workers' compensation laws?
10. How do task characteristics affect accident probability?
11. What is the current view on an accident causation theory?
12. Discuss unsafe conditions.
13. Discuss unsafe acts.

14. Why is machine guarding necessary?
15. List and discuss the three effects of electricity on living tissue.
16. What is byssinosis?
17. What is the major hazard in underground mining?
18. How could unsafe acts be minimized?
19. Discuss the frequency rate with an example.
20. What is disabling injury severity rate?
21. What is incidence rate?
22. Define *occupational illnesses.*
23. What are TLVs and PELs?
24. What are the three primary routes of entry of toxicants into the body?
25. Discuss biological monitoring of toxicants.
26. When is personal protective equipment used? What is the major problem in their use?

QUESTIONS

1. Discuss the activities of the OSHRC in detail.
2. Discuss the significance of the *general duty* of the employer under OSHAct.
3. What do you recommend for a significant reduction in accidents in industry?
4. Go to a local industry and review their accident records. Summarize your observations. Take two cases in particular. Discuss how these accidents could have been avoided.
5. The following data are given on the accident history of Company ABC:

Years	Number of disabling injuries	Worker hours of job exposure
1982	78	1,075,000
1983	71	1,050,000
1984	65	938,000
1985	67	953,000
1986	64	915,000
1987	51	898,000
1988	45	905,000
1989	43	903,000
1990	40	895,000

(a) Predict the disabling injury frequency rate for 1991.
(b) Do you observe any significance in the data trend? Discuss possible reasons.

REFERENCES

1. Hammer, W. 1985. *Occupational Safety Management and Engineering,* 3rd Ed. Prentice Hall, Englewood Cliffs, NJ.

2. National Safety Council. 1981. *Accident Prevention Manual for Industrial Operations: Administration and Programs*, 8th ed. McElroy, F. E. (Ed.). NSC, Chicago.

3. Kolb, J., and Ross, S. S. 1980. *Product Safety and Liability: A Desk Reference*. McGraw-Hill, New York.

4. Firenze, R. J. 1973. *Guide to Occupational Safety and Health Management*. Kendall/Hunt, Dubuque, IA.

5. Hannaford, E. S. 1976. *Supervisor's Guide to Human Relations*, 2nd Ed. National Safety Council, Chicago.

6. Denton, K. D. 1982. *Safety Management: Improving Performance*, McGraw-Hill, New York.

7. Hambly, W. D., and Bedford, T. 1972. *Preliminary Notes on Atmospheric Conditions in Boot and Shoe Factories*. No. 11. Industrial Fatigue Research Board, London.

8. Tolman, W. H., and Kendall, L. B. 1913. *Safety: Methods for Preventing Occupational and Other Accidents and Diseases*. Harper & Brothers, New York.

9. Vernon, H. M. 1940. An Experience of Munition Factories during the Great War. *Occupational Psychology,* 14(1), pp. 1–13.

10. Kerr, W. A. 1950. Accident Proneness of Factory Departments. *Journal of Applied Psychology,* 34(3), pp. 167–170.

11. Smith, M., Cohen, H. H., and Cleveland, R. J. 1978. Characteristics of Successful Safety Programs. *Journal of Safety Research,* 10(1), pp. 5–15.

12. Uusitalo, T., and Mattila, M. 1989. Evaluation of Industrial Safety Practices in Five Industries. In *Advances in Industrial Ergonomics and Safety I*. Mital, A. (Ed.). Taylor & Francis, London, pp. 353–358.

13. Khaleque, A., and Karim, M. M. 1990. Causes and Preventive Strategies of Industrial Accidents As Perceived by Managers and Workers. In *Advances in Industrial Ergonomics and Safety II*. Das, B. (Ed.). Taylor & Francis, London, pp. 521–526.

14. Ramsey, J. 1985. Ergonomic Factors in Task Analysis for Consumer Product Safety. *Journal of Occupational Accidents*, 7, pp. 113–123.

15. Century Research Corporation. 1973. *Are Some People Accident Prone?* CRC, Arlington, VA.

16. Burgess, W. A. 1981. *Recognition of Health Hazards in Industry*. Wiley, New York.

17. Asfahl, C. R. 1984. *Industrial Safety and Health Management*. Prentice Hall, Englewood Cliffs, NJ.

18. McElroy, F. E. (Ed.). 1980. *Accident Prevention Manual for Industrial Operations: Engineering and Technology*, 8th ed. National Safety Council, Chicago.

19. Boothroyd, G. 1975. *Fundamentals of Metal Machining and Machine Tools*. Scripta/McGraw-Hill, Washington, DC.

20. Waldron, H. A. 1977. *Journal of Society of Occupational Medicine*, 27, pp. 45.

21. McGuire, W. J. 1968. The Nature of Attitudes and Attitude Change. In *The Handbook of Social Psychology,* 2nd ed., Vol. 3. Lindzey, G., and Aronson, E. (Eds.), Addison-Wesley, Reading, MA.

22. Insco, C. A., and Schopler, J. 1972. *Experimental Social Psychology*. Academic Press, New York.

23. Grimaldi, J. V., and Simonds, R. H. 1975. *Safety Management*. Richard D. Irwin, Homewood, IL.

24. Anderson, K., and Scott, R. 1981. *Fundamentals of Industrial Toxicology*. Ann Arbor Science, Ann Arbor, MI.

25. Gardner, W., and Taylor, P. 1975. *Health at Work*. Halsted/Wiley, New York.

26. Baselt, R. C. 1980. *Biological Monitoring Methods for Industrial Chemicals.* Biomedical Publications, Davis, CA.

27. Waritz, R. S. 1979. Biological Indicators of Chemical Dosage and Burden. In *Patty's Industrial Hygiene and Toxicology,* Vol. 3. Cralley, L. V., and Cralley, L. J. (Eds.). Wiley, New York. pp. 257–318.

28. Moran, J. B., and Ronk, R. M. 1987. Personal Protective Equipment. In *Handbook of Human Factors.* Salvendy, G. (Ed.). Wiley-Interscience, New York.

_____CHAPTER 14

A Work Analysis Checklist

At this point, the reader has been exposed to the building blocks of industrial ergonomics, including theory and design. This chapter provides a summary of the main concepts in a format that will be of practical use to the reader. Specifically, we provide an evaluation tool to the designer in the form of a checklist.

As discussed earlier, ergonomics is most effective if applied during the design stage. Routine audits of jobs effectively bring into light creeping changes since design. Such audits are effective in identifying jobs that need ergonomic attention. The symptoms of ergonomic problems discussed in Chapter 2 will identify jobs on which to focus. Once a job is selected for further investigation, a checklist may be utilized to identify the specific problems for correction. A sample checklist is given below [1–4].

14.1 WORKPLACE CHARACTERISTICS

_____ There are extended lateral or forward reaches, beyond normal arm reach.

_____ There is inadequate clearance for handling and maintenance tasks.

_____ There are inaccessible workplaces for use of material-handling equipment.

379

_____ Chairs are difficult to adjust, with inadequate back support and no footrest.

_____ Dials and other displays are difficult to read.

_____ The workplace layout leads to inefficient motions.

_____ Awkward postures are required.

_____ There is inadequate space for temporary storage.

_____ The workplace has no built-in adjustability.

_____ The work surface appears to be too high or too low.

_____ Workers frequently sit on the front edge of their chairs for adequate reach.

_____ There is demand for continuous foot pedal operation while standing.

_____ Workers frequently adjust their chairs by adding cushions or pads.

_____ Workers are required to hold their arms or hands without the assistance of armrests.

_____ The workplace seems to be unnecessarily cluttered.

_____ The workplace frequently requires the worker to be engaged in static holding work.

_____ There is no room to move about.

_____ There is insufficient room for the hand while grasping all handles.

_____ There is insufficient kick space (for feet) for all standing operations.

_____ Most important and frequently used displays are not in close proximity to the normal operator vision.

_____ Writing surfaces are limited, less than 30 cm (12 in.) in depth.

_____ Maintenance instruments are not inconspicuous to production operators.

_____ Displays and controls are easily confused.

_____ There is inadequate space for legs.

_____ The work surfaces provided do not seem sufficient.

_____ Working surfaces do not seem to be of sufficient size.

_____ Seat surface is not padded.

_____ Chair adjustments cannot be made easily at a seated posture.

_____ Clearances are not free from obstruction.

14.2 PERCEPTUAL LOAD

_____ Auditory signals are difficult to hear or excessively loud.

_____ There are small and difficult-to-see defects.

_____ Fine color differences are to be discriminated.

_____ Critical distance judgments are necessary.

_____ There is an excessive or unacceptable need to discriminate parts by touch.

_____ There is demand for excessive eye movement.

_____ Numbers, words, symbols, and scale divisions are of inadequate size to suit the reading distance.

_____ Critical displays are not within the ± 15 degrees of NLS.

_____ Instruments, components, and labels are not within the visual field.

_____ Display-control compatibility is violated.

_____ There are inconsistencies as to the on/off positions of controls.

_____ There is insufficient target-to-background contrast.

_____ Vigilance is impaired by noise or other people's activities.

_____ The worker has insufficient time to sense and respond to signals.

_____ Trade names and other unnecessary information are not deleted.

_____ Labels obscure other necessary information.

_____ Magnifying devices do not exist on jobs that require them.

_____ The number "0" and the letter "O" can not be clearly distinguished.

14.3 MENTAL LOAD

_____ There is a need to keep track of multiple events simultaneously, especially in the case of operating complex machinery or multiple machines.

_____ Critical task elements exist where errors are not tolerated.

_____ High demand exists for short-term memory, such as working with nine-digit codes.

_____ Labels do not clearly indicate the function displayed or controlled.

_____ Some of the names used are highly similar.

_____ There are uncommon abbreviations.

_____ Interpolations are necessary.

_____ The system requires projections to the future without job aids.

_____ Coding schemes are not consistent.

14.4 ENVIRONMENT

_____ There is excessive noise that is annoying and distracting.

_____ Speech intelligibility is affected by noise.

_____ Process noise is possibly loud enough to cause hearing loss.

_____ There are direct or reflected glare sources in the area.

_____ The amount of illumination is not sufficient for the task.

_____ Lights shine on moving elements of machinery to produce distracting flashes.

_____ The visual field presents excessive contrasts.

_____ Eyes have to move between light and dark areas periodically.

_____ The thermal environment is not comfortable.

_____ There is vibration that is annoying and hazardous to health.

_____ Ventilation is not adequate.

_____ The worker is exposed to unacceptable thermal or visual environmental changes.

_____ Floors are not even.

_____ Floors are slippery.

_____ Housekeeping is poor.

_____ Hot surfaces present burn hazard.

_____ Air in the room contains toxic, flammable, explosive substances.

_____ There is a dermatitis hazard.

_____ Process dust settles on equipment and displays impairing visual performance.

_____ There is no local or direct lighting for equipment that is not properly lighted by general lighting.

_____ Air conditioning seems to be inadequate.

_____ There are places where the noise level exceeds 150 dB(A).

_____ There is insufficient clothing for people working in cold.

14.5 SPECIFIC EQUIPMENT

_____ Dials and controls are poorly labeled.

_____ Displays are not adequately lit.

_____ Pinch points are not adequately guarded.

_____ Electrical equipment is not properly grounded.

_____ Excessive strength requirements exist for operating equipment.

_____ Hand tools present injury potential.

_____ Maintenance manuals are missing or not up to date.

_____ Sharp edges exist in the work area.

_____ Warning signs and labels are not very apparent.

_____ Protection against accidental activation of controls is inadequate.

_____ Trip hazards exist in the work area.

_____ All numbers are not upright on stationary displays.

_____ On moving scales, numbers are not upright at reading position.

_____ Special materials and locations (toxic and flammable substances, covered electrical outlets, fuse boxes, etc.) are not color coded.

_____ Increase in numerical progression on scales does not read clockwise or from left to right.

_____ Pointers cover numerical graduations or scale marks.

_____ Emergency conditions lack support by flashing lights or audible warnings.

_____ Keyboards cannot be used in detached form from the screen consoles.

14.6 PHYSICAL DEMANDS

_____ There is frequent lifting of heavy objects.

_____ There is occasional lifting of very heavy objects.

_____ There is constant handling of material with very little variety.

_____ The task requires handling of difficult-to-grasp objects.

_____ The task requires exertion of forces in awkward postures.

_____ The task requires constant bending.

_____ The task requires constant standing.

_____ Frequent ladder or stair climbing is required.

_____ There is static muscle loading.

_____ Thin edges exert high pressure on the hands.

_____ There is a short-duration heavy-force exertion requirement.

_____ The worker is required to push or pull carts, boxes, and so on, that involve large breakaway forces to get started.

_____ Workers complain that fatigue allowances are insufficient.

_____ Heart rate and oxygen consumption are above the recommended maximum.

_____ The task requires sudden jerking motions.

14.7 WORK METHODS

_____ Motion range requirements are anatomically unacceptable.

_____ There are high-precision motion requirements.

_____ Sudden movements are necessary during handling.

_____ Machine pacing is not compatible with human capabilities.

_____ One motion pattern is repeated at a high frequency.

_____ Hand tools are used in incorrect hand positions.

_____ Visual control of manual movements is necessary.

_____ There are unnecessary moves.

_____ There are straight-line motions involving sudden and sharp changes in direction.

_____ There are nonsymmetrical motions.

_____ There are twisting motions with the elbows straight.

_____ Work between body members is not balanced.

_____ Muscle groups involved are not adequate for the job.

_____ There is frequent forceful application at joint extremes.

_____ Motion patterns frequently require work with elevated hands.

_____ Elbows are not close to the body.

14.8 MAINTAINABILITY

_____ Rapid and easy removal of devices is not possible.

_____ It is not obvious when a cover is in place but not secure.

_____ Openings and workspaces do not allow the appropriate body part through.

_____ Check points, lubrication, and adjustment points are not easily accessible.

_____ Connectors are not located far enough apart so that they can be grasped firmly for connection or disconnection.

_____ Connecting plugs are not color, shape, or otherwise coded.

_____ Lubrication points do not specify the type of lubricant and frequency of lubrication.

_____ Primary test points are not close to controls and displays used in the adjustment.

_____ Cables are routed in such a manner that they are not accessible for inspection and repair.

_____ Large parts prevent access to other smaller parts.

_____ Sensitive adjustments are not guarded. They are subject to disturbance.

_____ Internal controls are located close to dangerous voltages.

14.9 OTHER

_____ Work hours and breaks are not properly organized.

_____ Job performance aids are inadequate.

_____ Motivation is lacking.

_____ There is poor supervision.

_____ Supplies are inadequate.

_____ Feedback is inadequate or nonexistent.

_____ Skill demands are incompatible with skills possessed.

_____ Shift work is not properly organized.

_____ Boredom seems to be a problem.

_____ There are extensive vigilance demands in terms of work duration and complexity.

_____ Training seems to be insufficient.

_____ Supervisor–employee relationships are not pleasant.

_____ There is no freedom on the job for occasional personal needs.

_____ There are no allowances for retraining.

Static checklists such as the one above must be used with caution. Since the observations are categorized into predefined slots with no *degree* specifications, the user may have to take notes next to each item to define finer details. These notes must be considered in making improvements on the job. For example, elbows may not be close to the body in only three out of 15 job elements. In this case the designer will consider only three elements for improvement with respect to this item.

REFERENCES

1. Pulat, B. M. 1985. Summary. In *Industrial Ergonomics: A Practitioner's Guide*. Alexander, D. C., and Pulat, B. M. (Eds.). Industrial Engineering and Management Press, Atlanta, GA.

2. Grandjean, E. 1980. *Fitting the Task to the Man*. Taylor & Francis, London.

3. Woodson, W. E. 1981. *Human Factors Design Handbook*. McGraw-Hill, New York.

4. Cakir, A., Hart, D. J., and Stewart, T. F. M. 1980. *Visual Display Terminals*. Wiley, Chichester, West Sussex, England.

_____APPENDIX A

Ergonomic Characteristics of Common Controls

In this section we present attributes of commonly used controls that are ergonomically acceptable. Such information can be used in designing controls. Another and more important use is in screening controls for use in industrial environments. The data have been compiled from Refs. [1–6].

A.1 FOOT PUSHBUTTON

Although the feet are not as sensitive, as accurate, and as fast as the hands, they can be used when the hands are in danger of becoming overburdened. A foot pushbutton or switch is useful in a two-state-system response control. It also allows for standing operation as long as it is not used frequently. Design requirements include:

1. The foot contact surface should be threaded to avoid slip.
2. Resistance should be low at the start of activation. It should build up rapidly and drop when the control is activated.

Other design recommendations are shown in Table A.1.

TABLE A.1 FOOT PUSHBUTTON CHARACTERISTICS

	Diameter, L (mm)	Activation displacement, M (mm)	Resistance (foot resting) (newtons)
Minimum	12	12	44.5
Maximum	80	75	89

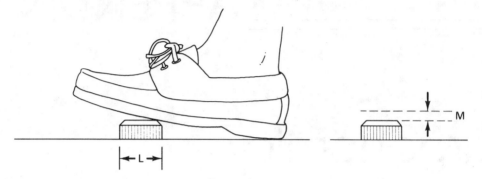

Figure A–1 Foot push button characteristics.

A.2 FOOT PEDAL

A foot pedal is useful for continuous control with the feet. They should be used primarily in a seated posture. Many assembly workers operate foot pedals for machine or conveyor control. Design requirements include:

1. Pedal shape is not important as long as it provides adequate area of contact with the shoe.
2. Optimum pedal angle varies with the specific location (vertical and fore–aft) of the pedal. A comfortable ankle angle must be maintained (within 25 degrees around the resting position of ankle). Maximum accuracy is obtained at knee angles between 95 and 135 degrees.
3. Elastic resistance should return pedal back to the null position when foot is removed.
4. A nonskid surface prevents foot slip during use.
5. The base of pedal should stand some distance (within 2.5 cm) above the floor to give the operator cues as to the foot being in contact with it.

Several other design considerations are given in Table A.2.

TABLE A.2 FOOT PEDAL CHARACTERISTICS

| | Width (mm) | | Length (mm) | | Displacement range (mm) | | Resistance (kg) | |
							Ankle flexion	total leg movement
	A	W	B	L	V	M		
Minimum	75	108	25	230	25	12	3	4
Maximum	—	51	305	300	180	65	10	100

Figure A–2   Foot pedal characteristics.

A.3 DETENT THUMBWHEEL

A detent thumbwheel may be used for setting a system to discrete values. The settings are visible on the control itself. Design recommendations include:

1. Downward movement of control should correspond to a decrement in numeric values.
2. Numbered surfaces should be separated by high friction areas raised at least 2 mm from the adjacent surfaces.
3. Numbered surfaces should be slightly concave.
4. Good contrast between the numbers and the background is recommended.
5. Numbered surfaces should be protected against dirt accumulation.

Table A.3 gives additional design details.

TABLE A.3 DETENT THUMBWHEEL CHARACTERISTICS

	Diameter, d (mm)	Width, w (mm)	Protrusion p (mm)	Separation, s (mm)	Resistance (kg)
Minimum	38	7	3	7	0.11
Maximum	63	14	6	20	0.34

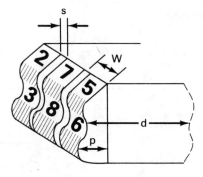

Figure A–3 Detent thumbwheel characteristics.

A.4 ROTARY SELECTOR SWITCH

Rotary selector switches are used in applications where a limited number of selections are necessary, mostly between 3 and 24. They can be bar or knob type. Additional space is necessary around such controls, compared to toggle switches, for the operating hand. Several design recommendations include:

1. Provide stops at the start and end of available selections. This allows for blind operation using feel.
2. If possible, limit the selection range to 180 degrees.
3. Minimize frictional and inertial resistance. Elastic resistance should be utilized to build up counterpressure as each position (or detent) is approached. Resistance should become zero while control is set to any position.
4. Bar-type rotary selector switch performs best in most situations.
5. If few control settings are required (fewer than eight), separation between adjacent settings should be 30 degrees.
6. A "click" at each setting would give additional (apart from visual and kinesthetic) cues as to accomplishment of the task.
7. There should be positive cues on the bar (tapered point, flat back, one end color coded, etc.) as to the direction of indication.
8. A clockwise turn of the switch should be associated with increasing setting values.
9. The numerical scale and the pointer should be close to each other to minimize visual parallax.
10. Index numbers should be visible while the hand is on the control.

Table A.4 provides additional design details.

TABLE A.4 ROTARY SELECTOR SWITCH CHARACTERISTICS

Bar Type

	Length, K (mm)	Width, W (mm)	Depth, D (mm)	Displacement, α (visual positioning)	Resistance (kg)
Minimum	25	13	12	15°	0.25
Maximum	100	25	75	45°	1.4

Round Type

	Diameter, K (mm)	Depth, D (mm)	Displacement, α (visual positioning)	Resistance (kg)
Minimum	25	12	15°	0.25
Maximum	100	75	45°	1.4

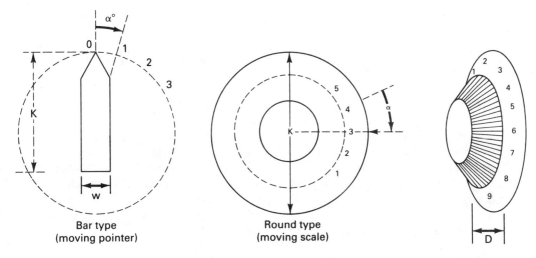

Bar type
(moving pointer)

Round type
(moving scale)

Figure A–4 Rotary selector switch characteristics.

A.5 FINGER PUSHBUTTON

A finger pushbutton is a discrete adjustment control that does not require much force to operate. Design requirements include:

1. The pressure face should preferably be concave.
2. Surface finish should provide sufficient frictional resistance to prevent finger from slipping during use.
3. An audible "click" indicating activation is a desired property.

4. Elastic resistance should be built into the pushbutton that gradually increases and then suddenly drops to indicate activation.

5. For emergency use, the size may be enlarged for finger or hand use.

Table A.5 gives additional design details.

TABLE A.5 FINGER-OPERATED PUSHBUTTON CHARACTERISTICS

	Diameter, d (mm)	Activation displacement, v (mm)	Resistance (kg)
Minimum	9	3	0.25
Maximum	19	38	1.1

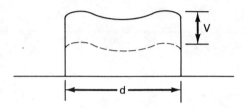

Figure A–5 Finger-operated push buttons.

A.6 LEGEND SWITCH

A legend switch is a variation of the finger-operated pushbutton. It may be used for the same purposes. The difference is that it is illuminated when activated with an appropriate label on. Since a label is placed on a legend switch, its size is normally bigger than that of a finger-operated pushbutton. It is used for two-state system response. Table A.6 gives additional details.

TABLE A.6 LEGEND SWITCH CHARACTERISTICS

	Size, K (mm)	Separators (mm) S_1	Separators (mm) S_2	Displacement activated/nonactivated (mm)	Resistance (kg)
Minimum	19	5	3	3	0.28
Maximum	38	7	7	7	1.27

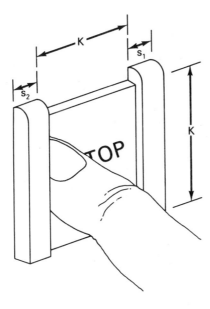

A.7 TOGGLE SWITCH

A toggle switch is another discrete adjustment control that does not require much activation force. However, as opposed to the finger pushbutton or legend switch, it can be set at three positions if required. Having more than three settings is not recommended. Specific design recommendations include:

1. Elastic resistance should build up as control is activated and suddenly drop as the desired position is approached. This snap action prevents the control from stopping between setting positions.
2. Activation should be accompanied by an audible click.
3. If a three-position toggle switch is required, adjacent positions should be separated by at least 30 degrees.
4. Frictional and inertial resistances should be minimum.
5. If toggle switches are required for momentary contacts, the control arm should be spring loaded to return to the null position after release.

Table A.7 gives additional design details.

TABLE A.7 TOGGLE SWITCH CHARACTERISTICS

	Displacement, D	Arm tip diameter, d (mm)	Arm length, L (mm)	Resistance (kg)
Minimum	30°	3	12	0.28
Maximum	120°	25	50	1.1

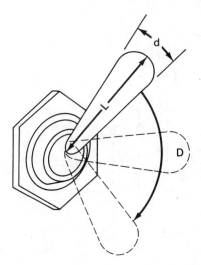

Figure A–7 Toggle switch.

A.8 ROCKER SWITCHES

These are two-state, discrete setting controls functionally similar to finger-operated buttons or legend switches. They can be located within close proximity of each other for effective relative setting review. Design recommendations include:

1. Switch resistance should range between 230 and 340 grams (8 to 12 oz).
2. They should snap into position with an audible click.
3. The two faces may be color coded for enhanced position identification.
4. The slope of the handle from the nominal plane should be 30 degrees.
5. The switch mounting plane should be at least 0.32 cm (0.125 in.) above the nominal plane.

Table A.8 gives additional design details.

TABLE A.8 ROCKER SWITCH CHARACTERISTICS

	Length, L (mm)	Width, W (mm)	Displacement, D
Minimum	13	5	30°
Maximum	—	—	30°

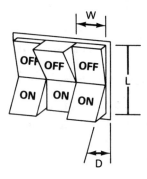

Figure A–8 Rocker switches.

A.9 JOYSTICKS

Joysticks are multidimensional controls. They can also be set at a multitude of values between applicable ranges. Hence they are classified as continuous adjustment controls. Design recommendations include:

1. The operating hand should be supported for extensive operation periods.
2. The pivot point should be recessed below the surface on which the wrist rests.
3. The surface of the handle should have frictional resistance.
4. These controls should be provided with elastic resistance.
5. Control should return to null position when released.

Additional design details are provided by Table A.9.

TABLE A.9 JOYSTICK CHARACTERISTICS

	Handle diameter, d (mm)	Displacement, D	Resistance (kg)
Minimum	5	—	0.34
Maximum	76	120°	0.9

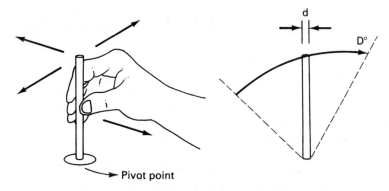

Pivot point

Figure A–9 Joy stick characteristics.

A.10 LEVERS

Levers are continuous adjustment controls that allow for multiple control operation at the same time. Pushbuttons on levers are examples. Long levers require relatively small force for operation. Design recommendations include:

1. For a large operating angle (displacement), a longer handle may be desired.
2. Use a ball grip or T-handle for large displacement. For small displacement (less than 30 degrees), use a straight grip.
3. The range of movement should not exceed arm reach.
4. The handle should be provided with frictional resistance.
5. Minimum separation between settings should be 50 mm for levers that are used as discrete controls and with lever arm length exceeding 150 mm.
6. The desirable resistance for levers is elastic resistance with spring loading.

Table A.10 gives additional design details.

TABLE A.10 LEVER CHARACTERISTICS

	Handle diameter, d (mm)	Grasp area height, h (mm)	Displacement, D (mm)		Resistance (kg)	
			Fore/aft	Right/left	Fore/aft	Right/left
Minimum	38	76	—	—	0.9	0.9
Maximum	70	—	350	950	13.5	9

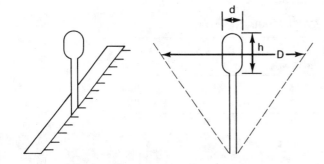

Figure A–10 Levers.

A.11 ROUND KNOBS

These are continuous adjustment controls that can accept multiple rotations. They can be operated by fingers, the palm, or with the thumb and finger circling. Design recommendations and requirements are:

1. A clockwise rotation of the knob should correspond to increasing values.
2. The scale on the knob skirt should be visible when operator's fingers are on the knob.

3. The grip surface should be provided with frictional resistance.

4. For palm-grasp knobs, a minimum of 22 mm clearance should be allowed between the knob and the surface on which it is mounted.

5. Only two concentric (on the same shaft) knobs are recommended, with a maximum total height of 44 mm.

6. If the knob is not for multirotation use, start and end stops should be provided. The gap between these stops should be greater than the separation between consecutive index marks.

Table A.11 gives additional design details.

TABLE A.11   ROUND KNOB CHARACTERISTICS

	Fingertip grasp		Thumb and finger encircled		Palm Grasp	
	Height, h (mm)	Diameter, d (mm)	Height, h (mm)	Diameter, d (mm)	Height, h (mm)	Diameter, d (mm)
Minimum	12	10	12	25	15	35
Maximum	25	100	25	75	—	75

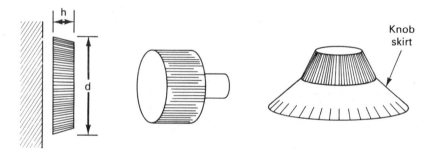

Figure A–11   Round knobs.

A.12 CRANKS

Cranks are also continuous setting controls with a wide range of adjustments. They can be used for fine or gross adjustments. Cranks take up a large amount of space. Several design recommendations are:

1. For fine adjustment, the handle should not rotate.

2. For gross adjustment, the handle should rotate.

3. Large slewing movements and fine adjustments can be attained through mounting a crank handle on a knob or handwheel. The crank handle may be utilized for large slewing movements.

4. The handle surface should have frictional resistance.

5. In general, clockwise rotation should be associated with increasing values, and vice versa.
6. Cranks should not be mounted directly in front of the operator but to either side.

Table A.12 gives additional details.

TABLE A.12 CRANK CHARACTERISTICS

	Handle			
	Diameter, d (mm)	Length, K (mm)	Radius, R (mm)	Resistance (kg)
Light loads ≤ 2.3 kg				
Minimum	9.5	25.4	12.7	0.9
Maximum	15.9	76.2	127	2.5
Heavy loads > 2.3 kg				
Minimum	25.4	76.2	127	1
Maximum	76.2	—	508	4

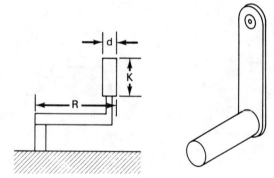

Figure A–12 Cranks.

A.13 HANDWHEELS

Handwheels are continuous controls. They can be used to apply large forces with two hands. They allow for greater than 360-degree rotation to make adjustments over a wide range of values. Specific design requirements are:

1. The rim may be slightly recessed to provide better grip on the wheel and more force to be applied. However, the small or large hand should not be at discomfort during use.
2. Inertial resistance should be minimized.

3. Except for valves, handwheels should turn clockwise for increasing values. The opposite direction should be associated with decreasing values. This relationship is the opposite of that for valves.

4. If a handwheel is expected to have one full revolution or less as maximum, the part that is not needed for grasp may be cut away. This improves visual performance.

5. Most effective use of handwheels can be realized when the total displacement is not greater than ±60 degrees. Large displacements are also possible with a crank handle placed on the rim to improve slewing efficiency.

Table A.13 gives additional design details.

TABLE A.13 HANDWHEEL CHARACTERISTICS

	Wheel diameter, K (mm)	Rim diameter (mm)	Resistance (kg)
Minimum	178	19	2.4
Maximum	520	50	24

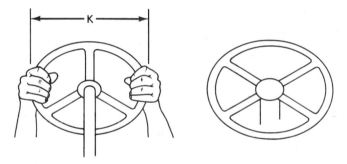

Figure A–13 Handwheels.

A.14 CONTINUOUS ADJUSTMENT THUMBWHEELS

As the name implies, these are continuous controls. They are not used if the operator wears gloves. Settings over a wide range are possible with thumbwheels. Specific design recommendations include:

1. At least 25 mm of the wheel should protrude from the panel surface.
2. Frictional resistance on the surface helps prevent finger slip during use.
3. Inertial resistance helps make smooth control movements.
4. Movement of the wheel up, forward, or to the right should be associated with increasing setting values.
5. When less than one revolution is required, a visible feature on the periphery should indicate the null position.

For further design details, see Table A.14.

TABLE A.14 CONTINUOUS ADJUSTMENT THUMBWHEEL
CHARACTERISTICS

	Diameter, D (mm)	Width, T (mm)	Protrusion, P (mm)	Resistance (torque) (cm–kg)
Minimum	38	6	3.2	1
Maximum	63	13	6.4	3

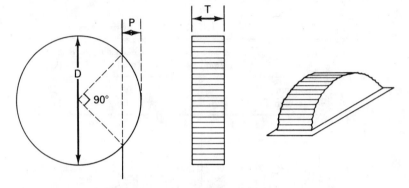

Figure A–14 Continuous-adjustment thumbwheels.

REFERENCES

1. *AFSC Design Handbook*. 1977. *DH1-3*, Personnel Subsystem, January. Air Force Systems Command. Andrews Air Force Base, D.C.

2. Van Cott, H. P., and Kinkade, R. G. (Eds.) 1972. *Human Engineering Guide to Equipment Design*. U.S. Government Printing Office, Washington, DC.

3. Woodson, W. E. 1981. *Human Factors Design Handbook*. McGraw-Hill, New York.

4. Diffrient, N., Tilley, A. R., and Harman, D. 1983. *Humanscale 4/5/6*. MIT Press, Cambridge, MA.

5. Eastman Kodak Company. 1983. *Ergonomic Design for People at Work*, Vol. 1, Lifetime Learning, Belmont, CA.

6. Clark, T. S., and Corlett, E. N. 1984. *The Ergonomics of Workspaces and Machines: A Design Manual*. Taylor & Francis, London.

_____APPENDIX B

Additional Anthropometric Data

B.1 DEFINITIONS

In this section we present additional anthropometric data. Each measure is defined and a short glossary follows. Specific data compiled from Refs. [1,2,3] are listed in Tables B.1 to B.24.

1. *Acromial (shoulder) height:* distance from standing surface to the most lateral point of the acromial process of the scapula
2. *Waist height:* distance from standing surface to waist landmark
3. *Crotch height:* distance from standing surface up into the crotch until contact
4. *Tibiale height:* distance from standing surface to proximal medial margin of the tibia
5. *Ankle height:* distance from standing surface to level of minimum circumference of ankle
6. *Elbow height:* distance from standing surface to depression at elbow between humerus and radius

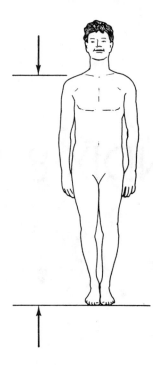

TABLE B.1   ACROMIAL (SHOULDER) HEIGHT [CM (IN.)]

	Percentile		
Subjects	5th	50th	95th
Females			
U.S. Air Force women	123 (48.9)	131.9 (51.9)	141.1 (55.6)
Males			
U.S. Air Force fliers	135.7 (53.4)	145.2 (57.2)	154.8 (60.9)
Italian military	129.4 (50.9)	138.9 (54.7)	148.2 (58.3)

Source: Adapted from Ref. [2].

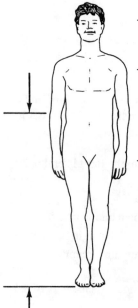

TABLE B.2   WAIST HEIGHT [CM (IN.)]

	Percentile		
Subjects	5th	50th	95th
Females			
U.S. Air Force women	93.1 (36.7)	100.3 (39.5)	107.9 (42.5)
Swedish civilians	98.2 (38.7)	91.5 (36.0)	104.8 (41.3)
Males			
U.S. Air Force flying personnel	98.7 (38.9)	106.5 (41.9)	114.3 (45.0)
Italian military	93.0 (36.6)	101.3 (39.9)	109.2 (43.0)

Source: Adapted from Ref. [2].

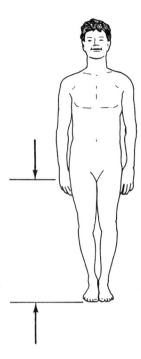

TABLE B.3 CROTCH HEIGHT [CM (IN.)]

Subjects	Percentile		
	5th	50th	95th
Females			
U.S. Air Force women	68.1 (26.8)	74.5 (29.3)	81.4 (32.0)
Males			
U.S. Air Force flying personnel	78.3 (30.8)	85.1 (33.5)	92.0 (36.2)
Italian military	73.6 (29.0)	80.7 (31.8)	87.6 (34.5)

Source: Adapted from Ref. [2].

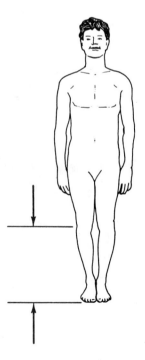

TABLE B.4 TIBIALE HEIGHT [CM (IN.)]

Subjects	Percentile		
	5th	50th	95th
Females			
U.S. Air Force women	38.2 (15.0)	42.0 (16.5)	46.0 (18.4)
Swedish civilians	36.4 (14.3)	43.9 (17.3)	51.4 (20.2)
Males			
NASA astronauts	43.8 (17.2)	46.6 (18.3)	49.4 (19.4)
French fliers	42.8 (16.9)	46.2 (18.2)	49.0 (19.3)

Source: Adapted from Ref. [2].

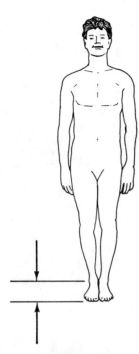

TABLE B.5 ANKLE HEIGHT [CM (IN.)]

	Percentile		
Subjects	5th	50th	95th
Females			
U.S. Air Force women	9.2 (3.6)	11.2 (4.4)	13.6 (5.4)
Males			
U.S. Air Force flying personnel	12.0 (4.7)	13.7 (5.4)	15.8 (6.2)
Italian military	11.9 (4.7)	12.9 (5.1)	13.9 (5.5)

Source: Adapted from Ref. [2].

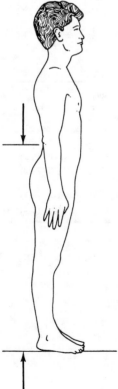

TABLE B.6 ELBOW HEIGHT [CM (IN.)]

	Percentile		
Subjects	5th	50th	95th
Females			
U.S. civilians	95 (37.4)[a]	100.6 (39.6)	109.2 (43)[a]
Males			
U.S. Air Force flying personnel	104.8 (41.3)	112.3 (44.2)	120.0 (47.2)
Italian military	98.5 (38.8)	106.1 (41.8)	113.7 (44.8)

Source: Adapted from Refs. [2] and [3].
[a]2.5th and 97.5th percentile values.

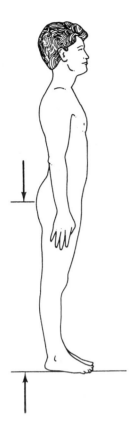

TABLE B.7 WRIST HEIGHT [CM (IN.)]

Subjects	Percentile		
	5th	50th	95th
Females			
U.S. civilians	75.7 (29.8)[a]	79.7 (31.4)	87.1 (34.3)[a]
Males			
U.S. Air Force fliers	80.2 (31.6)	86.6 (34.1)	93.3 (36.7)
Italian military	75.4 (29.7)	81.5 (32.1)	87.6 (34.5)

Source: Adapted from Refs. [2] and [3].
[a]2.5th and 97.5th percentile values.

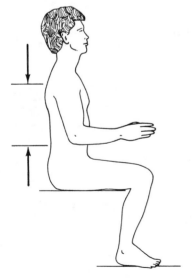

TABLE B.8 SHOULDER–ELBOW LENGTH [CM (IN.)]

Subjects	Percentile		
	5th	50th	95th
Females			
U.S. civilians	31 (12.2)[a]	34.5 (13.6)	37.4 (14.5)[a]
Males			
U.S. Air Force fliers	33.2 (13.1)	36.0 (14.2)	38.8 (15.3)
French fliers	30 (11.8)	32.2 (12.7)	34.7 (13.7)

Source: Adapted from Refs. [2] and [3].
[a]2.5th and 97.5th percentile values.

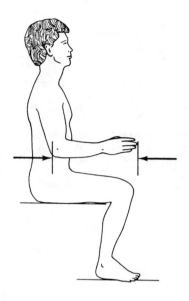

TABLE B.9 FOREARM–HAND LENGTH [CM (IN.)]

Subjects	Percentile		
	5th	50th	95th
Females			
Swedish civilians	40.2 (15.8)	44.2 (17.4)	48.2 (19.0)
Males			
NASA astronauts	44.3 (17.4)	47.6 (18.7)	50.9 (20)

Source: Adapted from Ref. [2].

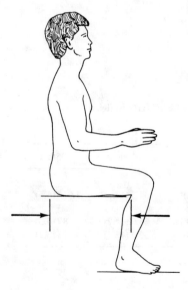

TABLE B.10 BUTTOCK–POPLITEAL LENGTH [CM (IN.)]

Subjects	Percentile		
	5th	50th	95th
Females			
U.S. Air Force women	43.5 (17.1)	47.7 (18.8)	52.6 (20.7)
Males			
U.S. Air Force fliers	46.1 (18.1)	50.4 (19.8)	54.6 (21.5)
German Air Force	44.8 (17.6)	48.9 (19.3)	53.0 (20.9)

Source: Adapted from Ref. [2].

TABLE B.11   BUTTOCK–KNEE LENGTH [CM (IN.)]

	Percentile		
Subjects	5th	50th	95th
Females			
U.S. Air Force women	53.2 (20.9)	57.4 (22.6)	61.9 (24.4)
Males			
U.S. Air Force fliers	56.1 (22.1)	60.4 (23.8)	65 (25.6)
French fliers	56.3 (22.2)	59.5 (23.4)	63.1 (24.8)

Source: Adapted from Ref. [2].

TABLE B.12   THUMB–TIP REACH [CM (IN.)]

	Percentile		
Subjects	5th	50th	95th
Females			
U.S. Air Force women	67.7 (26.7)	74.1 (29.2)	80.5 (31.7)
Males			
U.S. Air Force fliers	73.9 (29.1)	80.3 (31.6)	87.0 (34.3)
RAF fliers	74.4 (29.3)	80.2 (31.6)	85.1 (33.5)

Source: Adapted from Ref. [2].

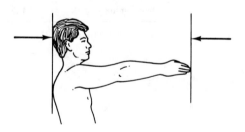

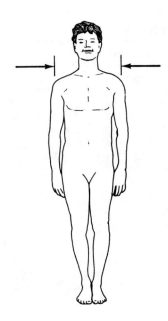

TABLE B.13 BIACROMIAL BREADTH [CM (IN.)]

| | Percentile | | |
Subjects	5th	50th	95th
Females			
U.S. Air Force women	33.2 (13.1)	35.8 (14.1)	38.6 (15.2)
Swedish civilians	32.9 (13.0)	35.4 (13.9)	37.8 (14.9)
Males			
U.S. Air Force fliers	37.5 (14.8)	40.7 (16.0)	43.8 (17.2)
Italian military	36.8 (14.5)	39.8 (15.7)	42.8 (16.9)

Source: Adapted from Ref. [2].

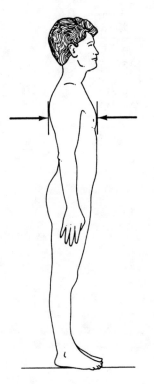

TABLE B.14 CHEST DEPTH [CM (IN.)]

| | Percentile | | |
Subjects	5th	50th	95th
Females			
U.S. Air Force women	20.9 (8.2)	23.6 (9.3)	27.2 (10.7)
Males			
U.S. Air Force fliers	21.3 (8.4)	24.5 (9.6)	27.7 (10.9)
Italian military	21.1 (8.3)	23.8 (9.4)	26.8 (10.6)

Source: Adapted from Ref. [2].

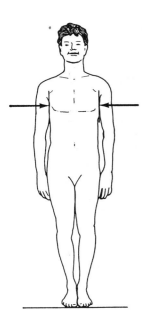

TABLE B.15   CHEST BREADTH [CM (IN.)]

Subjects	Percentile		
	5th	50th	95th
Females			
U.S. Air Force women	25.1 (9.9)	28.0 (11.0)	31.4 (12.4)
Males			
U.S. Air Force fliers	29.5 (11.6)	32.8 (12.9)	36.5 (14.4)
French fliers	29.0 (11.4)	32.1 (12.6)	35.7 (14.1)

Source: Adapted from Ref. [2].

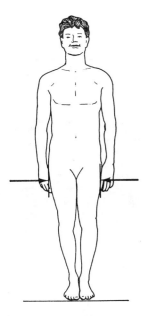

TABLE B.16   HIP BREADTH (STANDING) [CM (IN.)]

Subjects	Percentile		
	5th	50th	95th
Females			
U.S. Air Force women	31.6 (12.4)	35 (13.8)	38.8 (15.3)
Males			
U.S. Air Force flying personnel	32.3 (12.7)	35.3 (13.9)	38.5 (15.2)
Italian military	31.5 (12.4)	34.2 (13.5)	37.1 (14.6)
German Air Force	32.3 (12.7)	35.2 (13.9)	38.3 (15.1)

Source: Adapted from Ref. [2].

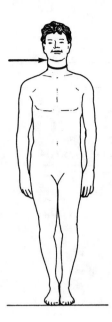

TABLE B.17 NECK CIRCUMFERENCE [CM (IN.)]

Subjects	Percentile		
	5th	50th	95th
Females			
U.S. Air Force women	31.1 (12.2)	33.8 (13.3)	36.7 (14.4)
British civilians	35.3 (13.9)	38.4 (15.1)	41.7 (16.4)
Males			
U.S. Air Force fliers	35.4 (13.9)	38.3 (15.1)	41.7 (16.4)
Japanese civilians	32.9 (13.0)	36.0 (14.2)	39.1 (15.4)

Source: Adapted from Ref. [2].

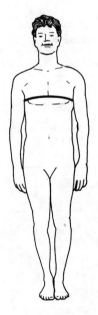

TABLE B.18 CHEST CIRCUMFERENCE [CM (IN.)]

Subjects	Percentile		
	5th	50th	95th
Females			
U.S. Air Force women	81.6 (32.1)	89.7 (35.3)	100.2 (39.4)
British civilians	81.5 (32.1)	92.7 (36.5)	109.6 (43.1)
Males			
U.S. Air Force fliers	88.6 (34.9)	98.6 (38.8)	109.4 (43.1)
German Air Force	84.7 (33.3)	94.7 (37.3)	105.3 (41.5)
Japanese civilians	79.4 (31.3)	88.1 (34.7)	96.8 (38.1)

Source: Adapted from Ref. [2].

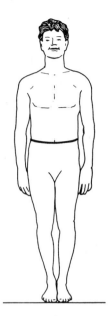

TABLE B.19 WAIST CIRCUMFERENCE [CM (IN.)]

Subjects	Percentile		
	5th	50th	95th
Females			
U.S. Air Force women	59.5 (23.4)	67.2 (26.5)	77.2 (30.4)
Swedish civilians	60.8 (23.9)	67.7 (26.7)	74.6 (29.4)
Males			
U.S. Air Force fliers	75.7 (29.8)	87.6 (34.5)	100.1 (39.4)
Japanese civilians	63.5 (25)	76.5 (30.1)	89.5 (35.2)

Source: Adapted from Ref. [2].

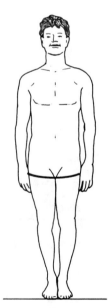

TABLE B.20 BUTTOCK CIRCUMFERENCE [CM (IN.)]

Subjects	Percentile		
	5th	50th	95th
Females			
U.S. Air Force women	85.5 (33.8)	95.3 (37.5)	105.6 (41.6)
Swedish civilians	78.1 (30.7)	88.1 (34.7)	98.0 (38.6)
Males			
U.S. Air Force fliers	89.7 (35.3)	98.6 (38.8)	107.9 (42.5)
Italian military	87.3 (34.4)	95.1 (37.4)	103.4 (40.7)
Japanese civilians	81.7 (32.2)	90.3 (35.6)	98.9 (38.9)

Source: Adapted from Ref. [2].

TABLE B.21 HEAD BREADTH [CM (IN.)]

Subjects	Percentile		
	5th	50th	95th
Females			
U.S. Air Force women	13.5 (5.3)	14.5 (5.7)	15.5 (6.1)
Males			
U.S. Air Force fliers	14.7 (5.8)	15.6 (6.1)	16.5 (6.5)
Italian military	14.6 (5.7)	15.5 (6.1)	16.5 (6.5)
German Air Force	14.7 (5.8)	15.7 (6.2)	16.7 (6.6)

Source: Adapted from Ref. [2].

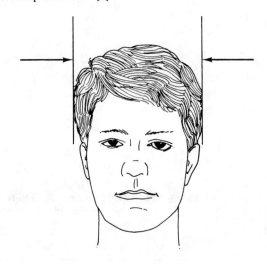

TABLE B.22 HEAD CIRCUMFERENCE [CM (IN.)]

Subjects	Percentile		
	5th	50th	95th
Females			
U.S. Air Force women	52.3 (20.6)	54.9 (21.6)	57.6 (22.7)
Japanese civilians	52.2 (20.6)	54.5 (21.5)	56.8 (22.4)
Males			
U.S. Air Force fliers	55.2 (21.7)	57.5 (22.6)	59.9 (23.6)
French fliers	54.5 (21.5)	56.8 (22.4)	59.2 (23.3)
Japanese civilians	54.0 (21.3)	56.5 (22.2)	

Source: Adapted from Ref. [2].

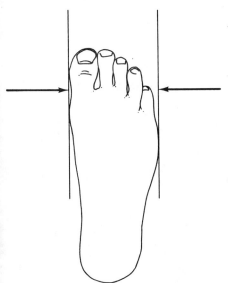

TABLE B.23 HAND BREADTH [CM (IN.)]

Subjects	Percentile		
	5th	50th	95th
Females			
U.S. Air Force women	6.9 (2.7)	7.6 (3.0)	8.2 (3.2)
Swedish civilians	7.1 (2.8)	7.7 (3.0)	8.3 (3.3)
Males			
U.S. Air Force fliers	8.2 (3.2)	8.9 (3.5)	9.6 (3.8)
French fliers	8.1 (3.2)	8.7 (3.4)	9.4 (3.7)

Source: Adapted from Ref. [2].

TABLE B.24 FOOT BREADTH [CM (IN.)]

Subjects	Percentile		
	5th	50th	95th
Females			
U.S. Air Force women	8.0 (3.1)	8.9 (3.5)	9.8 (3.9)
Swedish civilians	8.4 (3.3)	9.5 (3.7)	10.5 (4.1)
Males			
U.S. Air Force fliers	9.0 (3.5)	9.8 (3.9)	10.6 (4.2)
Italian military	9.4 (3.7)	10.2 (4.0)	11.0 (4.3)
German Air Force	9.2 (3.6)	10.1 (4.0)	11.0 (4.3)

Source: Adapted from Ref. [2].

7. *Wrist height:* distance from standing surface to most distal point of ulna

8. *Shoulder-elbow length:* distance from top of acromion process to bottom of elbow

9. *Forearm-hand length:* distance from tip of elbow to tip of largest finger

10. *Buttock-popliteal height:* distance from most posterior point of right buttock to back of lower leg at knee

11. *Buttock-knee length:* distance from most posterior point of right buttock to most anterior point of right kneecap

12. *Thumb-tip reach:* distance from wall to tip of finger, measured with subject's back against wall, arm extended forward, and index finger touching tip of thumb.

13. *Biacromial breadth:* distance across body between acromial landmarks

14. *Chest depth:* horizontal depth of trunk at level of nipples

15. *Chest breadth:* horizontal distance across trunk at level of nipples

16. *Hip breadth:* maximum horizontal distance across hips

17. *Neck circumference:* maximum circumference of neck at a point just inferior to bulge of thyroid cartilage

18. *Chest circumference:* horizontal circumference of chest at level of nipples

19. *Waist circumference:* horizontal circumference of trunk at waist-level landmarks

20. *Buttock circumference:* circumference of hips at level of maximum posterior protrusion of buttocks

21. *Head breadth:* maximum horizontal breadth of head above ears

22. *Head circumference:* maximum circumference of head passing above brow ridges

23. *Hand breadth:* breadth of hand between metacarpal–phalangeal joints II and V

24. *Foot breadth:* maximum horizontal distance across foot at right angles to long axis

B.2 GLOSSARY

1. *Acromion:* outer extremity of shoulder blade

2. *Anterior:* front, forward

3. *Brow ridge:* crest of brow

4. *Humerus:* bone of upper arm

5. *Inferior:* lower, low

6. *Lateral:* toward the side

7. *Medial:* toward the median axis (center) of body

8. *Metacarpal–phalangeal:* bones of hand

9. *Posterior:* toward rear, behind

10. *Proximal:* near center of body or near point of attachment of muscle, joint, etc.

11. *Radius:* smaller of two bones between elbow and wrist

12. *Scapula:* shoulder blade

13. *Thyroid cartilage:* the principal cartilage of the larynx

14. *Tibia:* inner and thicker of two bones between knee and ankle

15. *Ulna:* larger of two bones between elbow and wrist

REFERENCES

1. Hertzberg, H. T. E. 1972. Engineering Anthropology. In *Human Engineering Guide to Equipment Design.* Van Cott, H. P., and Kinkade, R. G. (Eds.). U.S. Government Printing Office, Washington, DC.

2. NASA. 1978. *Anthropometric Source Book.* Vol. I: *Anthropometry for Designers.* NASA Scientific and Technical Information Service. Yellow Springs, OH.

3. Diffrient, N., Tilley, A. R., and Harman, D. 1981. *Humanscale 7/8/9.* MIT Press, Cambridge, MA.

APPENDIX C

Further Reading

C.1 GENERAL

ABEYSEKERA, J. D. A. 1990. Ergonomics of Technology Transfer. *International Journal of Industrial Ergonomics,* 5(2), pp. 181–184.

AMICK, B. C. 1987. The Politics of the Quality of Work Life in Automated Offices in the USA. *Behavior and Information Technology,* 6(4), pp. 467–482.

AYOUB, M. M., and MITAL, A. 1989. *Manual Materials Handling.* Taylor & Francis, London.

BASS, L., and DEWAN, P. (Eds.). 1993. *User Interface Software.* John Wiley and Sons, Chichester.

BILLINGSLEY, K. 1994. Standardized Medical Icons May Be Beneficial to Your Health. *SIGCHI Bulletin,* 26(1), pp. 13–16.

BOUMA, H. 1988. Human Factors in Industry. In *Proceedings of the 12th International Symposium on the Human Factors in Telecommunication,* The Hague, The Netherlands, May 24–27, 5 pp.

BOUSSENNA, M., HORTON, D., and DAVIES, B. T. 1983. Ergonomics Approach Applied to the Problems of Two Disabled People. *Applied Ergonomics,* 14(4), pp. 285–290.

CAMPBELL, R. D. 1987. *Flight Safety in General Aviation.* Collins Professional Books, London.

416

CRAVEN, F. W. 1988. An Overview of Advanced Manufacturing. *Applied Ergonomics*, 19(1), pp. 9–16.

CURTIS, B. (Ed). 1986. *Human Factors in Software Development*. IEEE Computer Society Press, Washington, DC.

DHILLON, B. S. 1987. *Reliability in Computer System Design*. Ablex, Norwood, NJ.

DINMAN, B. D. 1987. Impact of the International Labor Organization on Occupational Health and Safety Laws and Practice. *Journal of Occupational Medicine*, 29(4), pp. 345–352.

ELSENNAWY, A. K., LEE, C. H., and HINES, M. I. 1989. Ergonomic Issues in Quality Control. *Computers and Industrial Engineering*, 17(1–4), pp. 514–518.

FRASER, T. M. 1989. *The Worker at Work: A Textbook Concerned with Men and Women in the Workplace*. Taylor & Francis, London.

GRAFSTEIN, O., and LEVINE, S. P. 1993. An Expert System for Guiding Selection of Direct-Reading Air Monitoring Instruments. *Applied Occupational and Environmental Hygiene*, 8(3), pp. 157–167.

GRAVES, R. J. 1989. Ergonomics in Practice. *Annals of Occupational Hygiene*, 33(3), pp. 401–410.

HALE, A. R., and GLENDON, A. I. 1987. *Individual Behavior in the Control of Danger*. Elsevier, Amsterdam.

HANSSON, J. E. 1988. Implementation of Ergonomics at the Workplace. *Scandinavian Journal of Work, Environment and Health*, 14(Suppl. 1), pp. 113–115.

HAWKINS, F. H. 1987. *Human Factors in Flight*. Gower Technical Press, Aldershot, Hampshire, England.

HELANDER, M. 1995. *A Guide to the Ergonomics of Manufacturing*. Taylor & Francis, London.

HORTON, W. K. 1990. *Designing and Writing Online Documentation: Help Files to Hypertext*. Wiley, New York.

JOHNSON, S. L. 1989. Manufacturing Ergonomics: A Historical Perspective. *Human Factors Society Bulletin*, 32(2), pp. 6–8.

KIDD, P. T. 1991. Organization, People, and Technology in European Manufacturing: Interdisciplinary Research for the 1990s. *International Journal of Human Factors in Manufacturing*, 1(3), pp. 257–279.

KILBOM, A. 1988. Intervention Programs for Work-Related Neck and Upper Limb Disorders: Strategies and Evaluations. *Ergonomics*, 31(5), pp. 735–747.

KONZ, S. 1989. The Rise of Ergonomics. In *Advances in Industrial Ergonomics and Safety I*. Mital, A. (Ed.). Taylor & Francis, London, pp. 13–18.

KUORINKA, I. 1987. Education and Training in Ergonomics. In *Ergonomics in Developing Countries*. International Labor Office, Geneva, pp. 481–483.

LANSDALE, M. W., and ORMEROD, T. C. 1994. *Understanding Interfaces: A Handbook of Human-Computer Dialogue*. Academic Press, London. 289 pp.

McCORMICK, E. J., and ILGREN, D. R. 1987. *Industrial and Organizational Psychology*, 8th ed. Allen & Unwin, London.

MEESE, G. B., and HILL, B. K. 1989. Ergonomics in Southern Africa. *International Journal of Industrial Ergonomics*, 4(2), pp. 177–184.

MULLINS, L. J. 1996. *Management and Organizational Behaviour*. Pitman Publishing, London, 324 pp.

O'BRIEN, T. G., and CHARLTON, S. G. (Eds.). 1996. *Handbook of Human Factors Testing and Evaluation.* Lawrence Erlbaum, Mahwah, NJ. 359 pp.

O'NEAL, D. H., and HASLEGRAVE, C. M. 1990. The Application of Ergonomics in Industrially Developing Countries. In *Contemporary Ergonomics.* Lovesey, E. J. (Ed.). Taylor & Francis, London, pp. 418–423.

PARSONS, H. M. 1986. Human Factors in Industrial Robot Safety. *Journal of Occupational Accidents,* 8(1–2), pp. 25–47.

RAHIMI, M., and KARWOWSKI, W. (Eds.). 1992. *Human-Robot Interaction.* Taylor & Francis, London.

RASMUSSEN, J. 1986. *Information Processing and Human-Machine Interaction.* North-Holland, New York.

REILLY, T., and USSHER, M. 1988. Sports Leisure and Ergonomics. *Ergonomics,* 31(11), pp. 1497–1500.

SANDERS, M. S., and McCORMICK, E. J. 1987. *Human Factors in Engineering and Design,* 6th ed. McGraw-Hill, New York.

SHERIDAN, T. B. 1989. Telerobotics. *Automatica,* 25(4), pp. 487–507.

SLOVAK, A. J. M., and TREVERS, C. 1988. Solving Workplace Problems Associated with VDTs. *Applied Ergonomics,* 19(2), pp. 99–102.

SNEL, J., and CREMER, R. (Eds.). 1994. *Work and Aging: A European Perspective.* Taylor & Francis, London, 417 pp.

THURMAN, M. T., ALEXANDER, D. C., and SMITH, L. A. 1994. OSHA's Proposed Ergonomics Standard: A Summary of Responses to OSHA's Advanced Notice of Proposed Rulemaking. *Professional Safety,* 39(12), pp. 18–23.

WARR, P. (Ed) 1987. *Psychology at Work.* Penguin Books, Harmondsworth, Middlesex, England.

WEIMER, J. 1995. *Research Techniques in Human Engineering.* Prentice Hall, Englewood Cliffs, NJ.

WICKENS, C. D. 1992. *Engineering Psychology and Human Performance,* 2nd ed. HarperCollins, New York.

WILSON, J. R. 1991. Critical Human Factors Contributions in Modern Manufacturing. *International Journal of Human Factors in Manufacturing,* 1/3, pp. 281–297.

WILSON, J. R., and CORLETT, E. N. (Eds.). 1990. *Evaluation of Human Work: A Practical Ergonomics Methodology.* Taylor & Francis, London.

C.2 EQUIPMENT DESIGN

AAGAARD-HANSEN, J., and STORR-PAULSEN, A. 1995. A Comparative Study of Three Different Kinds of School Furniture. *Ergonomics,* 38(5), pp. 1025–1035.

BOBJER, O. 1989. Ergonomic Knives. In *Advances in Industrial Ergonomics and Safety I.* Mital, A. (Ed.). Taylor & Francis, London, pp. 291–298.

BOHLEMONN, J., KLUTH, K., KOTZBAUER, K., and STRASSER, H. 1994. Ergonomic Assessment of Handle Design by Means of Electromyography and Subjective Rating. *Applied Ergonomics,* 25(6), pp. 346–354.

BRIDGER, R. S. 1988. Postural Adaptations to a Sloping Chair and Work Surface. *Human Factors,* 30(2), pp. 237–247.

BRUCE, V. 1989. Human Factors in the Design of Coins. *Psychologist,* 2(12), pp. 524–527.

CHI, C. F., and DRURY, C. G. 1988. Cross Validation of Measures of Handle/Human Fit. *Applied Ergonomics,* 19(4), pp. 309–314.

CHI, C. F., and DRURY, C. G. 1988. A Further Note on Psychophysical Testing of Handles. *Applied Ergonomics,* 19(4), pp. 315–318.

CHU, Z., KNOBLOCK, G., and CORNELL, P. 1987. Providing Dynamic Support in Office Seating: The Steelcase Sensor Chair. In *Proceedings of the 5th Symposium on Human Factors and Industrial Design in Consumer Products,* Rochester, NY, May 13–15.

CISNEROS, C. M., and ARMSTRONG, T. J. 1994. Diameter Preference for Cylindrical Handles for In Line Tools. In *Proceedings of the 12th Triennial Congress of the International Ergonomics Association,* Toronto, Canada, August 15–19. Vol. 2, pp. 78–79.

CORLETT, E. N. 1990. The Evolution of Industrial Seating. In *Evaluation of Human Work: A Practical Ergonomics Methodology.* Wilson, J. R., and Corlett, E. N. (Eds.). Taylor & Francis, New York.

EVANS, W. A., COURTNEY, A. J., and FOK, K. F. 1988. The Design of School Furniture for Hong Kong School Children. *Applied Ergonomics,* 19(2), pp. 122–134.

FAILEY, T. E. 1990. Predicting the Transmissibility of a Suspension Seat. *Ergonomics,* 33(2), pp. 121–135.

FOGLEMAN, M. T., FREIVALDS, A., and GOLDBERG, J. H. 1993. An Ergonomic Evaluation of Knives for Two Poultry Cutting Tasks. *International Journal of Industrial Ergonomics,* 11(3), pp. 257–265.

FRANK, A. S., KELLER, D., and RUBINI, D. 1987. Human Factors in Fishing Reel Design. In *Proceedings of the 5th Symposium on Human Factors and Industrial Design in Consumer Products,* Rochester, NY, May 13–15.

FREIVALDS, A. 1986. The Ergonomics of Shoveling and Shovel Design: A Review of the Literature. *Ergonomics,* 29(1), pp. 3–18.

GOODWIN, D. M. 1987. A Custom Wheelchair Which Allows for Leg-Assisted Propulsion. In *Proceedings of the 10th Annual Conference on Rehabilitation Technology,* San Jose, CA, June 19–23, pp. 489–91.

GOSWAMI, A., GANGULI, S., BOSE, K. S., and CHATTERJEE, B. B. 1986. Anthropometric Analysis of Tricycle Design. *Applied Ergonomics,* 17(1), pp. 25–29.

HAMMER, W., and SCHMALZ, U. 1992. Human Behaviour When Climbing Ladders with Varying Inclinations. *Safety Science,* 15(1), pp. 21–38.

HANAI, Y., and NAGASHIMA, H. 1988. Car Seat System of the Future as Seen in "Arc-X." In *Proceedings of the 32nd Annual Meeting of the Human Factors Society,* Anaheim, CA, October 24–28, pp. 588–592.

HOLDEN, J. M., FERNIE, G., and LUNAU, K. 1988. Chairs for the Elderly: Design Considerations. *Applied Ergonomics,* 19(4), pp. 281–288.

HSIAO, S. 1994. Fuzzy Set Theory on Car-Color Design. *Color Research and Application,* 19(3), pp. 202–213.

HSU, S. H., and WU, S. P. 1991. An Investigation for Determining the Optimum Length of Chopsticks. *Applied Ergonomics,* 22(6), pp. 395–400.

JIANGHONG, Z., and LONG, T. 1994. An Evaluation of Comfort of a Bus Seat. *Ergonomics*, 25(6), pp. 386–392.

JOHNSON, S. L. 1988. Evaluation of Powered Screwdriver Characteristics. *Human Factors*, 30(1), pp. 61–69.

JOHNSON, S. L., and CHILDRESS, L. J. 1988. Powered Screwdriver Design and Use: Tool, Task and Operator Effects. *International Journal of Industrial Ergonomics*, 2(3), pp. 183–191.

KANIS, H. 1988. Industrial Design of Consumer Products for Future Users. In *Proceedings of the 10th Congress of the International Ergonomics Association.* Adams, A. S.. Hall., R. R., McPhel, B. J., and Oxenburg, M. S. (Eds.). Taylor & Francis, London.

KARIS, D., and ZEIGLER, B. L. 1989. Evaluation of Mobile Telecommunication Systems. In *Perspectives: Proceedings of the Human Factors Society 33rd Annual Meeting,* Denver, CO. Human Factors Society, Santa Monica, CA, Vol. 1, pp. 205–209.

KEIL, J. 1987. Ergonomic Considerations in the Design and Operation of Solar Heaters in Nepal. In *Proceedings of the International Symposium on Ergonomics in Developing Countries.* International Labor Office, Geneva, pp. 385–394.

KONZ, S., and RAVISHANKAR, H. 1989. Knurls on Pop Bottle Lids. In *Perspectives: Proceedings of the Human Factors Society 33rd Annual Meeting,* Denver, CO. Human Factors Society, Santa Monica, CA, Vol. 1, pp. 483–485.

KREIFELDT, J. 1992. Ergonomics of Product Design. In *Handbook of Industrial Engineering*, 2nd ed., Salverdy, G. (Ed.). John Wiley, New York, pp. 1145–1163.

KROEMER, K. H. E. 1988. Ergonomic Seats for Computer Workstations. In *Trends in Ergonomics/Human Factors V.* Aghazadeh, E. (Ed.). North-Holland, Amsterdam, pp. 313–320.

KUMAR, S., and CHENG, C. 1990. Spinal Stresses in Simulated Raking with Various Rake Handles. *Ergonomics,* 33(1), pp. 1–11.

KUMAR, S., CHENG, C. K., and MAGEE, D. J. 1987. Comparison of Two Rake Handles. In *Trends in Ergonomics/Human Factors IV.* Asfour, S. S. (Ed.). North-Holland, Amsterdam.

LA BUDA, D. R. 1989. Aging and Design: The Issues and the Background. *SOMA: Engineering for the Human Body,* 3(1), pp. 7–13.

LEE, K., WAIKAR, A., AGHAZADEH, F., and CHEN, F. 1987. *Proceedings of the 31st Annual Meeting of the Human Factors Society,* New York, October 19–23, pp. 210–213.

LINDQVIST, B. 1993. Torque Reaction in Angled Nutrunners. *Applied Ergonomics*, 24(3), pp. 174–180.

MANDAL, A. C. 1987. The Influence of Furniture Height on Back Pain. *Behavior and Information Technology,* 6(3), pp. 347–352.

MITAL, A., GENAIDY, A. M., and FORD, H. F. 1989. Portability of Three Consumer Products. *Applied Ergonomics,* 20(4), pp. 301–306.

MITCHELL, J. W., and ADAMS, J. E. 1989. A Survey of U.S. Army Aeromedical Equipment. *Aviation, Space and Environmental Medicine,* 60(8), pp. 807–810.

NOCHOLLS, H. R., ROWLAND, J. J., and SHARP, K. A. I. 1989. Virtual Devices and Intelligent Gripper Control in Robotics. *Robotica,* 7(3), pp. 199–204.

PARK, D. YIN, M. H., and FREIVALDS, A. 1991. Knife Placement Studies at an Automobile Carpet Manufacturing Plant. In *Visions. Proceedings of the Human Factors Society 35th Annual Meeting,* San Francisco, CA, Sept. 2–6. The Human Factors Society, Santa Monica, CA, Vol. I, pp. 848–852.

PARSONS, C. ATKINSON, G. DOGGART, L., LEES, A., and REILLY, T. 1994. Evaluation of New Mail Delivery Bag Designs. In *Contemporary Ergonomics*. Robertson, S. A. (Ed.). Taylor & Francis, London, pp. 236–240.

RAUKO, M., HERRANEN, S., and VUORI, M. 1988. Ergonomics of Powered Hand Tools on Assembly Line Work. *Trends in Ergonomics/Human Factors, V.* Aghazadeh, F. (Ed.). North-Holland, Amsterdam, pp. 211–217.

RHOADES, T. P., and MILLER, J. M. 1987. Towards the Development of a Consensual Design Standard for Casual Furniture Chairs. In *Proceedings of the 5th Symposium on Human Factors and Industrial Design in Consumer Products,* Rochester, NY, May 13–15.

UDO, H., BHEEMA, R., and KONZ, S. 1995. Pen Grips Versus Grip Electromyograms and Pen Point Pressure. *Journal of Science and Labour*, 71(3), Pt. 2, pp. 1–9.

VELLING, A. R. 1987. A User Designed Terminal Table System. *Behavior and Information Technology*, 6(3), pp. 353–361.

YOGAMI, H., and NORO, K. 1987. Optimum Height of a VDT Work Table. *Japanese Journal of Ergonomics*, 23(3), pp. 155–162.

YUSUF, M. 1987. Improvement in Productivity through Changes in Workbenches. In *Proceedings of the International Symposium on Ergonomics in Developing Countries.* International Labor Office, Geneva, pp. 385–394.

C.3 JOB DESIGN

ASFOUR, S. S., WALY, S. M., GENAIDY, A. M., and GONZALEZ, R. M. 1988. Physiological Stresses Associated with Television Camera Operators. *Applied Ergonomics,* 19(4), pp. 275–280.

ASHFORD, S. J. 1993. The Feedback Environment: An Exploratory Study of Cue Use. *Journal of Organizational Behavior*, 14(3), pp. 201–224.

AYOUB, M. M. 1987. The Problem of Manual Material Handling. *Trends in Ergonomics/Human Factors IV.* Asfour, S. S. (Ed.). North-Holland, Amsterdam, pp. 901–907.

BAO, S., MATHIASSEN, S. E., and WINKEL, J. 1996. Ergonomic Effects of Management Based Rationalization in Assembly Work. *Applied Ergonomics*, 27(2), pp. 89–92.

CAVESTRO, W. 1986. Automation, Work Organization and Skills. *Automatica*, 22(6), pp. 739–743.

CHAKRAVARTY, A. K. 1988. Line Balancing with Task Learning Effects. *IIE Transactions*, 20(2), pp. 186–193.

CZAJA, S. J., and DRURY, C. G. 1988. An Ergonomic Evaluation of Traditional and Modern Offices. In *Proceedings of the 10th Congress of the International Ergonomics Association.* Adams, A. S., Hall, R. R., McPhee, B. J., and Oxengurg, M. S. (Eds.). Taylor & Francis, London, pp. 132–133.

CAMPION, M. A., and MEDSKER, G. J. 1992. Job Design. In *Handbook of Industrial Engineering*, 2nd ed., Salvendy, G. (Ed.). John Wiley, New York, pp. 845–881.

CHRISTMANSSON, M. 1994. Repetitive and Manual Jobs-Content and Effects in Terms of Physical Stress and Work-Related Musculoskeletal Disorders. *International Journal of Human Factors in Manufacturing*, 4(3), pp. 281–292.

CLARK, J. 1991. Skill Changes in Maintenance Work in British Telecom: An Alternative View. *New Technology, Work and Employment*, 6(2), pp. 138–143.

COLARELLI, S. M., and BOOS, A. L. 1992. Sociometric and Ability-Based Assignment to Work Groups: Some Implications for Personnel Selection. *Journal of Organizational Behavior*, 13(1), pp. 187–196.

DAVID, F. R., PEARCE, J. A., and RANDOLPH, W. A. 1989. Linking Technology and Structure to Enhance Group Performance. *Journal of Applied Psychology*, 74(2), pp. 233–241.

EKLUND, J. A. E. 1995. Relationships Between Ergonomics and Quality in Assembly Work. *Applied Ergonomics*, 26(1), pp. 15–20.

ELIZUR, D. 1987. Work and Nonwork Relations: A Facet Analysis. *Journal of General Psychology*, 114(1), pp. 47–55.

FAFF, J., and TUTAK, T. 1989. Physiological Responses to Working with Firefighting Equipment in the Heat in Relation to Subjective Fatigue. *Ergonomics*, 32(6), pp. 629–638.

GAGNON, M., CHEHADE, A., KEMP, F., and LORTIE, M. 1987. Lumbo-Sacral Loads and Selected Muscle Activity while Turning Patients in Bed. *Ergonomics*, 30(7), pp. 1013–1032.

GAGNON, M., CHEHADE, A., KEMP, F., and LORTIE, M. 1987. Mechanical Work and Energy Transfers while Turning Patients in Bed. *Ergonomics*, 30(11), pp. 1515–1530.

GAGNON, M., ROY, D., LORTIE, M., and ROY, R. 1988. Examination of Biomechanical Principles in a Patient Handling Task. *International Journal of Industrial Ergonomics*, 3(1), pp. 29–40.

GERHART, B. 1988. Sources of Variance in Incumbent Perceptions of Job Complexity. *Journal of Applied Psychology* 73(2), pp. 154–162.

HAGAN, J. 1988. Skills and Job Commitment in High Technology Industries. *New Technology Work, and Employment*, 3(2), pp. 112–124.

HAYASHI, Y., and KOSUGO, R. 1987. A Study of the Mental Workload of Software Engineers. *Journal of Science and Labor*, 63(7), pp. 351–359.

HELLER, F. 1989. On Humanizing Technology. *Applied Psychology: An International Review*, 38(1), pp. 15–28.

ILGEN, D. R., and MOORE, C. F. 1987. Types and Choices of Performance Feedback. *Journal of Applied Psychology*, 72(3), pp. 401–406.

INGELGARD, A., KARLSSON, H. NONAS, K., and ORTENGREN, R., 1996. Psychosocial and Physical Work Environment Factors at Three Workplaces Dealing with Materials Handling. *International Journal of Industrial Ergonomics*, 17(3), pp. 209–220.

JOHANSSON, G. 1989. Stress, Autonomy and the Maintenance of Skill in Supervisory Control of Automated Systems. *Applied Psychology: An International Review*, 38(1), pp. 45–56.

JOHNSSON, B. 1988. Electromyographic Studies of Job Rotation. *Scandinavian Journal of Work, Environment and Health*, 14 (Suppl. 1), pp. 108–109.

JUDGE, T. A., and WELBOURNE, T. M. 1994. A Confirmatory Investigation of the Dimensionality of the Pay Satisfaction Questionnaire. *Journal of Applied Psychology*, 79(3), pp. 461–466.

KELLER, K. 1994. Conditions for Computer-Supported Cooperative Work: The Significance of Psychosocial Work Environment. *Technology Studies*, 1(2), pp. 242–269.

KLEIN, J. A. 1991. A Reexamination of Autonomy in Light of New Manufacturing Practices. *Human Relations*, 44(1), pp. 21–38.

LEE, K. S., WAIKAR, A. M., and WU, L. 1988. Physical Stress Evaluation of Microscope Work Using Objective and Subjective Methods. *International Journal of Industrial Ergonomics*, 2(3), pp. 203–209.

LJUNBERG, A. S., KILBOM, A., and HAAG, G. M. 1989. Occupational Lifting by Nursing Aides and Warehouse Workers. *Ergonomics*, 32(1), pp. 59–78.

LOUHEVAARA, V., PERASLINNA, P., PIIRILAS, R. SALMIO, S., and ILMARINEN, J. 1988. Physiological Responses during and after Intermittent Sorting of Postal Parcels. *Ergonomics*, 31(8), pp. 1165–1175.

LURA, S., LOUHEVAARA, V., and KIHNUNEN, K. 1994. Are the Job Demands on Physical Work Capacity Equal for Young and Aging Firefighters? *Journal of Occupational Medicine*, 36(1), pp. 70–74.

MACK, K., and HASLEGRAVE, C. 1990. Evaluation of the Use of a Hand Pump. In *Contemporary Ergonomics*. E. J. Lovesey (Ed.). Taylor & Francis, London.

NEMETH, G., ARBORELIUS, U. P., SVENSSON, O. K., and NISEL, R. 1990. The Load on the Low Back and Hips and Muscular Activity during Machine Milking. *International Journal of Industrial Ergonomics*, 5(2), pp. 115–123.

PUFFER, S. M. 1989. Task Completion Schedules: Determinants and Consequences of Performance. *Human Relations*, 42(10), pp. 937–955.

SAUTER, S. L. 1989. Moderating Effects of Job Control on Health Complaints in Office Work. In *Job Control and Worker Health*. Sauter, S. L., Hurrell, J. J., and Cooper, C. L. (Eds.). Wiley, Chichester, West Sussex, England.

SAVERY, L. K. 1988. The Influence of Social Support on the Reaction of an Employee. *Journal of Managerial Psychology*, 3(1), pp. 27–31.

SCHNAUBER, H. 1988. Optimum Stresses and Strains Represented by Examples from Shop Practice. In *Ergonomics of Hybrid Automated Systems*. Karwowski, W., Parsai, H. R., and Wilhelm, M. R. (Eds.). Elsevier, Amsterdam, pp. 671–687.

SKOKO, M. 1989. The Man-Machine Environment in the Textile Industry. *Organiczcija Rada*, 39(11–12), pp. 998–1000.

SMITH, J. 1988. Cardiac Measures of Stress in British Prison Officers. *Work and Stress*, 2(4), pp. 301–308.

TEMMYO, Y., and SAKAI, K. 1987. An Ergonomic Study on Workload in a Slaughterhouse. In *Ergonomics in Developing Countries: An International Symposium*. International Labor Office, Geneva, pp. 375–384.

VOJTECKY, M. A., HARBER, P., SAYRE, J. W., BILLETT, E., and SHIMOZAKI, S. 1987. The Use of Assistance while Lifting. *Journal of Safety Research*, 18(2), pp. 49–56.

WALL, T. D. 1987. New Technology and Job Design. In *Psychology at Work*. Warr, P. (Ed.). Penguin Books, Harmondsworth, England, pp. 270–290.

WALL, T. D., and MARTIN, R. 1987. Job and Work Design. *International Review of Industrial and Organizational Psychology*, pp. 61–91.

WALSH, J. P. 1989. Technology Change and the Division of Labor: The Case of Retail Meatcutters. *Work and Occupations*, 16(2), pp. 165–183.

WAND, S. E., and YASSI, A. 1993. Modernization of a Laundry Processing Plant: Is it Really an Improvement? *Applied Ergonomics*, 24(6), pp. 387–396.

WEBER, R. 1988. Computer Technology and Jobs. *Communications of the ACM,* 31(1), pp. 68–77.

WELLS, R., BERUBE, D., and MOORE, A. 1994. Hand, Arm, and Shoulder Loads and Physical Characteristics of MIG Welding Guns. In *Proceedings of the 12th Triennial Congress of the International Ergonomics Association,* Toronto, Canada, Aug. 15–19, Vol. 2, pp. 75–77.

WINKEL, J., and GARD, G. 1988. An EMG Study of Work Methods and Equipment in Crane Coupling as a Basis for Job Redesign. *Applied Ergonomics,* 19(3), pp. 178–184.

ZICKLING, G. 1987. Numerical Control Machining and the Issue of Deskilling: An Empirical View. *Work and Occupations,* 14(3), pp. 452–466.

C.4 WORKPLACE DESIGN

ANDRES, R. 1995. Adjustable Workstations: Flexible Concepts Ease Strain on Factory Workers. *Workplace Ergonomics,* 1(1), pp. 28–29, 31–32, 45.

BENDIX, T. 1986. Chair and Table Adjustments for Seated Work. In *The Ergonomics of Working Postures: Models, Methods and Cases.* Corlett, N., Wilson, J., and Manenica, I. (Eds.). Taylor & Francis, London, pp. 355–362.

COLES, S. 1994. Practical Issues in the Redesign of a Complex Production Process Layout: A Case Study. In *Contemporary Ergonomics.* Robertson, S. A. (Ed.). Taylor & Francis, London, pp. 394–399.

COMBS, R. B., and AGHAZADEH, F. 1988. Ergonomically Designed Chemical Plant Control Room. In *Trends in Ergonomics/Human Factors V.* Aghazadeh, F. (Ed.). North-Holland, Amsterdam, pp. 357–364.

COURTNEY, A. J., and EVANS, W. A. 1987. A Preliminary Investigation of Bus Cab Design for Cantonese Drivers. *Journal of Human Ergology* 16(2), pp. 163–171.

DeGREEVE, T. B., and AYOUB, M. M. 1987. A Workplace Design Expert System. *International Journal of Industrial Ergonomics,* 2(1), pp. 37–48.

FLOHRER, U., and WEIKINNIS, H. 1988. Man-Machine Aspects of an Office Videophone. In *Proceedings of the 12th International Symposium on the Human Factors in Telecommunication,* The Hague, The Netherlands, 10 pp.

FOOD MARKETING INSTITUTE. 1992. *Suggestions for Ergonomic Improvement of Scanning Checkstand Designs.* Washington, D.C., 35 pp.

GILAD, I., and MESSER, E. 1992. Ergonomic Design of the Diamond Polishing Workstation. *International Journal of Industrial Ergonomics,* (9)1, pp. 53–63.

GUPTA, R., and SHARIT, J. 1988. Human Computer Interaction in Facilities Layout. In *Handbook of Human-Computer Interaction.* Helander, M. (Ed.). North-Holland, Amsterdam, pp. 729–736.

HADLEY, M. A. 1988. Present Trends in Naval Bridge Design and Integrated Navigation. *Journal of Navigation,* 41(2), pp. 276–287.

HELLA, F., TISSERAND, M., and SCHOULLER, J. F. 1988. Visibility Requirements for the Driver's Seat of Lift Trucks: Experimental Study of Driver's Lateral Head Movements. *Applied Ergonomics,* 19(3), pp. 225–232.

HORNICK, R. J. 1988. Workstation Design. In *Automotive Engineering and Litigation,* Vol. 2. Peters, G. A. and Peters, B. J. (Eds.). Garland, New York.

IVEGARD, T. 1989. *Handbook of Control Room Design and Ergonomics*. Taylor & Francis, London.

KERN, R. and BAUER, W. 1988. Computer Aided Workplace Design: Stage of Development and Application Fields. In *Proceedings of the 10th Congress of the International Ergonomics Association*. Adams, A. S., Hall, R. R., McPhee, B. J., and Oxenburg, M. S. (Eds.). Taylor & Francis, London, pp. 99–101.

KROEMER, K. H. E. 1988. VDT Workstation Design. In *Handbook of Human-Computer Interaction*. Helander, M. (Ed.): North-Holland, Amsterdam, pp. 521–539.

LANNERSTEN, L., and HARMS-RINGDAHL, K. 1990. Neck and Shoulder Muscle Activity during Work with Different Cash Registers. *Ergonomics,* 33(1), pp. 49–65.

LECLERCQ, S., TISSERAND, M., and SAULNIER, M. 1995. Tribological Concepts Involved in Slipping Accident Analysis. *Ergonomics*, 38(2), pp. 197–208.

LUEDER, R. 1986. Workstation Design. In *The Ergonomic Payoff: Designing the Electronic Office*. Lueder, R. (Ed.). Holt, Rinehart and Winston of Canada, Toronto, pp. 142–183.

MACDONALD, G. A. H. 1980. Implications of Male/Female Differences for Workplace Design. In *Proceedings of the 13th Annual Meeting of the Human Factors Society of Canada*, Lake of Bays, Ontario. Stager, P. (Ed.), pp. 9–10.

MACIEL, R. H. 1987. An Ergonomic Study of a Data Processing Bureau. In *Ergonomics in Developing Countries: An International Symposium*. International Labor Office, Geneva, pp. 283–289.

MCCROBIE, D. 1991. Integrating Personnel-Computer Interaction Requirements into Console Design. *Human Factors Society Bulletin*, 34(10), pp. 1–3.

MILL, P. A. D., HARTKOPF, V., and LOFTNESS, V. 1986. Evaluating the Quality of the Workplace. In *The Ergonomic Payoff: Designing the Electronic Office*. Lueder, R. (Ed.). Holt, Rinehart and Winston of Canada, Toronto, pp. 295–343.

MITAL, A, 1991. Workspace Clearance and Access Dimensions and Design Guidelines. In *Workspace, Equipment and Tool Design*, Mital, A., and Karwowski, W. (Eds.). Elsevier, Amsterdam.

MITAL, A., FARD, H. F., and KHALEDI, H. 1987. A Biomechanical Evaluation of Staircase Riser Heights and Tread Depths during Stair Climbing. *Clinical Biomechanics,* 2(3), pp. 162–164.

MITAL, A., and MOTORWALA, A. 1995. An Ergonomic Evaluation of Steel and Composite Access Covers. *International Journal of Industrial Ergonomics*, 15(4), pp. 285–296.

NOWAK, E. 1989. Workplace for Disabled People. *Ergonomics,* 32(9), pp. 1077–1088.

PEKKARINEN, A., and ANTTONEN, H. 1988. The Effect of Working Height on the Loading of the Muscular and Skeletal Systems in the Kitchens of Workplace Canteens. *Applied Ergonomics,* 19(4), pp. 306–308.

RENDLE, J., and WATT, J. 1993. Floor Choices. *Australian Safety News*, 64(8), pp. 48–49.

RAYFIELD, J. K. 1994. *The Office Interior Design Guide: An Introduction for Facilities Managers and Designers.* John Wiley and Sons, New York, 249 pp.

REDFERN, M. S., and CHAFFIN, D. B. 1988. The Effects of Floor Types on Standing Tolerance in Industry. In *Trends in Ergonomics/Human Factors V*. Aghazadeh, F. (Ed.). North-Holland, Amsterdam, pp. 401–405.

RODRIGUES, C. C. 1989. A Technical Analysis of Supermarket Laser Scanning Workstation. In *Advances in Industrial Ergonomics and Safety I*. Mital, A. (Ed.). Taylor & Francis, London, pp. 953–959.

ROE, R. W. 1993. Occupant Packaging. In *Automotive Ergonomics*, Peacock, B. and Korwowski, W. (Eds.). Taylor & Francis, London.

ROMBACH, V., and LAURIG, W. 1988. Ergon-Expert: A Knowledge Based Approach to the Design of Workplaces. In *Trends in Ergonomics/Human Factors V*. Aghazadeh, F. (Ed.). North-Holland, Amsterdam, pp. 53–61.

RYS, M., and KONZ, S. 1989. An Evaluation of Floor Surfaces. In *Perspectives: Proceedings of the Human Factors Society 33rd Annual Meeting*, Denver, CO. Human Factors Society, Santa Monica, CA, Vol. 1, pp. 517–520.

STRASSER, H. 1990. Evaluation of a Supermarket Twin Checkout involving Forward and Backward Operation. *International Journal of Industrial Ergonomics*, 5(1), pp. 7–14.

TOMPKINS, J. A. 1992. Facilities Layout. In *Handbook of Industrial Engineering*, 2nd ed. Salverdy, G. (Ed.). John Wiley, New York, pp. 1777–1813.

ULICH, E., SCHUPBACH, H., SCHILLING, A., and KUARK, J. K. 1990. Concepts and Procedures of Work Psychology for the Analysis, Evaluation and Design of Advanced Manufacturing Systems: A Case Study. *International Journal of Industrial Ergonomics*, 5(1), pp. 47–57.

VISCHER, J. C. 1989. *Environmental Quality in Offices*. Van Nostrand Reinhold. New York.

WOOD, J. 1986. Designing Control Rooms. In *Proceedings* of the *Ergonomics Society's 1986 Conference*. Oborne, D. J. (Ed.). Taylor & Francis, London, pp. 158–162.

YUST, B. L., and OLSON, W. W. 1987. Microwave Cooking Appliance Placement in Residential Kitchens. *Home Economics Research Journal*, 16(1), pp. 70–78.

ZHANG, L., HELANDER, M. G., and DRURY, C. G. 1996. Identifying Factors of Comfort and Discomfort in Sitting. *Human Factors*, 38(3), pp. 377–389.

C.5 ENVIRONMENT DESIGN

AKBAR-KHANZADEH, F. 1993. Hearing Impairment Study to Stimulate Management Support for a Hearing Conservation Program. *Applied Occupational and Environmental Hygiene*, 8(5), pp. 472–478.

ALEXANDER, D. C., and SMITH, L. A. 1988. An Efficient Method of Verifying Heat Stress Problems in Industry. *Trends in Ergonomics/Human Factors V*. Aghazadeh, F. (Ed.). North-Holland, Amsterdam, pp. 471–478.

BARON, R. A. 1987. Effect of Negative Ions on Cognitive Performance. *Journal of Applied Psychology*, 72(1), pp. 131–137.

BERGLUND, B., HARDER, K., and PREIS, A. 1994. Annoyance Perception of Sound and Information Extraction. *Journal of the Acoustical Society of America*, 95(3), pp. 1501–1509.

BILLMEYER, F. W. 1988. Quantifying Color Appearance Visually and Instrumentally. *Color Research and Application*, 13(3), pp. 140–145.

BRITTON, L. A., and DELAY, E. R. 1989. Effects of Noise on a Simple Visual Attentional Task. *Perceptual and Motor Skills*, 68(3), pp. 875–878.

CHAD, K. E., and BROWN, J. M. M. 1995. Climatic Stress in the Workplace: Its Effect on Thermoregulatory Resources and Muscle Fatigue in Female Workers. *Applied Ergonomics*, 26(1), pp. 29–34.

CHANT, R. 1986. Air Quality. In *The Ergonomics Payoff: Designing the Electronic Office.* Lueder, R. (Ed.). Holt, Rinehart, and Winston of Canada, Toronto, pp. 272–294.

COMMITTEE ON AIRLINER CABIN AIR QUALITY. 1986. *The Airliner Cabin Environment: Air Quality and Safety.* National Academy Press, Washington, DC.

CORBRIDGE, C., and GRIFFIN, M. J. 1986. Vibration and Comfort: Vertical and Lateral Motion in the Range 0.5 to 5.0 Hz. *Ergonomics,* 29(2), pp. 249–272.

CORNELIUS, K. M., REDFERN, M. S., and STEINER, L. J. 1994. Postural Stability after Whole-Body Vibration Exposure. *International Journal of Industrial Ergonomics*, 13(4), pp. 343–351.

DAVISON, G. 1993. Is Your Computer Killing You? *Australian Safety News*, 64(3), pp. 43–46.

EMBRECHTS, J. J. 1988. Combination of Illuminants and Its Effect on Color Rendering. *Lighting Research and Technology,* 20(1), pp. 1–10.

EMERY, A. F. 1986. Thermal Comfort. In *The Ergonomics Payoff: Designing the Electronic Office.* Lueder, R. (Ed.). Holt, Rinehart, and Winston of Canada, Toronto, pp. 249–271.

ENANDER, A. 1987. Effects of Moderate Cold on Performance of Psychomotor and Cognitive Tasks. *Ergonomics,* 30(10), pp. 1431–1445.

FOXCROFT, W. J., and ADAMS, W. C. 1986. Effects of Ozone Exposure on Four Consecutive Days on Work Performance and $V_{O_2}max$. *Journal of Applied Physiology,* 61(3), pp. 960–966.

FRANDSEN, S. 1987. The Scale of Light. *International Lighting Review,* 38(3), pp. 108–112.

GOODMAN, S. L. 1990. Noise and the Petrochemical Industry. *Health and Safety at Work,* 12(4), pp. 26–27.

HAFEZ, H. A., and BESHIR, M. Y. 1987. Effects of Thermostat Temperature on Human Performance. In *Trends in Ergonomics/Human Factors IV.* Asfour, S. S. (Ed.). North-Holland, Amsterdam, pp. 352–332.

HANCOCK, P. A., and VERCRUYSSEN, M. 1988. Limits of Behavioral Efficiency for Workers in Heat Stress. *International Journal of Industrial Ergonomics,* 3(2), pp. 149–158.

HAYMES, E. M., and WELLS, C. L. 1986. *Environment and Human Performance.* Human Kinetics Publishers, Champaign, IL.

HEUS, R., DAANEN, H. A. M., and HAVENITH, G. 1995. Physiological Criteria for Functioning of Hands in the Cold: A Review. *Applied Ergonomics*, 26(1), pp. 5–13.

JACKSON, C. G. R., and SHARKEY, B. J. 1988. Altitude, Training and Human Performance. *Sports Medicine,* 6(5), pp. 279–284.

KANAYA, S., and MIYAMAE, A. 1989. The Interior Visual Environment for Aged People. *Japanese Journal of Ergonomics*, 25(3), pp. 163–167.

KJELLBERG, A., and SKOLDSTROM, B. 1991. Noise Annoyance During the Performance of Different Nonauditory Tasks. *Perceptual and Motor Skills*, 73(1), pp. 39–49.

LAIDEBEUR, A. 1987. Lighting, Pedestrians and the City. *International Lighting Review,* 2, pp. 50–53.

LAWTHER, A., and GRIFFIN, M. J. 1988. Motion Sickness and Motion Characteristics of Vessels at Sea. *Ergonomics,* 31(10), pp. 1373–1394.

LEFTHERIOTIS, G., SAVOUREY, G., SAUMET, J. L., and BITTEL, J. 1990. Finger and Forearm Vasodilatory Changes after Local Cold Acclimation. *European Journal of Applied Physiology and Occupational Physiology,* 60(1), pp. 49–53.

MCLELLAN, T. M., MEUNIER, P., and LIVINGSTONE, S. 1992. Influence of New Vapor Protective Clothing Layer on Physical Work Tolerance Times at 40°C. *Aviation, Space, and Environmental Medicine,* 63(2), pp. 107–113.

MELNICK, W. 1987. Noise Standards for the Work Place. *Noise Control Engineering Journal,* 29(1), pp. 13–17.

MUSSON, Y., BURDORF, A., and VAN DRIMMELEN, D. 1989. Exposure to Shock and Vibration and Symptoms in Workers Using Impact Power Tools. *Ergonomics,* 33(1), pp. 85–96.

NIELSEN, R. 1986. Clothing and Thermal Environments. *Applied Ergonomics,* 17(1), pp. 47–57.

OULETTE, M. J., and REA, M. S. 1989. Illuminance Requirements for Emergency Lighting. *Journal of the Illuminating Engineering Society,* 18(1), pp. 37–42.

PALIN, S. L. 1994. Does Classical Music Damage the Hearing of Musicians? A Review of the Literature. *Occupational Medicine,* 44(3), pp. 130–136.

PERRY, M. J. 1988. Fundamental Vision: A Glaring Case. *Lighting Research and Technology* 20(4), pp. 161–165.

POPE, M. H., BROMAN, H., and HANSSON, T. 1989. The Dynamic Response of a Subject Seated on Various Cushions. *Ergonomics,* 32(10), pp. 1155–1166.

RAMSEY, J. D. 1987. Practical Evaluation of Hot Working Areas. *Professional Safety,* 32(2), pp. 42–48.

SAGAWA, S., SHIRAKI, K., YOUSEF, M. K., and MIKI, K. 1988. Sweating and Cardiovascular Responses of Aged Men to Heat Exposure. *Journal of Gerontology: Medical Sciences,* 43(1), p. Ml–8.

SAITO, K., HOSOKAWA, T., INUZUKA, S., and ITOH, T. 1993. Evaluation of the Combined Strain of Sound and Physical Exercise by Measurement of Mental Activities and Catecholamines. *Archives of Complex Environmental Studies,* 5(1–2), pp. 85–90.

SALAME, P., and BADDELEY, A. 1987. Noise, Unattended Speech and Short Term Memory. *Ergonomics,* 30(8), pp. 1185–1194.

SCHULTZ, U. 1988. Color-Distance Judgement and the Influence of the Background in Color Reproduction. *Color Research and Application,* 13(2), pp. 99–105.

SEROUSSI, R. E., WILDER, D. G., and POPE, M. H. 1989. Trunk Muscle Electromyography and Whole Body Vibration. *Journal of Biomechanics,* 22(3), pp. 219–229.

SHAMSSAIN, M. H., THOMPSON, J., and OGSTON, S. A. 1988. Effects of Cement Dust on Lung Function in Libyans. *Ergonomics,* 31(9), pp. 1299–1303.

SHERWOOD, N. and GRIFFIN, M. J. 1991. Evidence of Impaired Learning During Whole-Body Vibration. *Journal of Sound and Vibration,* (152)2, pp. 219–225.

SHOENBERGER, R. W. 1988. Intensity Judgements of Vibrations in the X Axis, Z Axis and X plus Z Axis. *Aviation Space and Environmental Medicine,* 59(8), pp. 749–753.

SIVAK, M., and OLSON, P. L. 1988. Toward the Development of a Field Methodology for Evaluating Discomfort Glare from Automobile Headlights. *Journal of Safety Research,* 19(3), pp. 135–143.

SMITH, A. 1989. A Review of the Effects of Noise on Human Performance. *Scandinavian Journal of Psychology,* 30(3), pp. 185–256.

SORENSEN, S., and BRUNNSTROM, G. 1995. Quality of Light and Quality of Life: An Intervention Study among Older People. *International Journal of Lighting Research and Technology,* 27(2), pp. 113–118.

THIBAULT, T. P. 1987. Direct/Indirect Lighting for Display Screen Operators. *Proceedings of the 33rd IES National Convention and Expo,* Melbourne.

C.6 WORKER CHARACTERISTICS

ASPDEN, R. D. 1987. Intra-abdominal Pressure and Its Role in Spinal Mechanics. *Clinical Biomechanics,* 2(3), pp. 168–174.

BONNEY, R. 1988. Some Effects on the Spine from Driving. *Clinical Biomechanics,* 3(4), pp. 236–240.

BURTON, A. K. 1986. Spinal Strain from Shopping Bags with and without Handles. *Applied Ergonomics,* 17(1), pp. 19–23.

COHEN, H. H., and COHEN, D. M. 1991. Human Error: Myths about Mistakes. *Professional Safety,* 36(10), pp. 32–36.

CRAIG, A. 1988. Self-Control over Performance in Situations That Demand Vigilance. In *Vigilance: Methods, Models and Regulation.* Leonard, J. P. (Ed.). Verlag Peter Lang, Frankfurt, pp. 237–246.

DAMOS, D. L. 1988. Individual Differences in Subjective Estimates of Workload. In *Human Mental Workload.* Hancock, P. A., and Meshkati, N. (Ed.). North-Holland, Amsterdam, pp. 321–237.

DIVITA, J. C., and Hanna, T. E. 1992. Human Efficiency for Visual Detection of Targets on Cathode Ray Tube Displays Using a Two-Level Multiple-Channel Time History Format. *Journal of the Acoustical Society of America*, 91(3), pp. 1552–1564.

DONDERI, D. C. 1994. Visual Acuity, Color Vision, and Visual Search Performance at Sea. *Human Factors*, 36(1), pp. 129–144.

EBERTS, R. E. 1987. Internal Models, Tracking Strategies, and Dual-Task Performance. *Human Factors,* 29(4), pp. 407–419.

EGAN, D. E. 1988. Individual Differences in Human-Computer Interaction. In *Handbook of Human Computer Interaction.* Helander, M. (Ed.). North-Holland, Amsterdam, pp. 543–568.

FIBIGER, W., CHRISTENSEN, F., SINGER, G., and KAUFMANN, H. 1986. Mental and Physical Components of Sawmill Operatives' Workload. *Ergonomics,* 29(3), pp. 363–375.

FOLKARD, S. 1987. Circadian Rhythms and Hours of Work. In *Psychology at Work.* Warr, P. (Ed.). Penguin Books, London, pp. 30–52.

FRANSSON-HALL, C., and KILBOM, A. 1993. Sensitivity of the Hand to Surface Pressure. *Applied Ergonomics*, 24(3), pp. 181–189.

GAWRON, V. J., and RANNEY, Y. A. 1988. The Effects of Alcohol Dosing on Driving Performance on a Closed Course and in a Driving Simulator. *Ergonomics,* 31(9), pp. 1219–1244.

GENAIDY, A. M., and ASFOUR, S. S. 1989. Effects of Frequency and Load of Lift on Endurance Time. *Ergonomics,* 32(1), pp. 51–57.

Gilbert, D. K., and ROGERS, W. A. 1996. Age-Related Differences in Perceptual Learning. *Human Factors*, 38(3), pp. 417–424.

HALLBACK, M. S. 1994. Flexion and Extension Forces Generated by Wrist-Dedicated Muscles Over the Range of Motion. *Applied Ergonomics*, 25(6), pp. 379–385.

HALPERN, D. L., and Blake, R. R. 1988. How Contrast Affects Stereoacuity. *Perception,* 17(4), pp. 483–495.

HARRINGTON, D. L., and HAALAND, K. Y. 1987. Programming Sequences of Hand Postures. *Journal of Motor Behavior,* 19(1), pp. 77–95.

HOFFMAN, E. R., and SHEIKH, I. H. 1994. Effect of Varying Target Height in a Fitts' Movement Task. *Ergonomics*, 37(6), pp. 1071–1088.

KLAPP, S. T., and NETICK, A. 1988. Multiple Resources for Processing and Storage in Short Term Working Memory. *Human Factors,* 30(5), pp. 617–632.

KOLODNER, J. L., and KOLODNER, R. M. 1987. Using Experience in Clinical Problem Solving. *IEEE Transactions on Systems, Man, and Cybernetics,* SMC-17(3), pp. 420–431.

KUMAR, S. 1994. Lumbosacral Compression in Maximal Lifting Efforts in Sagittal Plane with Varying Mechanical Disadvantage in Isometric and Isokinetic Modes. *Ergonomics*, 37(12), pp. 1975–1983.

LATHAM, K., and WHITAKER, K. 1996. Relative Roles of Spatial Interference in Foveal and Peripheral Vision. *Ophthalmic and Physiological Optics*, 16(1), pp. 49–57.

LI, S. ZHU, Z., and ADAMS, A. S. 1995. An Exploratory Study of Arm-Reach Reaction Time and Eye-Hand Coordination. *Ergonomics*, 38(4), pp. 637–650.

LINTON, S. J., and KAMWENDS, K. 1989. Risk Factors in the Psychosocial Work Environment for Neck and Shoulder Pain in Secretaries. *Journal of Occupational Medicine,* 31(7), pp. 609–613.

LOVELACE, E. A. 1989. Vision and Kinesthesis in Accuracy of Hand Movement. *Perceptual and Motor Skills*, 68(3), pp. 707–714.

MALATERRE, G., FERNANDEZ, F., KLEURY, D., and LECHNER, D. 1988. Decision Making in Emergency Situations. *Ergonomics,* 31(4), pp. 643–655.

MCEVOY, G. M., and CASCIO, W. F. 1989. Cumulative Evidence of the Relationship between Employee Age and Job Performance. *Journal of Applied Psychology,* 74(1), pp. 11–17.

MESTRE, D. 1988. Visual Control of Displacement at Slow Speeds. *Human Factors,* 30(6), pp. 663–675.

MICALIZZI, J., and GOLDBERG, J. H. 1989. Knowledge of Results in Visual Inspection Decisions: Sensitivity or Criterion Effect? *International Journal of Industrial Ergonomics,* 4(3), pp. 225–235.

MISHRA, B., and SILVER, N. 1989. Some Discussions of Static Gripping and Its Stability. *IEEE Transactions on Systems, Man and Cybernetics,* 19(4), pp. 783–796.

MORRIS, R. G., CRAIK, F. I. M., and GICK, M. L. 1990. Age Differences in Working Memory Tasks: The Role of Secondary Memory and the Central Executive System. *Quarterly Journal of Experimental Psychology,* 42A(1), pp. 67–86.

NAGEL, D. C. 1988. Human Error in Aviation Operations. In *Human Factors in Aviation.* Wiener, E. L., and Nagel, D. C. (Eds.). Academic Press, San Diego, CA, pp. 263–303.

ONG, T. C., and SOTHY, S. P. 1986. Exercise and Cardiorespiratory Fitness. *Ergonomics,* 29(2), pp. 273–280.

PAPCUN, G., KREIMAN, J., and DAVIS, A. 1989. Long-Term Memory for Unfamiliar Voices. *Journal of the Acoustical Society of America*, 85(2), pp. 913–925.

PARASURANAM, R. 1987. Human-Computer Monitoring. *Human Factors,* 29(6), pp. 695–706.

PIERCE, J. R. 1991. Periodicity and Pitch Perception. *Journal of the Acoustical Society of America*, 90(4), Pt. 1, pp. 1889–1893.

POYNTER, D. 1988. Variability in Brightness Matching of Colored Lights. *Human Factors*, 30(2), pp. 143–151.

RAAIJMAKERS, J. G. W., and VERDUYN, W. W. 1996. Individual Differences and the Effects of an Information Aid in Performance of a Fault Diagnosis Task. *Ergonomics*, 39(7), pp. 966–974.

RAFTOPOULOS, D. D., RAFCO, M. C., GREEN, M., and SCHULTZ, A. B. 1988. Relaxation Phenomenon in Lumbar Trunk Muscles during Lateral Bending. *Clinical Biomechanics*, 3(3), pp. 166–172.

REID, D. C., OEDEKOVEN, G., KRAMER, J. F., and SABOLE, L. A. 1989. Isokinetic Muscle Strength Parameters for Shoulder Movements. *Clinical Biomechanics*, 4(2), pp. 97–104.

RIPOLL, H., and FLEURANCE, P. 1988. What Does Keeping One's Eye on the Ball Mean? *Ergonomics*, 31(11), pp. 1647–1654.

RIZZO, A., BAGNARA, S., and VISCIOLA, M. 1987. Human Error Detection Process. *International Journal of Man-Machine Studies*, 27(5–6), pp. 555–570.

ROSENBAUM, D. A., and Jorgensen, M. J. 1992. Planning Macroscopic Aspects of Manual Control. *Human Movement Science*, 11(1–2), pp. 61–69.

SHEPHERD, M., and MULLER, H. J. 1989. Movement versus Focusing of Visual Attention. *Perception and Psychophysics*, 46(2), pp. 146–154.

SHIOMI, T. 1994. Effects of Different Patterns of Stairclimbing on Physiological Cost and Motor Efficiency. *Journal of Human Ergology*, 23(2), pp. 111–120.

SIMS, M. T., and GRAVELING, R. A. 1988. Manual Handling of Supplies in Free and Restricted Headroom. *Applied Ergonomics*, 19(4), pp. 289–292.

SKOLDSTROM, B. 1987. Physiological Responses of Fire Fighters to Workload and Thermal Stress. *Ergonomics*, 30(11), pp. 1589–1597.

SNAPE, J. 1988. Stress Factors among Lecturers in a College of Further Education. *Work and Stress*, 2(4), pp. 327–331.

STATON, N. A., and BOOTH, R. T. 1990. The Psychology of Alarms. In *Contemporary Ergonomics*. Lovesey, E. J., (Ed.). Taylor & Francis, London, pp. 378–383.

SWELLER, J. 1988. Cognitive Load during Problem Solving. *Cognitive Science*, 12(2), pp. 257–285.

THOMPSON, D. 1989. Reach Distance and Safety Standards. *Ergonomics*, 32(9), pp. 1061–1076.

WAGNER, C. H. 1988. The Pianist's Hand: Anthropometry and Biomechanics. *Ergonomics*. 31(1), pp. 97–131.

WELFORD, A. T. 1987. On Rates of Improvement with Practice. *Journal of Motor Behavior*, 19(3), pp. 401–415.

WICKERS, G. N. 1988. Knowledge Structures of Expert-Novice Gymnasts. *Human Movement Science*, 7(1), pp. 47–72.

WYNN, V. T. 1993. Accuracy and Consistency of Absolute Pitch. *Perception*, 22(1), pp. 113–121.

Index